Mukesh Yadav, Nirmala Sehrawat
Functional Foods and Gut Microbiome

Also of Interest

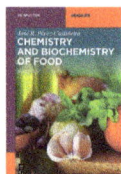

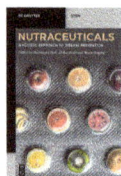

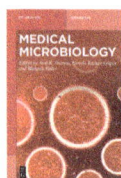

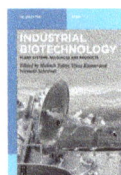

Functional Foods and Gut Microbiome

Probiotics, Prebiotics, and Synbiotics

Edited by
Mukesh Yadav and Nirmala Sehrawat

DE GRUYTER

Editors
Dr. Mukesh Yadav
Associate Professor
Department of Bio-Sciences and Technology, M.M.E.C.,
Maharishi Markandeshwar (Deemed to be University), Mullana-Ambala,
Haryana-133207, India
Email: mukeshyadav7@gmail.com

Dr. Nirmala Sehrawat
Associate Professor
Department of Bio-Sciences and Technology, M.M.E.C.,
Maharishi Markandeshwar (Deemed to be University), Mullana-Ambala,
Haryana-133207, India
Email: nirmalasehrawat@gmail.com

ISBN 978-3-11-914795-8
e-ISBN (PDF) 978-3-11-220515-0
e-ISBN (EPUB) 978-3-11-220537-2

Library of Congress Control Number: 2026930557

Bibliographic information published by the Deutsche Nationalbibliothek
The Deutsche Nationalbibliothek lists this publication in the Deutsche Nationalbibliografie;
detailed bibliographic data are available on the Internet at http://dnb.dnb.de.

© 2026 Walter de Gruyter GmbH, Berlin/Boston, Genthiner Straße 13, 10785 Berlin
Cover image: Eoneren/E+/gettyimages
Typesetting: Integra Software Services Pvt. Ltd.

www.degruyterbrill.com
Questions about General Product Safety Regulation:
productsafety@degruyterbrill.com

Book Summary/Preface

– Currently, a significant part of human population is consuming various probiotics in form of traditional fermented food or commercial supplements. The use of probiotics along with prebiotics and synbiotics has increased in recent times due to increase in public awareness on functional foods. It has been found that probiotics and prebiotics may be useful in maintaining the gut microbiome and therefore, good health. Probiotics are living microbes having potential to exert health benefits when administered in the host in sufficient amounts. Prebiotics are nondigestible food substances that can be utilized by selective host microbes and have potential to exert significant health benefits. Synbiotics are specific combinations of prebiotics and probiotics which collectively provides significant health benefits to the host by affecting the gut microbiota. Appropriate combination of prebiotic and probiotic as a single product ensure greater effects in comparison to the individual effect of prebiotic or probiotic as therapeutics. The microbial community inside the gut and their products influence the host functions in the gastrointestinal tract both locally as well as systemically. Gut microbes also affects the nutritional status of the host by several mechanisms. Probiotics (and also prebiotics and synbiotics) exerts health benefits through improved digestion, stimulating the metabolism, supporting growth of beneficial microbes, production of specific metabolites having therapeutic potential, maintaining favorable intestinal environment and promoting overall host health. The research suggests that the use of suitable prebiotics, probiotics and synbiotics might play significant role in human life as part of their food.
– This book covers conceptual knowledge of probiotics, prebiotics, synbiotics and related aspects. The book covers comprehensive and systamatic information on potential of probiotics, prebiotics as well as synbiotics as functional foods.
– Recent advances, managemnt of disorders/ diseases by using probiotics have been covered. The effect of gut-microbiome on various organs, gut-organ axis has been discussed.
– Mechanism of actions for probiotics, prebiotics and synbiotics have been included. This book will be very useful for the readers.
– Information has been presented in the form of reader friendly text, figures and tables.
– It will be useful for UG, PG students, PhD scholars as core subject and text book information. Researchers, academiicnas and dieticians, biochemists, microbiologists, biotechnologists and other life science disciplines will find it useful for thier concept, update and recent information.
– This book covers the concept of probiotics, prebiotics and synbiotics along with their potential as functional food for maintaining a good health.

https://doi.org/10.1515/9783112205150-202

- In recent times, the researchers have focused on gut microbiome, use of probiotics and related aspects to manage human health. This book provides useful insights into the area of gut microbiome, and probiotics as functional food.
- This book is suitable for UG, PG, PhD students as well as academicians working in the area of biotechnology, microbiology, foods, diets, nutrition, gut microbiome, probiotics as well as synbiotics.
- **Keywords:** Probiotics, gut health, gut microbiome, prebiotics, synbiotics, human health, disease management

Contents

List of Contributing Authors

Mukesh Yadav
Department of Bio-Sciences and Technology
MMEC
Maharishi Markandeshwar (Deemed to be
University)
Mullana-Ambala 133207, Haryana
India

Sunil Kumar
Department of Microbiology
Graphic Era University
Dehradun, Uttarakhand
India

Sachin Kumar
Department of Bioinformatics
Janta Vedic College
Baghpat, Baraut, Uttar Pradesh
India

Manoj Singh
Department of Bio-Sciences and Technology
MMEC
Maharishi Markandeshwar (Deemed to be
University)
Mullana-Ambala 133207, Haryana
India

Nirmala Sehrawat
Department of Bio-Sciences and Technology
MMEC
Maharishi Markandeshwar (Deemed to be
University)
Mullana-Ambala 133207, Haryana
India
nirmalasehrawat@gmail.com

Nijendra Pratap Singh
Department of Biotechnology
Sharda School of Bio-Sciences and Technology
Sharda University
Greater Noida, Uttar Pradesh
India

Shivam Singh
Sharda School of Bio-Sciences and Technology
Sharda University
Greater Noida, Uttar Pradesh
India

Ravindra Kumar Jain
Department of Biotechnology
Swami Vivekanad Subharti University
Meerut, Uttar Pradesh
India

Amit Kumar
Sharda School of Bio-Sciences and Technology
Sharda University
Greater Noida, Uttar Pradesh
India
amit.kumar23@sharda.ac.in; baliyaniitr@gmail.com

Harshada Joshi
Department of Biotechnology
Mohanlal Sukhadia University
Udaipur 313001, Rajasthan
India

Jyoti
Department of Microbiology
Mohanlal Sukhadia University
Udaipur 313001, Rajasthan
India

Vidhi Jain
Department of Microbiology
Mohanlal Sukhadia University
Udaipur 313001, Rajasthan
India

Tamanna Ajmera
Department of Biotechnology
Mohanlal Sukhadia University
Udaipur 313001, Rajasthan
India

https://doi.org/10.1515/9783112205150-204

Moomal Acharya
Department of Biotechnology
Mohanlal Sukhadia University
Udaipur 313001, Rajasthan
India

Namita Ashish Singh
Department of Microbiology
Mohanlal Sukhadia University
Udaipur 313001, Rajasthan
India
namita.singh@mlsu.ac.in

Priti
KL Mehta Dayanand College for Women
Faridabad, Haryana
India

Kumkum Verma
IARI
Regional Station
Karnal, Haryana
India

Soniya Goyal
Department of Biosciences and Technology
Maharishi Markandeshwar (Deemed to be
University)
Mullana-Ambala 133207, Haryana
India
soni.goyal48@gmail.com

Praveen
Faculty of Agriculture
Guru Kashi University
Talwandi Sabo
Bathinda, Punjab
India

Charulata
ICAR-Indian Institute of Wheat and Barley
Research
Shimla, Himachal Pradesh
India

Bhupesh Gupta
Department of Computer Sciences
MMDU University
Ambala, Haryana
India

Poonam Bansal
Department of Biosciences and Technology
MMDU University
Ambala, Haryana
India

Mahiti Gupta
Department of Biosciences and Technology
Maharishi Markandeshwar (Deemed to be
University)
Mullana-Ambala 133207, Haryana
India

Sushil Kumar Upadhyay
Department of Bio-Sciences and Technology
MMEC
Maharishi Markandeshwar (Deemed to be
University)
Mullana-Ambala 133207, Haryana
India

Nafees Ahmed
Department of Microbiology
Mohanlal Sukhadia University
Udaipur 313001, Rajasthan
India

Harsh M. Prajapati
Department of Microbiology and Biotechnology
University School of Sciences
Gujarat University
Ahmedabad 380009, Gujarat
India

Mohit R. Chauhan
Department of Microbiology and Biotechnology
University School of Sciences
Gujarat University
Ahmedabad 380009, Gujarat
India

Vikram H. Raval
Department of Microbiology and Biotechnology
University School of Sciences
Gujarat University
Ahmedabad 380009, Gujarat
India

Kalidoss A.
Department of Microbiology
SRM Arts and Science College
Kattankulathur
Chengalpattu 603203, Tamil Nadu
India
kalidassgene@gmail.com

Susmitha B.
Department of Microbiology
St. Pious X Degree & PG College for Women
(Autonomous)
Nacharam
Hyderabad 500076, Telangana
India

Ashwitha Kodaparthi
Department of Microbiology
MNR PG College
Kukatpally
Hyderabad 500085, Telangana
India

Yogananth N.
Department of Biotechnology
Mohamed Sathak College of Arts and Science
Sholinganallur
Chennai 600119, Tamil Nadu
India

R. Nithyatharani
Department of Microbiology
Shrimati Indira Gandhi College
Tiruchirappalli, Tamil Nadu
India
nithyatharanir@gmail.com

R. Manikandan
Department of Biochemistry
Shrimati Indira Gandhi College
Tiruchirappalli, Tamil Nadu
India

M. Vinoth
Department of Microbiology
Government Arts and Science College
Perambalur, Tamil Nadu
India

R. Malathi
Department of Biochemistry
Enathi Rajappa Arts and Science College
Pattukkottai, Tamil Nadu
India

Sachin Kumar
Department of Bioinformatics
Janta Vedic College
Baraut, Baghpat, Uttar Pradesh
India
sachinsuryan@gmail.com

Ashwani Kumar
Department of Biotechnology
KVSCOS
Swami Vivekanand Subharti University
Meerut, Uttar Pradesh
India

Ashish Kumar Singh
Biomolecular Toxicology Lab
CSIR-Indian Institute of Toxicology Research
Lucknow 206001, Uttar Pradesh
India

P. Saranraj
PG and Research Department of Microbiology
Sacred Heart College (Autonomous)
Tirupattur, Tamil Nadu
India
microsaranraj@gmail.com

K. Gayathri
PG and Research Department of Microbiology
Sacred Heart College (Autonomous)
Tirupattur, Tamil Nadu
India

K. Kesavardhini
PG and Research Department of Microbiology
Sacred Heart College (Autonomous)
Tirupattur, Tamil Nadu
India

B. Lokeshwari
PG and Research Department of Microbiology
Sacred Heart College (Autonomous)
Tirupattur, Tamil Nadu
India

U. Subhalakshmi
PG and Research Department of Microbiology
Sacred Heart College (Autonomous)
Tirupattur, Tamil Nadu
India

L. Charlie Jelura
PG and Research Department of Microbiology
Sacred Heart College (Autonomous)
Tirupattur, Tamil Nadu
India

About the Editors

Dr. Mukesh Yadav: Currently, I am working as an Associate Professor at Maharishi Markandeshwar (Deemed to be University), Mullana-Ambala, India. I accomplished my B.Sc. and M.Sc. Degree in Biotechnology from M.D. University, Rohtak, India. I completed my Ph.D. degree in Microbial Biotechnology from Punjabi University, Patiala, India. I am working in the area of microbial biotechnology specifically emphasizing on fermentation technology, food biotechnology, industrial enzymes and microbial metabolites of industrial importance. I have published more than 100 articles in various reputed Journals. I have also edited and authored six books on Microbial Biotechnology related aspects (Industrial Biotechnology, Microbial Enzymes for Food Industries, Medical Microbiology, Therapeutic Potential of Flavonoids and Instrumentation). I am serving as member of editorial board of various International Journals. I wish to contribute to the development of science and technology in the developing world.

Dr. Nirmala Sehrawat: Currently, I am working as an Associate Professor at Maharishi Markandeshwar (Deemed to be University), Mullana-Ambala, India. I accomplished my M.Sc. and Ph.D. Degree in Biotechnology from M.D. University, Rohtak, India. I am working in the area of functional foods, probiotics, legumes production, plant metabolites of industrial importance and plant biotechnology. I have published more than 120 articles and book chapters. I have also edited and authored books in my profile (Industrial Biotechnology, Microbial Enzymes for Food Industries, Therapeutic Potential of Flavonoids and Instrumentation). I am working as editor and reviewer for various International Journals. I regularly connect with scientific peoples through various professional bodies and societies related to biotechnology.

https://doi.org/10.1515/9783112205150-205

Mukesh Yadav, Sunil Kumar, Sachin Kumar, Manoj Singh,
Nirmala Sehrawat*

Chapter 1
Gut Microbiome, Probiotics, and Gut Health: Concept, Recent Developments, and Future Insights

Abstract: Gut microbiota constitutes a complex diversity of microbes involved in regulation of host metabolism and gut-associated axis to maintain good host health. An imbalance of pathogenic and commensal bacteria, as well as the generation of microbial metabolites and antigens, causes dysbiosis, which in turn causes serious disorders in the host. Active microbes known as probiotics have a variety of uses, including enhancing host immunity and reestablishing the gut microbiota's makeup and overall health by preventing the development of gut-related disease or disorders. The control of gut flora by probiotics to enhance host immunity has garnered a lot of attention lately. In addition to recent developments and challenges that aid in the future treatment of gut-related diseases, this succinct review gives a general overview of the significance of gut microbiota in human health and the potential of probiotics in restoring gut microbiota and managing diseases.

Keywords: gut microbiota, dysbiosis, probiotics, gut homeostasis, disease management

Acknowledgment: The authors are thankful to the head of the Department of Bio-Sciences and Technology, MMEC, Maharishi Markandeshwar (Deemed to be University), Mullana-Ambala (Haryana), India, for providing necessary support and facilities.

*Corresponding author: Dr Nirmala Sehrawat, Department of Bio-Sciences and Technology, MMEC, Maharishi Markandeshwar (Deemed to be University), Mullana-Ambala 133207, Haryana, India,
e-mail: nirmalasehrawat@gmail.com
Mukesh Yadav, Manoj Singh, Department of Bio-Sciences and Technology, MMEC, Maharishi Markandeshwar (Deemed to be University), Mullana-Ambala, 133207 Haryana, India
Sunil Kumar, Department of Microbiology, Graphic Era University, Dehradun, Uttarakhand, India
Sachin Kumar, Department of Bioinformatics, Janta Vedic College, Baraut, Baghpat, Uttar Pradesh, India

https://doi.org/10.1515/9783112205150-001

1.1 Gut Microbiota and Its Importance

The gut microbiota is made up of around 100 trillion microbial cells that control host metabolism and general health [1]. Humans and gut bacteria have a mutualistic or commensal relationship. The five main phyla of bacteria that predominate in the human gut are Verrucomicrobia, Actinobacteria, Proteobacteria, Bacteroidetes, and Firmicutes. The microbiome population in the colon is the largest, whereas the stomach and small intestine only contain a few species of bacteria. The microbial makeup of the various parts of the gastrointestinal (GI) system varies (Table 1.1). From the stomach to the jejunum to the colon, the commensal microbiota load in the human GI system progressively rises, as indicated by colony-forming units (cfu)/mL [2]. The human gut microbiota has the ability to restrict the colonization of pathogenic microbes through various mechanisms, and the healthy gut microbiome is known to increase the immunity of host [3]. The gut microbiota of a neonate is generally acquired immediately after birth [3–5]. Various factors influence the initial development of an individual's microbiota. During the first 3 years of life, the gut microbiota undergoes a significant transformation toward an adult-like composition. Gut dysbiosis develops in early infancy but may persist until maturity. Consequently, it is crucial to create a beneficial gut microbiota throughout infancy [6, 7]. Significant effects of a variety of factors on children's gut microbiota establishment were examined in studies. Maternal microbiota from the bowel and vagina, delivery method, feeding style, use of antibiotics and other medications, gestational age, siblings, pets, and regional variations, such as dietary practices and hygienic conditions, are some of these determinants [8–10].

In equilibrium, the microbiome has a warm and nutrient-rich environment that favors various biochemical activities. Bacterial diversity and richness in the GI microbiota are crucial for regulating the body's natural immunological and cellular metabolic processes, especially through the production of different metabolites and bioactive compounds that are involved in different metabolic pathways and the enzyme pathways that break down complex foods [11, 12]. The breakdown of complex carbohydrates generated from plants and the generation of short-chain fatty acids (SCFAs), such as acetate, butyrate, and propionate, are carried out by beneficial microorganisms in the GI tract. These microbes also regulate the diversity, growth, and composition of gut microbes [13]. Additionally, the gut microbiota supplies the branched-chain amino acids needed for the production of glutathione, a crucial intracellular antioxidant and detoxifying agent that is essential for numerous host biological processes. Gut microbiota play significant role in intestinal development, homeostasis, strengthening gut barrier, and protection against pathogenic bacteria. These microbes also regulate cellular processes, physiological functions, and different gut-associated axes in the host. These regulations control the onset and progression of gut-associated diseases or disorders and other chronic health issues [14, 15].

Table 1.1: Microbiota load and diversity present in gastrointestinal (GI) tract.

GI tract	Microbiota load (cfu/mL)	Microbial species
Stomach	10^1–10^3	*Lactobacillus*, Enterobacteriaceae, *Streptococcus*, *Staphylococcus*, yeast
Small intestine		
– Duodenum	10^1–10^3	*Lactobacillus*, Enterobacteriaceae, *Streptococcus*, *Staphylococcus*, yeast
– Jejunum/ileum	10^4–10^7	*Bacteroides*, Fusobacteriota, *Bifidobacterium*, *Lactobacillus*, *Streptococcus*, Enterobacteriaceae
– Cecum, colon, rectum, and anus	10^{10}–10^{12}	Yeast, Protozoa, *Bacteroides*, Clostridia, Eubacteria, *Bifidobacterium*, *Lactobacillus*, *Streptococcus*, *Peptostreptococcus*, *Veillonella*, Enterobacteriaceae, Fusobacteria, *Proteus*, *Pseudomonas*

Gut bacteria and the immune system variably interact with each other via different metabolic and signaling pathways helping in maintaining the overall metabolism [3, 16]. Dietary habits, lifestyle choices, antibiotic use, food scarcity or excess, environmental pollution, stress, and dairy products (sugars, coffee, tea, and alcohol), all have an impact on the stability and activity of the gut microbiota. These factors also have a significant impact on the intestinal mucous membrane [3], gut barrier, and therefore, immune system. The lifestyle, food, and dietary factors not only affect the gut microbiome, intestinal barrier, and gut mucosa but also regulate the homeostasis.

1.2 Gut Dysbiosis and Host Health

Dysbiosis of the gut microbiota is caused by an imbalance in the composition and activities of the gut microbiota. The host phenotype may change as a result of a metabolic shift brought on by changes in its composition. Other organ malfunctions or diseases have also been intimately linked to an imbalance of the gut microbial communities and their metabolites. This results in development of metabolic and intestinal diseases or disorders, neurodegenerative diseases, and associated chronic inflammation-related diseases [13, 17]. Eubiosis, or the balance of the physiological microbiota community, in the gut is essential for maintaining a person's health. Dysbiosis can have a detrimental effect on organ function and can be caused by a number of factors, including a poor diet, prolonged fasting, alcohol use, smoking, physical or psychological stress, chronic inflammation, and abuse of antibiotics. Imbalances and anomalies in the gut microbiota can cause a wide range of diseases that affect every organ. The possible interaction between metabolic disorders and host-gut microbiome dysbiosis was revised by Belizário et al. [11]. Probiotics, prebiotics, synbiotics, nutrients, fecal treatment, and small-molecule inhibitors of metabolic pathway enzymes, which act as preventive and therapeutic measures for metabolic disorders, were also compiled by Belizário et al. [11]. Nowadays, gut dysbiosis is a common issue since intestine microbial communities are changeable and can be influenced by a number of factors, including the host's lifestyle, food habits, bacterial infections, and usage of antibiotics [14, 18].

1.3 Probiotics

Probiotics are the microbial strains (generally including viable bacteria or yeast), which provides various health benefits to host when administered appropriately [19, 20]. Probiotics have been found to play an important role in human gut health and therefore helps in strengthening the immunity through balanced gut microbiome [3, 21]. Probiotics in intact living form, non-viable form (dead probiotic cells), and also their metabolites have been reported for various biological activities [21–23]. Selection of probiotic culture is very important and crucial. The safety as well as functionality of probiotic cultures are assessed on the basis of various parameters.

Bacterial probiotic strains have been reported and studied extensively. Various bacterial strains have been reported for their probiotic attributes. Among these bacterial probiotics, strains of *Lactobacillus*, *Lactococcus*, *Bifidobacterium*, and *Enterococcus* are considered as common bacterial probiotics [3, 21–24]. Several yeasts have also been investigated for their probiotic potential [3, 21]. Among these, *Saccharomyces cerevisiae* var. *boulardii* has been studied most extensively and also found most promising [25–27]. Among bacteria, the most extensively studied and characterized probiotics belongs to the lactic acid bacteria group. Bacteria of the genera *Bifidobacterium* and *Faecalibacterium* are also studied in an elaborative way [28, 29]. Several bacteria have been used for a long time in several industrial processes for the production of fermented foods, such as cheese and yogurt, and they usually have probiotic properties [30].

1.4 Prebiotics

Prebiotics are nondigestible food substances that have the ability to be utilized by specific host microbes and therefore exert considerable health benefits [3, 21, 31–33]. According to Bindels et al. [34], the prebiotics refer to "selectively fermented nondigestible food ingredients or substances that specifically support the growth and/or activity of health-promoting bacteria that colonize the gastrointestinal tract." Appropriate combination of prebiotic and probiotics as a single product ensures greater effects than the individual effect of prebiotic or probiotic as therapeutics. In the present era, the gut microbiota is focused as a promising therapeutic target to treat several disorders or diseases [3, 21, 33]. The prebiotics emerged as an important tool along with probiotics to improve and maintain healthy gut microbiome, which is helpful in prevention as well as management of various diseases/disorders [3, 21, 33].

1.5 Role of Probiotics in Reestablishing Gut Microbiota Balance

Probiotics are "live microorganisms which when administered in adequate amounts confer a health benefit on the host" [35, 36]. Probiotics constantly maintain the gut homeostasis of internal microbiota by decreasing the harmful microbes and increase in beneficial microbes in the gut to maintain good human health and immunity [37]. By competing with pathogenic bacteria for resources and space, beneficial bacteria in the digestive system prevent their invasion and proliferation. Harmful microbes are unable to survive in acidic environment of gut whereas the beneficial ones grow well and proliferate to prevent gut dysbiosis [3, 21, 38]. Probiotics are capable to effectively modulate gut microbiota composition that seems beneficial to treat various organ-specific diseases via regulation of inter-organic axes. Probiotic consumption has been proposed to improve human immunity and general health. Furthermore, it has been demonstrated that taking the right probiotics can help control and improve the prognosis of illnesses linked to metabolism [2, 14, 39].

1.6 Gut Dysbiosis Regulation and Disease Management

Throughout the human GI system, a symbiotic consortium of bacteria, viruses, fungus, and host eukaryotic cells is necessary to maintain a healthy human metabolism. After receiving antibiotic treatment, probiotics are one of the most important ways to help patients to regain the balance of gut microbes, which strengthens the intestinal barrier, lowers inflammation, fends off infections, and resists pathogen attack [2]. Probiotics prevent infections or harmful microorganisms by promoting the function of the epithelial barrier, releasing antimicrobial substances, competitively rejecting pathogens through binding sites, and restricting their access to nutrients. Colonization resistance, organic acids, SCFA production, competitive exclusion of pathogens, normalization of altered microbiota abundance, intestinal transit regulation, direct antagonism, gut barrier reinforcement, and carcinogen neutralization are all considered to be the contributing factors to the positive effects of probiotics. Promising impacts of probiotics on gut microbiota and host health are presented in Figure 1.1. Many metabolic, intestinal, and cardiovascular disorders are caused by gut dysbiosis, and probiotics have demonstrated encouraging results as supplements or adjunct therapies in reducing the symptoms of these illnesses [40]. Previous articles outlined the latest research on probiotics in humans and animals, as well as their positive impacts on gut microbiota and host immunity. They also emphasized the probiotics' evidence-based health-promoting properties [2, 41, 42].

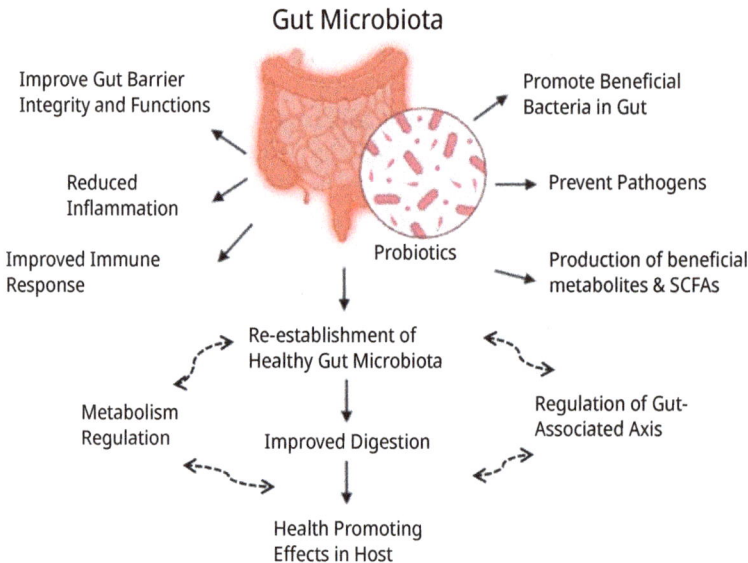

Figure 1.1: Beneficial effects of probiotics in maintaining gut microbiota and host health (Source: www.biorender.com).

Furthermore, a number of omics technologies, such as transcriptomics, metagenomics, metabolomics, and microbial genomics, have increased our understanding of the identification and selection of novel probiotic strains, the validation of their genuine probiotic effects, including molecular mechanisms, and the effects of probiotics on health in experimental studies. However, the use of proper amounts of probiotics for specific patients is required for therapeutic purpose [43, 44]. The host's gender, race, and genetic variations all affect how well probiotic supplements work. Because there are numerous variables that could affect a particular "probiotic's" biological activity, prior research shows that designating a probiotic can be difficult and complex. The ways that probiotics are delivered have drawn increased interest recently, i.e., encapsulation of probiotics with prebiotics, nanoparticles, and artificial enzymes. These strategies enhance the efficacy of probiotic-mediated modulatory effects, anti-oxidative effects, anti-inflammatory effect, and immunomodulatory effects to establish gut microbiota balance [45–48]. Beside probiotics, the prebiotics or synbiotic formulations (probiotic and suitable prebiotic) have also been reported to affect gut microbiota more significantly to manage human illnesses as compared to the probiotic alone [3, 21, 33]. In addition, the gut microbiota may be successfully analyzed through the application of artificial intelligence and systems biology techniques, leading to a better comprehension of the role of the microbial community and its metabolites in the etiology of specific illnesses. According to the current research, the newest technologies – single-cell omics, isotope tracking, and CRISPR/Cas – will likely reveal novel host delivery methods and probiotics' immunomodulatory effects [2, 18, 49].

1.7 Conclusions and Future Aspects

Gut microbiota is a potential target in current scenario to treat or manage human illnesses. Dysbiosis is a major concern to evoke severe health problem due to alteration in different gut-associated axes, which collectively affects host health. Probiotics have immense potential in reshaping host-microbiota interactions and modulatory effects to restore the gut microbiota to treat human illnesses. Probiotics are essential for controlling the gut microbiota, improving immunity in general, growing the number of good bacteria, and reducing the symptoms of a number of illnesses. Earlier studies demonstrated the positive impacts of probiotics in establishing gut microbiota balance in several clinical trials but deeper understanding and exploration of the mechanisms involved is still required.

Conflicts of Interest: The authors do not declare any conflicts of interest.

References

[1]   Ley RE, Peterson DA, Gordon JI. Ecological and evolutionary forces shaping microbial diversity in the human intestine. Cell. 2006;124:837–848.
[2]   Chandrasekaran P, Weiskirchen S, Weiskirchen R. Effects of probiotics on gut microbiota: An overview. International Journal of Molecular Sciences. 2024;25:6022.
[3]   Sehrawat N, Yadav M, Singh M, Kumar V, Sharma VR, Sharma AK. Probiotics in microbiome ecological balance providing a therapeutic window against cancer. Seminars in Cancer Biology. 2021;70:24–36.
[4]   Palmer C, Bik EM, DiGiulio DB, Relman DA, Brown PO. Development of the human infant intestinal microbiota. PLoS Biology. 2007 Jul;5(7):e177.
[5]   Vael C, Desager K. The importance of the development of the intestinal microbiota in infancy. Current Opinion in Pediatrics. 2009;21(6):794–800.
[6]   Sender R, Fuchs S, Milo R. Revised estimates for the number of human and bacteria cells in the body. PLoS Biology. 2016;14:e1002533.

[7] Webb CR, Koboziev I, Furr KL, Grisham MB. Protective and pro-inflammatory roles of intestinal bacteria. Pathophysiol. 2016;23:67–80.

[8] Thaiss CA, Zmora N, Levy M, Elinav E. The microbiome and innate immunity. Nature. 2016;535:65–74.

[9] Akagawa S, Tsuji S, Onuma C, Akagawa Y, Yamaguchi T, Yamagishi M, Yamanouchi S, Kimata T, Sekiya SI, Ohashi A, Hashiyada M, Akane A, Kaneko K. Effect of delivery mode and nutrition on gut microbiota in neonates. Annals of Nutrition and Metabolism. 2019;74:132–139.

[10] Akagawa S, Akagawa Y, Yamanouchi S, Kimata T, Tsuji S, Kaneko K. Development of the gut microbiota and dysbiosis in children. Bioscience of Microbiota, Food and Health. 2021;40:12–18.

[11] Belizário JE, Faintuch J, Garay-Malpartida M. Gut microbiome dysbiosis and immunometabolism: New frontiers for treatment of metabolic diseases. Mediators of Inflammation. 2018;2037838.

[12] Rowland I, Gibson G, Heinken A, Scott K, Swann J, Thiele I, Tuohy K. Gut Microbiota functions: Metabolism of nutrients and other food components. European Journal of Nutrition. 2018;57:1–24.

[13] Sharma VR, Singh M, Kumar V, Yadav M, Sehrawat N, Sharma DK, Sharma AK. Microbiome dysbiosis in cancer: Exploring therapeutic strategies to counter the disease. Seminars in Cancer Biology. 2021a;70:61–70.

[14] Kim SK, Guevarra RB, Kim YT, Kwon J, Kim H, Cho JH, Kim HB, Lee JH. Role of probiotics in human gut microbiome-associated diseases. Journal of Microbiology and Biotechnology. 2019;29:1335–1340.

[15] Sharma V, Sharma N, Sheikh I, Kumar V, Sehrawat N, Yadav M, Ram G, Sankhyan A, Sharma AK. Probiotics and prebiotics having broad spectrum anticancer therapeutic potential: Recent trends and future perspectives. Current Pharmacology Reports. 2021b;7:67–79.

[16] Cani PD, Van Hul M, Lefort C, Depommier C, Rastelli M, Everard A. Microbial regulation of organismal energy homeostasis. Nature Metabolism. 2019;1(1):34–46.

[17] Carding S, Verbeke K, Vipond DT, Corfe BM, Owen LJ. Dysbiosis of the gut microbiota in disease. Microbial Ecology in Health and Disease. 2015;26:26191.

[18] Wang Z, Li L, Wang S, Wei J, Qu L, Pan L, Xu K. The role of the gut microbiota and probiotics associated with microbial metabolisms in cancer prevention and therapy. Frontiers in Pharmacology. 2022;13:1025860.

[19] FAO. Guidelines for the Evaluation of Probiotics in Food. Report of a Joint FAO/WHO Working Group on Drafting Guidelines for the Evaluation of Probiotics in Food. 2002, London: *FAO*.

[20] WHO-FAO. World Health Organization-Food and Agricultural Organization. Probiotics in Food: Health and Nutritional Properties and Guidelines for Evaluation FAO Food and Nutritional Paper. 2006, Rome: *FAO/WHO*.

[21] Sehrawat N, Yadav M, Kumar S, Lata S, Kumar A, Kumar A, Singh M, Kumar V. Therapeutic potential, health benefits and mechanism of action of probiotics: A concise review. Bulletin of Environment, Pharmacology and Life Sciences. 2022;1(Special Issue):123–129.

[22] Adams CA. The probiotic paradox: Live and dead cells are biological response modifiers. Nutrition Research Reviews. 2010;23(1):37–46.

[23] Bedada TL, Feto TK, Awoke KS, Garedew AD, Yifat FT, Birri DJ. Probiotics for cancer alternative prevention and treatment. Biomedicine & Pharmacotherapy. 2020;129:110409.

[24] Georgiev K, Georgieva M, Iliev I, Peneva M, Alexandrov G. Antiproliferative effect of Bulgarian spring water probiotics (Lakter a Nature Probiotics®), against human colon carcinoma cell line. World Journal of Pharmacy and Pharmaceutical Sciences. 2015;4(6):130–136.

[25] Smith IM, Baker A, Arneborg N, Jespersen L. Non-Saccharomyces yeasts protect against epithelial cell barrier disruption induced by *Salmonella enterica* subsp. *enterica serovar Typhimurium*. Letters in Applied Microbiology. 2015 Nov 1;61(5):491–497.

[26] Ricci A, Allende A, Bolton D, Chemaly M, Davies R, Girones R, Koutsoumanis K, Herman L, Lindqvist R, Nørrung B. Update of the list of QPS-recommended biological agents intentionally added to food

or feed as notified to EFSA 5: Suitability of taxonomic units notified to EFSA until September 2016. EFSA Panel on Biological Hazards (BIOHAZ). EFSA Journal. 2017;15(3):e04663.

[27] Saber A, Alipour B, Faghfoori Z, Yari Khosroushahi A. Cellular and molecular effects of yeast probiotics on cancer. Critical Reviews in Microbiology. 2017;43(1):96–115.

[28] Pot B, Foligné B, Daniel C, Grangette C. Understanding immunomodulatory effects of probiotics. Nestlé Nutrition Institute Workshop Series. 2013;77:75–90. https://doi.org/10.1159/000351388.

[29] Chang CW, Liu CY, Lee HC, Huang YH, Li LH, Chiau JS, Wang TE, Chu CH, Shih SC, Tsai TH, Chen YJ. *Lactobacillus casei* variety rhamnosus probiotic preventively attenuates 5-fluorouracil/oxaliplatin-induced intestinal injury in a syngeneic colorectal cancer model. Frontiers in Microbiology. 2018;9:983.

[30] Plavec TV, Berlec A. Engineering of lactic acid bacteria for delivery of therapeutic proteins and peptides. Applied Microbiology and Biotechnology. 2019;103(5):2053–2066.

[31] Gibson GR, Roberfroid MB. Dietary modulation of the human colonic microbiota: Introducing the concept of prebiotics. Journal of Nutrition. 1995;125:1401–1412.

[32] Gibson GR, Hutkins R, Sanders ME, Prescott SL, Reimer RA, Salminen SJ, Scott K, Stanton C, Swanson KS, Cani PD, Verbeke K, Reid G. Expert consensus document: The International Scientific Association for Probiotics and Prebiotics (ISAPP) consensus statement on the definition and scope of prebiotics. Nature Reviews Gastroenterology & Hepatology. 2017;14:491–502.

[33] Yadav M, Sehrawat N, Sharma AK, Kumar S, Singh R, Kumar A, Kumar A. Synbiotics as potent functional food: Recent updates on therapeutic potential and mechanistic insight. Journal of Food Science and Technology. 2024;61:1–5.

[34] Bindels LB, Delzenne NM, Cani PD, Walter J. Towards a more comprehensive concept for prebiotics. Nature Reviews Gastroenterology & Hepatology. 2015;12(5):303–310.

[35] Singh R, Kumar M, Mittal A, Mehta PK. Microbial metabolites in nutrition, healthcare and agriculture. Biotech. 2017;7:15.

[36] Cremon C, Barbaro MR, Ventura M, Barbara G. Pre- and probiotic overview. Current Opinion in Pharmacology. 2018;43:87–92.

[37] Marteau P, Seksik P, Jian R. Probiotics and intestinal health effects: A clinical perspective. British Journal of Nutrition. 2002;88:S51–S57.

[38] Wang W, Xing W, Wei S, Gao Q, Wei X, Shi L, Kong Y, Su Z. Semi-rational screening of the probiotics from the fecal flora of healthy adults against DSS-induced colitis mice by enhancing anti-inflammatory activity and modulating the gut microbiota. Journal of Microbiology and Biotechnology. 2018;29:1478–1487.

[39] Agus A, Clément K, Sokol H. Gut microbiota-derived metabolites as central regulators in metabolic disorders. Gut. 2021;70:1174–1182.

[40] Hou K, Wu ZX, Chen XY, Wang JQ, Zhang D, Xiao C, Zhu D, Koya JB, Wei L, Li J, Chen ZS. Microbiota in health and diseases. Signal Transduction and Targeted Therapy. 2022;7:135.

[41] Wang X, Zhang P, Zhang X. Probiotics regulate gut microbiota: An effective method to improve immunity. Molecules. 2021;26:6076.

[42] Cao F, Jin L, Gao Y, Ding Y, Wen H, Qian Z, Zhang C, Hong L, Yang H, Zhang J, Tong Z. Artificial-enzymes-armed Bifidobacterium longum probiotics for alleviating intestinal inflammation and microbiota dysbiosis. Nature Nanotechnology. 2023;18:617–7.

[43] Varela-Trinidad GU, Domínguez-Díaz C, Solórzano-Castanedo K, Íñiguez-Gutiérrez L, Hernández-Flores TD, Fafutis-Morris M. Probiotics: Protecting our health from the gut. Microorganisms. 2022;10:1428.

[44] Gebrayel P, Nicco C, Al Khodor S, Bilinski J, Caselli E, Comelli EM, Egert M, Giaroni C, Karpinski TM, Loniewski I, Mulak A. Microbiota medicine: Towards clinical revolution. Journal of Translational Medicine. 2022;20:111.

[45] Zheng DW, Li RQ, An JX, Xie TQ, Han ZY, Xu R, Fang Y, Zhang XZ. Prebiotics-encapsulated probiotic spores regulate gut microbiota and suppress colon cancer. Advanced Materials. 2020;32:2004529.

[46] Kim CS, Jung MH, Choi EY, Shin DM. Probiotic supplementation has sex-dependent effects on immune responses in association with the gut microbiota in community-dwelling older adults: A randomized, double-blind, placebo-controlled, multicenter trial. Nutrition Research and Practice. 2023;17:883–898.

[47] Li X, Lin Y, Chen Y, Sui H, Chen J, Li J, Zhang G, Yan Y. The effects of race and probiotic supplementation on the intestinal microbiota of 10-km open-water swimmers. Heliyon. 2023;9: e22735.

[48] Fu J, Liu X, Cui Z, Zheng Y, Jiang H, Zhang Y, Li Z, Liang Y, Zhu S, Chu PK, Yeung KW. Probiotic-based nanoparticles for targeted microbiota modulation and immune restoration in bacterial pneumonia. National Science Review. 2023;10:nwac221.

[49] Kumar R, Sood U, Gupta V, Singh M, Scaria J, Lal R. Recent advancements in the development of modern probiotics for restoring human gut microbiome dysbiosis. Indian Journal of Microbiology. 2020;60:12–25.

Nijendra Pratap Singh, Shivam Singh, Ravindra Kumar Jain,
Mukesh Yadav, Amit Kumar*

Chapter 2
Comprehensive Insights into the Provoking Agents, Biological Ramifications, and Clinical Responses to Microbiome Dysbiosis

Abstract: Microbiome dysbiosis means an imbalance of microbes in the body, especially in the gut, skin, and other mucosal areas. This imbalance is linked to many diseases such as metabolic, heart, brain, immune, lung, and reproductive disorders. It is also associated with viral infections like COVID-19. This chapter explains how the microbiome is formed and how factors like environment and body changes affect it. It describes how dysbiosis can harm health by weakening immunity, causing inflammation, and damaging body barriers. Conditions such as obesity, diabetes, high blood pressure, gut inflammation, Alzheimer's disease, and infertility are discussed in detail. The chapter highlights the changes in specific microbes in these conditions. It also covers ways to bring back a healthy balance of microbes. These include the use of probiotics, prebiotics, synbiotics, fecal transplants, diet changes, and microbial products. New methods like engineered probiotics and probiotic-material combinations are also explained. This work stresses the need for a personalized, microbiome-focused approach to understand the causes, effects, and treatments of disease.

Keywords: microbiome dysbiosis, diseases, immunity, treatments

2.1 Introduction

Microbiota is a community of microorganisms themselves [1, 2]. A person with a weight of 70 kg has approximately 3.8×10^{13} bacteria. The number of bacteria is higher than the total number of cells in healthy individuals [3]. Disruption of the mi-

---

*Corresponding author: Amit Kumar, Department of Biotechnology, Sharda School of Bio-Sciences and Technology, Sharda University, Greater Noida, Uttar Pradesh, India,
e-mail: amit.kumar23@sharda.ac.in; baliyaniitr@gmail.com
Nijendra Pratap Singh, Shivam Singh, Department of Biotechnology, Sharda School of Bio-Sciences and Technology, Sharda University, Greater Noida, Uttar Pradesh, India
Ravindra Kumar Jain, Department of Biotechnology, Swami Vivekanad Subharti University, Meerut, Uttar Pradesh, India
Mukesh Yadav, Department of Bio-Sciences and Technology, MMEC, Maharishi Markandeshwar (Deemed to be University), Mullana-Ambala 133207, Haryana, India

https://doi.org/10.1515/9783112205150-002

crobiome that deviates from a normal state is known as microbial dysbiosis [4]. Microbial dysbiosis is characterized by alterations in the composition or functions of the microbiome.

Microbial dysbiosis is mainly divided into two types: (a) gut microbial dysbiosis and (b) non-gut microbial dysbiosis.

(a) Gut Microbial Dysbiosis

Gut microbial dysbiosis is an imbalance marked by increased pathogens, reduced beneficial microbes, and decreased microbial diversity, which disrupts gut function and environment [5]. Gut microbial dysbiosis plays important roles in systemic lupus erythematosus. Geographic location and dietary habits influence the microbial composition of a population [6].

Dysbiosis of intestinal microorganisms causes many systemic diseases. It also increases the likelihood of developing periodontitis [7].

(b) Non-gut Microbial Dysbiosis

Oral Dysbiosis: Oral microbiome dysbiosis is an imbalance in diversity and functioning of the oral microbial flora [8]. It can contribute to gum disease, tooth decay, and other oral health problems. Dental and oral diseases often arise due to imbalances in the oral microbiota. In recent research, it has been reported that oral microbiome dysbiosis has relevant effects on systemic health. Dysbiosis of the oral microbiomes may cause neurodegenerative, cardiovascular, and other human diseases [9].

Skin Dysbiosis: Due to an imbalance in the community of microbes that live on our skin, its intricate balance is disrupted. The skin microbiome is dominated by genus *Malassezia* and few other fungi under steady-state conditions. It coupled with skin diseases like atopic dermatitis [9–11].

2.2 Lung Dysbiosis

Imbalances in the lung microbial community are known to be an important factor in various respiratory diseases. It is associated with respiratory diseases such as asthma, chronic obstructive pulmonary disease (COPD), and lung cancer [12, 13]. Lung microbiota dysbiosis is essential for silica-induced pulmonary fibrosis [14]. Microbiota in the gut and lungs can influence both the inception and progression of asthma [15].

Vaginal Dysbiosis: An imbalance of the ovarian microflora is a major factor in a variety of health problems. These include bacterial vaginosis, vaginal candidiasis, and sexually transmitted infections. Cervical biopsies also show a loss of *Lactobacillus* dominance and an increase in microbial diversity, as well as increased gut permeability [16, 17]. It may be associated with yeast infections and bacterial vaginosis.

Microbial dysbiosis can disrupt gut barrier function, promote systemic inflammation, and affect distant organs through gut-skin axis, gut-brain axis (GBA), and gut-liver axis. Microbial dysbiosis disrupt intestinal barrier function, resulting in intestinal barrier dysfunction [18]. Pathology caused by *Faecalibacterium prausnitzii* is frequently seen in patients suffering from inflammatory bowel disease (IBD) [19–21]. Increased infection with opportunistic pathogens, such as *Clostridium difficile, Escherichia coli,* and *Enterococcus faecalis,* is associated with intestinal inflammation and infection. Similarly, overrepresentation of probiotics and Firmicutes is associated with bacteremia and metabolic disorders [22, 23]. Apart from this, there are also ovarian dysfunction, type 2 diabetes, heart disease, autism spectrum disorder, and even neurodegenerative diseases like Parkinson's and Alzheimer's [24–26].

Recent therapeutic measures aim to restore the microbiota balance using probiotics such as *Lactobacillus rhamnosus, Bifidobacterium longum,* dietary interventions, fecal microbiota transplantation (FMT), and prebiotics [27–29]. Understanding the role of specific microbial changes and their metabolism is essential to develop personalized treatments and maintain overall health.

2.3 Provoking Agents

COVID-19 infection causes changes in the makeup of the human gut microbiome, regardless of whether patients took medication or not. It has been reported that there is a correlation between gut microbiota composition, cytokine levels, and inflammation in COVID-19 patients [30, 31]. Gut bacteria with known immune-modulating potential are depleted in COVID-19 patients such as *F. prausnitzii, Eubacterium rectale,* and *Bifidobacterial* species [31, 32].

2.4 Ecological Drivers of Microbial Dysbiosis

Recent studies have indicated that elevated levels of host-derived respiratory electron acceptors represent a significant environmental factor contributing to intestinal dysbiosis by altering microbial community structure [33].

The production of reactive oxygen species (ROS) has increased due to air pollution. ROS are one of the main causes of damage from air pollution. The consequences of exposure to air pollution are epithelial barrier integrity. The effects of air pollution based on the pollutant composition, sources, exposure mode, and host health status [34, 35]. Oxidative stress reactions cause the production of ROS, inflammation, and cell death, which lead to microbial dysbiosis [34].

Air pollutant modifies the respiratory environment, pH, local oxidation, and local inflammation, resulting in oxidative stress [35]. Air pollutant alters interactions between microbial metabolites and it also activates the nuclear factor kappa-B pathway using pattern recognition receptors and pro-inflammatory cytokines [34]. Alveolar macrophages and T cells help fight bacteria, but too much activity can upset the immune system [34, 36].

2.5 Antibiotic-Induced Microbial Perturbation

Antibiotics change the microbiome, disrupt immune function, and making the body more prone to disease (Table 2.1). Reduced levels of *Lactobacillus* in both maternal vaginal and infant meconium microbiota have been linked to antibiotic exposure. Antibiotic-induced alterations in the microbiome negatively affect immune function, leading to a greater predisposition to diseases associated with these changes. Antibiotic exposure has been reported to reduce the abundance of *Lactobacillus* spp. in the vaginal meconium microbiome of mothers and newborns [37–39]. Exposure to antibiotics also plays a role in increasing the diversity of microbial composition and decreasing the diversity of microbial composition.

The use of antibiotics during pregnancy, especially antibiotics during labor, enriches the phyla Actinobacteria, Bacteroidetes, and Proteobacteria [40]. The genus *Lactobacillus* is more prevalent in newborns without exposure to antibiotics, in contrast [41, 42].

It has been reported that exposure to prophylactic antibiotics during pregnancy decreased the abundance of *Bifidobacteria* in normally born infants [43].

An imbalance is correlated with decreased levels of short-chain fatty acids (SCFAs), tryptophan, and increased levels of purines. It is also associated with food allergies [43, 45].

2.6 Diet-Induced Modulation of the Microbiome

Eating a high-calorie Western diet can lead to intestinal biopsy and intestinal inflammation [75]. Diet has a profound effect on the microbiome during colonization (Table 2.2). Breastfed infants had higher abundances of *Bifidobacteria* species, *Clostridium coccoides*, and *Lactobacillus* species [76].

It has been reported that when mice are switched to a "Western" diet after regularly consuming a low-fat diet rich in polysaccharides, the microbiota shifts toward more Firmicutes, including *Clostridium inoculum*, *Eubacterium dolichorum*, Bacteroidetes, and Myrtaceae. In mice, low-carbohydrate diets promote the growth of bacteria of the Bacteroidetes lineage and low-calorie diets promote the growth of *Clostridium coccoides*, *Lactobacillus* spp., and *Bifidobacteria* species. This inhibits the growth of

Table 2.1: Gut ecosystem perturbation due to antibiotics.

Antibiotics	Amplification of gut microbes or family	Microbial depletion	Associated effects	References
Amoxicillin	*Escherichia* *Enterobacter* Enterobacteriaceae *Parabacteroides distasonis*	*Bifidobacterium* *Bacteroides fragilis* *Clostridium* cluster XIVa *Roseburia* Rhodospirallales *Negativibacillus* *Akkermansia* *Harryflintia* Candidatus *Saccharimonas* Lachnospiraceae family Ruminococcaceae family	Reduced the amount of short-chain fatty acid	[46–49]
Clindamycin	*Bacteroides* *Clostridia* *Clostridioides difficile* *Lactobacillus*	*Bifidobacterium* *Citrobacter* *Enterobacter* *Enterococcus* *Klebsiella* *Lactobacillus*	Premature infants, reduced SCFA production (e.g., butyrate) leads to impaired epithelial barrier function and reduced permeability	[38, 49–53]
Vancomycin	Firmicutes *Pseudomonas* and *Klebsiella*	*Clostridium* Bacteroidetes Firmicutes	–	[54, 55]
Ciprofloxacin	*Bacteroides* *Enterococcus* *Roseburia, Faecalicatena* *Escherichia*	*Bifidobacterium* Enterobacteriaceae *Alistipes* *Faecalibacterium* *Oscillospira* *Ruminococcus* *E. coli* *Bifidobacterium*	Mutations in *gyrA/B* and *parC/E* concentrations, serotonin diminuta in the foetus	[38, 56–59]
Azithromycin	*Streptococcus*	*Bifidobacterium* *Actinobacteria* Veillonellaceae *Leptotrichia* *Fusobacterium* *Neisseria* *Haemophilus*	Enhances protein synthesis and immune response	[38, 60–62]

Table 2.1 (continued)

Antibiotics	Amplification of gut microbes or family	Microbial depletion	Associated effects	References
Cefprozil	*Lachnoclostridium bolteae* *Enterobacter cloacae*	*Bacteroides* enterotype	Decreased activities of CYP1A2, CYP2C19, and CYP3A Beta diversity increased and alpha diversity decreased	[63–65]
Ceftazidime	–	Enterobacteriaceae *Lactobacillus*	Short-chain fatty acids lower blood vessel leakiness and their levels in the colon are usually between 65 and 123 mM	[66, 67]
Ceftriaxone	*E. coli, Clostridium, Staphylococcus* sp. *Lactobacillus OTU*	*Lactobacillus murinus* and *Lactobacillus animalis*	Disseminated to the extraintestinal organs Disrupted colonization resistance and mucosal balance promote infection. Ultrastructural changes of kidneys Increases central nervous system protein expression of the glutamate transporters GLT-1	[68–70]
Levofloxacin	Firmicutes *Streptococcus, Haemophilus,* and *Neisseria*	Bacteroidetes *Staphylococcus* and *Bacillus*	Enhanced total length and supported embryonic growth Decreased gut microbiota diversity and richness	[71–73]
Meropenem	Enterococcaceae	*Roseburia, Lachnospiraceae, Ruminococcaceae*	Cytokine levels elevated Recapitulated in an immune-competent CAR-T mouse model	[73, 74]

Bifidobacteria, which are essential for butyric acid homeostasis throughout the colon [77–79]

Recent studies show that prebiotics and fiber-rich foods help increase *Faecalibacterium* and *Bacteroides*, while consuming more refined grains and gluten lowers microbial diversity. Intake of insoluble dietary fiber increases the relative abundance of beneficial gut bacteria and reduces harmful bacteria. Similarly, diets rich in animal fat can raise the levels of harmful bacteria like *Lachnoclostridium* and *Trichoderma* [80, 81]. Oat, psyllium, wheat, and apple contain high concentration of crude dietary fiber.

Glutamine supplementation can change the balance of gut bacteria in overweight people. A recent study found that glutamine supplementation reduced the *Bacillus/*Bacteroidetes ratio (0.85–0.57) [82, 83]. Higher incidence of the butyrate-producing *Anaerotruncus* bacterium, which is linked to obesity, was observed in individuals with higher intake of saturated fatty acids (SFAs) than the WHO recommendations, including those with overweight or metabolic disorders [84]. When women consume a lot of fatty acids, the number of bacteria increases and increasing their risk of developing inflammatory diseases associated with endometriosis [85, 86].

Table 2.2: Dietary patterns and modulation of intestinal microbiota.

Dietary intake	Influence on intestinal microbiota	
Bambusa vulgaris (fiber)	Boosts the growth of *Bifidobacterium adolescentis* and *Lactobacillus acidophilus*	[87]
Laminaria japonica (polysaccharide)	Improves the Bacillota/Bacteroidota ratio and boosting Bacteroidaceae and Tannerellaceae	[80]
Gracilaria fisheri	Promotes increased richness and diversity of *Roseburia* and *Faecalibacterium*	[88]
Moringa oleifera	Decreases the abundance of Pseudomonadota and boosts the growth of Bacteroidota	[89]
Linum usitatissimum (seed)	Negatively modulates the populations of Bacillota, Pseudomonadota, and Bacteroidota	[90]
Dictyopteris divaricata (crude polysaccharide)	Boosts the growth of Bacillota, Bacteroidota, and *Lactobacillus*	[91]
Fish ball, noodles, collagen egg, minced meat, and seafood-style meals	Stimulates the growth of *Roseburia* sp. CAG 182 *Faecalibacterium prausnitzii, Roseburia hominis, Prevotella* CAG 5226	[92]

2.7 E-Cigarette Usage Trends and Their Public Health Consequences

E-cigarette usage has significantly increased over the past decade, particularly among adolescents and young adults (Table 2.3). This trend is driven by perceptions of reduced harm compared to traditional cigarettes, flavored nicotine options, and aggressive marketing.

Table 2.3: E-cigarette consumption patterns and associated public health outcomes.

Organisms	Consequence	References
Gut microbiome	Abundance of *Prevotella* lowered *Bacteroides*	[93]
	Elevated levels of Erysipelotrichia-*Catenibacterium* lineage and Alphaproteobacteria	[94]
	Suppresses the abundance of the genera *Collinsena*, *Enterorhabdus*, and *Gordonibacter*	[35]
	Elevated levels of *Veillonella* and *Megasphaera* Suppresses the abundance of *Haemophilus*	[96]
Intestinal microbiome	Facilitates the proliferation of Firmicutes and Actinobacteria Diminishes the relative abundance of Bacteroidetes and Proteobacteria on phylum	[97]
Microbiota found in the lower airway in patients with COPD	Facilitates the proliferation of *Pseudomonas* Suppresses the abundance of Bacteroidetes (*Prevotella* spp.)	[98, 99]

Pathological conditions in the respiratory tract appear to be associated with decreased nasal microbial diversity. It is also implicated in the pathophysiology of many respiratory diseases.

2.8 Lung Dysbiosis

Lung dysbiosis, can be linked to a variety of respiratory conditions, including those affecting the ear, nose, and throat tract. Lung dysbiosis is associated with pulmonary fibrosis, asthma, idiopathic COPD, and idiopathic lung cancer through the gut-lung axis [100–102]. In addition, dysbiosis of the oral flora increases the infectious content of inflammatory microorganisms, allowing the entry of microorganisms into the respiratory tract and lungs.

2.9 Genitourinary Dysbiosis

Bacterial vaginosis involves an imbalance in the vaginal microbiota. It is marked by a decrease in *Lactobacillus* species and an overgrowth of bacteria like *Gardnerella vaginalis* [103, 104]. Dysbiosis of the female genital microflora can impact fertility through various mechanisms. These include altered vaginal pH, reduced sperm viability, motility, and inflammation that hinders endometrial receptivity [105, 106]. Comprehensive sexual health education serves as an essential lifestyle intervention. It empowers youth with the knowledge, skills, attitudes, and values needed to protect and manage their sexual and reproductive health.

2.10 Skin Dysbiosis

Skin dysbiosis refers to an imbalance in the skin's microbial flora, which can lead to conditions like acne, eczema, or dermatitis. Eye dysbiosis disrupts the ocular microbiota, increasing the risk of inflammation and eye infections.

The changes in the microbiota like *Staphylococcus epidermidis* cause opportunistic infections of symbiotic skin cells and lead to catheter infections, prosthetic valve endocarditis, and endophthalmitis [107–111]. It has been found that 12 different genera, such as *Pseudomonas, Methylobacterium, Propionibacterium, Corynebacterium, Acinetobacter,* staphylococci, *Aquabacterium, Brevundimonas, Sphingomonas, Streptococcus, Streptophyta,* and *Bradyrhizobium,* were forming the core of the conjunctival microbiota. Skin aging, disease, and injury can disrupt the skin barrier and cause microbial dysbiosis. This increases the risk of conditions such as acne, atopic dermatitis, psoriasis, and wound infections [112, 113]. Increased androgen levels during puberty can cause changes in skin cell activity, inflammation, etc. Acne can block hair follicles, leading to acne. *Cutibacterium acnes* is a long-term inflammatory condition of the hair follicles, which contain hair and sebum. It affects more than 85% of adolescents and young adults [114–116].

2.11 Atopic Dermatitis

Microbial changes associated with atopic dermatitis included *C. acnes, Corynebacterium, Dermatococcus,* many symbiotic bacteria, *Actinobacteria,* and phytopathogenic bacteria. Bacteroidetes accounted for the majority of the conjunctival microbiota [117–120].

2.12 Biological Ramifications

Patients with COVID-19 showed a reduction in gut bacteria with immunomodulatory and anti-inflammatory properties. Patients with COVID-19 have reduced levels of beneficial gut bacteria. These include *F. prausnitzii*, *E. rectale*, and several *Bifidobacterium* species [31, 32].

2.13 Gut Microbiota in Major Human Diseases

As a result, gut microbiome imbalance has been linked to a variety of human diseases, including anxiety, obesity, depression, IBD, hypertension, heart disease, diabetes, and cancer [121, 122].

2.14 Allergic Rhinitis

A long-term study followed gut microbiome changes from infancy to school age in children with immunoglobulin E (IgE)-related allergies like allergic rhinitis. It found increased levels of *Bifidobacteria* and decreased levels of Lachnospiraceae compared to children without IgE-related allergies [123–125].

2.15 Nonalcoholic Fatty Liver Diseases

An unbalanced diet can lead to an imbalance in the intestinal flora and increase intestinal permeability. This allows more substances to move from the intestine to the liver. As a result, it can lead to nonalcoholic fatty liver disease (NAFLD) and nonalcoholic steatohepatitis [75, 126].

2.16 Hypertension

It has been reported that the development of hypertension is coupled with dysbiosis of the intestinal flora in animals and humans. An increase in *Lactobacillus*, *Akkermansia*, *Coprococcus*, *Roseburia*, and *Bifidobacterium* is linked to lower blood pressure. In contrast, higher levels of *Blautia*, *Streptococcus*, and *Prevotella* are associated with higher blood pressure [127, 128].

2.17 Firmicutes/Bacteroidetes

An elevated Firmicutes/Bacteroidetes (F/B) ratio is a marker of gut microbial dysbiosis. Increased F/B ratios are observed in various animal models of hypertension, including spontaneously hypertensive rats, angiotensin II (Ang-II)-induced rats, pulmonary arterial hypertension rats, hypoxia-induced pulmonary hypertension rats, Dahl salt-sensitive rats, hypertensive heart failure rats, and *E. coli*-induced hypertensive monkeys [127–130].

People with high blood pressure show distinct gut microbiome features. They have more harmful bacteria like *Prevotella*, *Klebsiella*, and *Streptococcus*, and fewer beneficial SCFA-producing bacteria such as Bacteroidetes, *Roseburia*, and *Faecalibacterium* [131].

Hypertension was accompanying with higher abundance of Lachnospiraceae, *Ruminococcus*, and *Parasutterella* and lower abundance of *Lactobacter and Oscillibacter* [132, 133]. These microbial changes are associated with increased levels of SCFAs and Ang-II/trimethylamine-N-oxide (TMAO), which contribute to increased blood pressure. A fiber-rich diet supports gut health and has been associated with lower blood pressure by reducing microbes [134, 135].

2.18 Cardiovascular Diseases

Atherosclerosis is associated with an increase in *Lactobacillus* and a decrease in *Roseburia* and *Faecalibacterium* [136, 137]. These changes are allied with increased levels of TMAO, which is a cardiovascular risk factor [139–140].

Atrial fibrillation (AF) is associated with an imbalance of intestinal flora. These include increase in *Ruminococcus*, *Streptococcus*, and *Enterococcus* [141]. These changes correlate with increases in TMAO, choline, and SCFAs. The duration of persistent AF is influenced by these microbial changes. Adopting a Mediterranean diet (MD) helps restore gut health by increasing *Bifidobacteria*, improving gut health, and reducing inflammation associated with AF [142, 143].

Heart failure is associated with an increase in *E. coli*, *Shigella*, *Enterococcus*, and *Klebsiella*, as well as an increase in TMAO levels [144, 145]. Elevated TMAO is associated with left ventricular remodeling and poor outcomes [146, 147].

2.19 Diabetes Mellitus

A Western diet high in refined carbohydrates and fats but low in fiber is linked to type 2 diabetes, obesity, and other metabolic diseases [148, 149].

Gut microbiota help to digest nutrients, produce important compounds, support the immune system, and maintain gut health [150, 151]. *Lactobacillus, Subdoligranulum, Escherichia, Enterococcus, Shigella*, and *Fusobacteria* are linked to a higher risk of type 2 diabetes. In contrast, *Akkermansia, Roseburia, Facilibacterium, Bacteroides, Bifidobacterium*, and *Prevotella* are linked to a lower risk [150, 152]. Patients with type 2 diabetes mellitus (T2DM) have significantly lower diversity of gut bacteria and number of butyrate-producing bacteria compared to healthy individuals [153, 154].

Within the Firmicutes, *Clostridia* was found to be predominant, although this class is significantly less abundant in adult T2D patients, despite studies describing the contrary. In T2DM patients, there is a higher abundance of non-butyrate-producing and harmful Clostridiales. However, beneficial Clostridiaceae and Peptostreptococcaceae levels are reduced, and Betaproteobacteria levels are strongly increased [155].

2.20 Cancer

Helicobacter pylori infection is a major risk factor for gastric cancer. Certain strains of *E. coli* and *Bacteroides fragilis* are associated with colorectal cancer (CRC) by producing toxins that damage DNA and cause chronic inflammation [159, 160]. The gut microbiome includes *H. pylori, B. fragilis*, and *Fusobacterium nucleatum*, which are known to cause stomach cancer [158–160].

The oral microorganisms like *Porphyromonas gingivalis* and *Treponema denticola* may influence oral and systemic cancers [161, 162]. *Mycobacterium tuberculosis* and *Acidovorax temperans* are lung microbiomes that contribute to lung cancer [163, 164]. *S. aureus* and *Cutibacterium granulosum* on the skin are associated with inflammatory cancers [165]. Microorganisms present in the urinary tract, reproductive organs, and breast are *Lactobacillus, E. coli*, and *Staphylococcus*. These affect cancer risk through microbial imbalance and immunomodulation.

2.21 Gut Microbiota and Pregnancy Complications

Intestinal dysbiosis causes and contributes to various pregnancy complications in patients [165]. Gut microbiota metabolites are involved in several pathophysiological pathways that are closely linked to the pathogenesis of pregnancy complications [166, 167]. Metabolites of the gut microbiome are involved in several pathophysiological pathways associated with the pathogenesis of pregnancy complications [168]. Gut dysbiosis, i.e., changes in microbiome composition or metabolism, has adverse effects on the course of pregnancy and can lead to complications by disrupting maternal adaptation. In fact, pregnant women who experience pregnancy-related illnesses have a higher risk of developing gestational diabetes, which is a common condition [168, 169].

Bacteria like *Porphyromonas, Aggregatibacter*, streptococci, staphylococci, and *Candida* are more common in pregnant women. *Prevotella intermedia* is found in both pregnant and nonpregnant women. The high percentage of *F. nucleatum* and streptococci species parallels maternal intestinal dysbiosis through intestinal barrier disruption, inflammation, and hormonal changes in the offspring [170–172].

Inflammatory responses, cytokine release, and changes in oral bacteria may play a role in preterm birth and other pregnancy problems. These factors can also lead to low birth weight and preeclampsia. Good oral hygiene during pregnancy may reduce the transmission of cariogenic bacteria to the unborn child and reduce the risk of caries in future children.

2.22 Microbiota Dysbiosis and Alzheimer's Disease

The link between gut dysbiosis and neurological disorders is not fully clear. However, studies suggest that it may contribute to Alzheimer's disease by increasing amyloid-beta buildup, inflammation, oxidative stress, and insulin resistance [173].

2.23 Gut Microbiota and Female Infertility

The microbiota can influence fertility even before pregnancy. Infertile women showed evidence of gut dysbiosis, with increased levels of *Verrucomicrobia* and decreased levels of *Streptococcus, Stenotrophomonas*, and *Roseburia* compared with control subjects [106, 174, 175].

2.24 Gut Microbiota Regulates Steroid Hormones

Influence of gut microbiota on sex hormone production has been reported. *Prevotella* has been shown to be strongly associated with the hormone testosterone [176–178].

2.25 Gut Microbiota and Male Fertility Behavior

Alginate oligosaccharides are natural polymers derived from brown seaweed with desired effect-mediated modulation of gut microbiota that influences the sperm quality [179].

Oxidative stress significantly affects the development of male infertility. Increased levels of OS or DNA-damaged sperm increase the risk of infertility [180, 181]. One of the hallmarks of dysbiosis in the male genital microflora is a decrease in diversity and richness.

Dysbiosis in the male genital tract can increase harmful bacteria like *E. coli* and *Ureaplasma urealyticum*, which are linked to chronic prostatitis, inflammation, and reduced sperm quality and fertility [105, 182, 183].

2.26 Gut Microbiota and Female Fertility Behavior

It has been reported that the composition and function of gut microbiota affect women's reproductive health [184, 185]. Gut microbiota has been shown to regulate estrogen levels and found decreased estrogen levels after antimicrobial supplementation. β-Glucuronidase is an enzyme with the ability to deconjugate estrogen, binding it to its receptors with subsequent physiological effects. This protein is secreted by the intestinal microbiota [186, 187].

2.27 Antibiotic and Diarrhea

Antibiotic-associated diarrhea is caused by disrupted gut microbiota and the overgrowth of harmful or opportunistic bacteria after taking antibiotics. These organisms mostly include *Klebsiella pneumoniae*, *Staphylococcus aureus*, and *C. difficile* [187–191].

2.28 Ulcerative Colitis

Ulcerative colitis (UC) patients have been reported to have low levels of gut microbiota, due to low levels of beneficial bacteria such as *Bifidobacterium* and *Lactobacillus*. Particularly in patients with UC, it is believed that the intestinal flora, especially members of the Desulfovibrionaceae family, play an important role in the pathogenesis of the disease, including high levels of virulence factors [192, 193].

Bacteroidetes, *Bacillota*, *Clostridia*, *Eubacteriales*, Clostridiaceae, and *Clostridia* contribute significantly to the functions of the gut microbiota, but they are cytotoxic and persist in disease [193]. Shift in the microbiota affect mucosal integrity directly or indirectly through changes in the production of SCFAs and cytokines [193, 194].

Butyrate deficiency has a negative effect on intestinal balance. SCFs bind to G-protein-coupled receptors, whose signaling pathways have anti-inflammatory effects on IBD [45, 195]. Additionally, SCFAs promote regulatory T-cell differentiation by inhibiting histone deacetylase activity. It also promotes regulatory T-cells that produce protective cytokines [196, 197].

Cytokines act as signaling molecules of the immune system and it influences the onset and progression of IBD. Cytokinins associated with the pathophysiology of IBD are tumor necrosis factor-alpha (TNF-α), interferon-gamma (INF-γ), interleukin (IL)-1, IL-6, IL-4, IL-5, IL-10, transforming growth factor-beta (TGF-β), IL-13, IL-12, IL-18, and IL-23 [198, 199].

Rheumatoid arthritis is accompanying with dysbiosis of the gut and oral microbiota. Dysbiosis alters microbial metabolites like SCFAs, lipopolysaccharides, and peptidoglycans, which stimulate immune responses via TLR2/4 signaling. These pathways lead to Th17 cell activation and elevated pro-inflammatory cytokines (e.g., TNF-α and IL-6), resulting in immune dysregulation and chronic inflammation.

Microbial dysbiosis alters metabolites such as peptidoglycan, SCFAs, lipopolysaccharide, and SCFAs that activate immune responses via TLR2/4 signaling pathways. These processes lead to the production of Th17 cells and an increase in pro-inflammatory TNF-α, IL-6 cytokines, resulting in immunosuppression and chronic inflammation. The imbalance promotes T follicular helper cell activity and autoantibody production, exacerbating RA pathology. The microbiome-immune axis thus plays a pivotal role [200–202].

The incidence of type 1 diabetes mellitus (T1DM) is significantly reduced in the presence of *F. prausnitzii, Roseburia, Bifidobacteria, Lactobacillus*, and BC strains, which is associated with poor glycemic control. This is due to changes in temperature [203, 204]. This dysbiosis interferes with the production of small molecules such as butyrate, which are crucial for gut barrier integrity and immune regulation. Elevated lipopolysaccharide defense by gram-negative bacteria contributes to inflammation and gut permeability. Disruption of the intestinal barrier and loss of immunity provoke an aberrant immune response against pancreatic β-cells [205, 206]. Thus, microbiota-driven immune dysfunction plays a critical role in T1DM development and progression.

Multiple sclerosis (MS) is linked to gut microbiota imbalance, with an increase in pro-inflammatory bacteria such as *Akkermansia muciniphila, Methanobrevibacter, Ruminococcus*, and *Streptococcus*, and a decrease in beneficial bacteria such as *F. prausnitzii, Bifidobacterium, Prevotella*, and members of the Lachnospiraceae family. MS has been associated with an intestinal inflammatory response, including inflammatory cytokines, and members of the Lachnospiraceae family have been identified [207, 208]. The microbial dysbiosis alters SCFAs and tryptophan metabolites, affecting the integrity of the gut and blood-brain barrier. It promotes neuroinflammation through molecular transcription and activation of Th1, Th17 cells, and TLR2-mediated pathways, which increase cytokines such as TNF-α, IFN-γ, and IL-17 [209, 210].

Concurrently, Treg cell suppression exacerbates immune dysregulation, contributing to MS pathogenesis through persistent CNS inflammation.

2.29 Mechanisms of Dysbiosis-Induced Disease

Hypertension has been increasingly associated with gut microbiota dysbiosis, particularly through the overgrowth of bacteria such as *Prevotella* and *Klebsiella*. These bacteria are associated with inflammation. They also activate the immune system. Such immune activation can raise the blood pressure. Dysbiosis disrupts microbial balance [211]. It affects both beneficial and harmful species. Microbial metabolites are also altered. SCFAs are key among them. SCFAs help regulate vascular tone. They also reduce inflammation. Lower SCFAs may contribute to hypertension. Bacterial overgrowth harms the gut barrier. The barrier becomes more permeable [212]. Endotoxins can then enter the bloodstream. This condition is called endotoxemia that promotes vascular inflammation. It also causes endothelial dysfunction. Both are major contributors to hypertension. The immune system responds to this damage. Th17 cells increase in number. Pro-inflammatory cytokines are also elevated [213]. These immune changes impair vascular function. They also affect the kidneys. Dysbiosis changes bile acid metabolism. This impacts FXR and TGR5 signaling. Collectively, these findings highlight the gut microbiome's crucial role in modulating systemic vascular health and suggest that targeting microbial composition could be a novel therapeutic strategy for hypertension management [18].

2.29.1 Clinical Responses to Microbiome Dysbiosis

Various strategies help restore and maintain gut microbiota health. FMT introduces beneficial bacteria directly into the gut. Gut biotics include probiotics, prebiotics, and synbiotics. Probiotics supply live beneficial microbes. Prebiotics support the growth of these microbes. Synbiotics combine both probiotics and prebiotics. Dietary choices are crucial for microbiome health. Fermented foods and fiber-rich diets boost microbial diversity. The MD improves gut health with nutrient-dense, minimally processed foods. Ketogenic diets (KDs) influence microbiota through metabolic shifts. Postbiotics like SCFAs enhance gut barrier integrity and suppress pathogens [246].

Recent studies highlight the role of prebiotics, probiotics, and synbiotics in disease treatment through gut microbiota modulation (Table 2.4). Prebiotics like docosahexaenoic acid (DHA) help reduce hyperphosphorylation in Alzheimer's disease. DHA also promotes synapse formation in animal models. In CRC, prebiotics increase SCFAs with antitumor effects. Probiotics, especially *Lactobacillus* strains, aid bile acid metabolism. These effects help treat diarrhea, hypercholesterolemia, and NAFLD. The FXR-

Table 2.4: Disease, target organ system, pathogen, and therapeutic microorganism.

Disease	Organ	Causing organism	Curing organism	References
Irritable bowel syndrome	Large and small intestine	Elevated Firmicutes with diminished levels of Bacteroidetes, *Bifidobacterium*, and *Faecalibacterium* spp.	*Lactobacillus* strains, *Bifidobacterium*, and *Streptococcus*	[214]
Type I diabetes	Pancreas	Decreased relative abundance of *Streptococcus thermophilus* and *Lactococcus lactis*, enrichment of *Bifidobacterium pseudocatenulatum*, *Roseburia hominis*, and *Alistipes shahii*	Yet not to be stabilized	[215, 216]
Asthma	Lungs and airway	*Chlamydophila pneumonia*, *Streptococcus pneumoniae*, *Prevotella intermedia*, Veillonellaceae, Selenomonadales, and Negativicutes	Yet not to be stabilized	[217, 218]
Food poisoning	Stomach and intestine	*Campylobacter, Salmonella, Escherichia coli, Shigella*	*Lactobacillus rhamnosus* and *Lactobacillus reuteri*	[219, 220]
Malnutrition	Whole body	Overabundance of Enterobacteriaceae Abundance of *Bilophila wadsworthia* and *Clostridioides innocuum* combined with a low-nutrient diet enhanced malnutrition		[221]
Depression		*Prevotella, Eggerthella, Subdoligranulum, Coprococcus, Sellimonas, Lachnoclostridium, Hungatella,* Ruminococcaceae, Lachnospiraceae, *Eubacterium ventriosum, Ruminococcus gauvreauii* group, Ruminococcaceae	*Bifidobacterium breve* CCFM1025 *Lacticaseibacillus paracasei* YIT 9,029	[222–225]

Table 2.4 (continued)

Disease	Organ	Causing organism	Curing organism	References
Anxiety		Decreased *Bacteroides* and increased *Clostridium*, *Campylobacter jejuni*, *Paraprevotella*, Euryarchaeota, *Caldivirga*, Porphyromonadaceae, and Desulfovibrionales	*Lactobacillus plantarum* P	[226–229]
Hypertension	Cardiovascular	Elevated *Lactobacillus*, *Roseburia*, *Coprococcus*, *Akkermansia muciniphila*, and *Bifidobacterium* correlate with reduced BP, whereas increased *Streptococcus*, *Blautia*, and *Prevotella* are linked to elevated BP	*Lactobacillus plantarum* *Lactobacillus delbrueckii* Lb100 *and Lactococcus lactis*, *Lactiplantibacillus plantarum* NK181, and *Lactobacillus delbrueckii* KU200171 *Lactobacillus acidophilus* NCDC-15	[127, 230–233]
Dyslipidemia	Cardiovascular	High *Candida albicans*	*Lactobacillus paracasei* sup. *paracasei* TzISTR 2593 *Lactiplantibacillus plantarum* strains (CECT7527, CECT7528, and CECT7529)	[237, 238]
Atherosclerosis	Cardiovascular	*Chlamydia pneumoniae*, *Helicobacter pylori*, *Mycoplasma pneumoniae*, and *Candida albicans*	*Lactobacillus coryniformis* CECT5711	[234, 235]
Endocarditis	Cardiovascular	*Lactobacillus endocarditis*	*Lactobacillus rhamnosus*	[236, 237]
COPD		*Pseudomonas aeruginosa*	*Lactobacillus* and *Bifidobacterium*	[238]
Cystic fibrosis		*Clostridium difficile*, *Escherichia coli*, and *Pseudomonas aeruginosa*	*Lactobacillus rhamnosus* GG, *L. reuteri*	[239, 240]
Ventilator-associated pneumonia		*Staphylococcus aureus*, *Pseudomonas aeruginosa*, and *Acinetobacter* species	*L. rhamnosus* GG and *B. breve*	[241, 242]
Lung cancer		*Streptococcus pneumonia* *Veillonella parvula* *Cyanobacteria* *Acidovorax* *Haemophilus*	*Lactobacillus*, Eubacteriaceae, *Ruthenibacterium*, *Faecalicatena*, *Pseudobutyrivibrio*, and *Roseburia*	[243–245]

FGF15 signaling pathway mediates many probiotic actions. In psoriasis and arthritis models, inulin-type fructans and probiotics reduce inflammatory cytokines. These agents also lower disease incidence. Clinical trials confirm that prebiotics enhance immunity and beneficial microbes in T2DM, irritable bowel syndrome (IBS), and autism spectrum disorder (ASD). Probiotics alleviate depression and gastrointestinal issues in ASD. Synbiotics show mixed outcomes depending on the condition. They improve gut flora in NAFLD but not in liver fibrosis. In obesity, synbiotics lack clear synergistic effects. However, synbiotics benefit IBS symptoms and metabolic markers. These findings support microbiome-based therapies. Personalized medicine could leverage them for neurological and metabolic disorders. Modulating the gut microbiome presents a novel therapeutic approach for cardiovascular disease prevention and management through microbial and metabolic rebalancing [247, 248].

2.30 Targeted Genetic Modification of Probiotic Strains

Probiotic bacterial species *E. coli* Nessel 1917 (ECN) has a sophisticated toolbox for gene editing, and it is known to be effective for a wide range of genes [249–251]. ECNs were genetically engineered to transcript the immunomodulatory protein in extracellular vesicles, cystatin, in response to thiosulfate, a biomarker of inflammation [252]. Genetic engineering techniques involve the use of genetic templates to create biological products or biomaterials such as vaccines or cytokine delivery systems.

2.31 Hybrid Systems of Probiotics and Biomaterials

Oral administration of probiotics severely limits their therapeutic potential due to severe gastrointestinal conditions. Oral administration of probiotics can reduce their effectiveness because of harsh conditions in the gut [253, 254]. Many probiotics do not survive long enough to work properly. Using special coatings made from smart biomaterials can protect them. This helps prevent damage and allows them to stay and grow in the gut. Tannic acids is an ROS scavenger and anchor for mucin, thus enhancing intestine adhesiveness [255]. Most commonly used microbial probiotics are *Lactobacillus* species, *Bifidobacterium* species, *Saccharomyces* species, and *Bacillus* species [256, 257].

Probiotics can have one or more strains. Multi-strain probiotics often include a mix of *Lactobacillus* and *Bifidobacterium* [258].

2.32 Prebiotics

Prebiotics are carbohydrates that are not digested in the intestine. They are fermented by gut bacteria and help certain beneficial bacteria grow. Fermentation produces SCFAs, CO_2, and H_2. These compounds acidify the intestine, promoting *Bifidobacteria* and lactobacilli growth [81, 259].

Prebiotics aim to restore balance of gastrointestinal tract by increasing production of SCFAs, number of beneficial bacteria, and lowering the pH of the gut [133, 260]. The prebiotic oligofructose-rich inulin selectively altered the gut microbiota and significantly reduced body fat and body mass index in overweight/obese children [259, 261]. The use of probiotics has shown benefits in reducing the risk of cardiovascular diseases, such as dyslipidemia, hypercholesterolemia, hypertension, and chronic inflammation.

Synbiotics are preparations combining probiotics and prebiotics. Synbiotics combine probiotics and prebiotics for a stronger, more beneficial effect. Their combined action works better than using them alone. The combination of these, due to their synergistic effect, has a more positive effect on the recipient than using them separately [262, 263].

2.33 Dietary Fiber

Fiber-rich diets are known to be beneficial for gut health. Inflamed gut with dysbiotic bacterial community leads to changes in fiber metabolism based on the physicochemical properties of fiber. Resistant maltodextrin, wheat bran, high-methoxyl pectin, and inulin were selected for fermentation using UC-associated bacteria to assess the bacterial dysbiosis effect. UC-related gut microbiota had markedly lower α- and β-diversity than healthy microbiota [264].

Orange peels are rich in dietary fibers and polyphenols with little amounts of sorbitol [265]. Polyphenols exhibit prebiotic effects by promoting the growth of beneficial microbes. Large amounts of dietary fiber may reduce the risk of heart failure by promoting a healthy gut microbiota and by reducing TMAO production.

2.34 Microbial Metabolite

The effects of microbial dysbiosis can be treated with artificially engineered bacteria, which are increasingly being used in medicine for targeted therapy and drug production. Intravenous injection allows bacteria such as *Salmonella typhimurium* to reach the tumor via the bloodstream, while intertumoral injection delivers them directly to the tumor site, maximizing local efficacy. Oral drugs pose problems due to stomach acidity, but encapsulation offers a potential solution. Combining bacteria with chemotherapy or immunotherapy may increase the efficacy of treatment, as has been seen with *Listeria monocytogenes* in clinical trials with melanoma. Bacterial proteins such as HapQ target cancer cells and reduce metastasis. Additionally, *E. coli* K5 has been engineered to produce heparin, an important anticoagulant used in clinical treatment [266].

2.35 Antimicrobial Peptide Self-Assembly

A new multifunctional agent has three parts: AMP (P-113), C8, and 8DSS. It fights harmful bacteria like *Streptococcus mutans*. It resists acid attacks and removes dental biofilms. It also helps repair damaged tooth tissue by promoting mineralization [267].

An in vivo mastitis infection model in mice confirmed the therapeutic efficacy and promising biological safety of fetal fibronectin, which effectively alleviates mastitis caused by multidrug-resistant *E. coli* and *S. aureus*, and eliminates pathogenic bacteria [268].

2.36 Oral RNA Delivery System

miRNAs in outer membrane vesicles help shape the gut microbiota. They play an important role in human health and disease. These miRNAs can affect inflammation and gut microbial balance. Their changes may lead to dysbiosis and related health issues [269]. Microbiome modulation has been reported in patients with MS. Surprisingly, stool collected at the peak of disease from an experimental autoimmune encephalomyelitis (EAE) model of MS helped reduce symptoms in the host. This effect depended on the presence of specific miRNAs. Oral delivery of synthetic miR-30d improves EAE by increasing regulatory T cells. It works by regulating lactase in *A. muciniphila*, which boosts its growth in the gut [270].

2.37 Food and Food Products

Polyphenols, biologically active compounds, have been reported in barley, cocoa, and green tea. They play an important role in maintaining gut health. These compounds interact with the intestinal flora and produce biologically active metabolites that showed anti-inflammatory and antioxidant properties. The anthocyanins reported blueberries promote the growth of *A. muciniphila*, which lead to improved gut health and reduced systemic inflammation [271].

2.38 Mediterranean Diet

MD is widely known for its health benefits and it modifies the gut microbiota composition. MD has the potential to reduce the risk of metabolic, cardiovascular, and neurological diseases. It is characterized by being rich in plant foods, monounsaturated fats, and polyphenols, whose main component is extra virgin olive oil, which promotes the growth of beneficial intestinal bacteria such as *Bifidobacterium, Faecalibacterium*, and Rosaceae, forming a metabolic chain that improves intestinal integrity and reduces inflammation and stimulation of intestinal metabolism [272]. Adopting an MD can help restore gut health by enhancing *Bifidobacteria*, improving gut health and reducing inflammation associated with AF.

2.39 Microbiome-Based Therapies

Microbial metabolism-based therapies use compounds derived from gut microbes, such as SCFAs. They have the capability to directly modulate host signaling, enhance immune responses, and improve gut health. They provide targeted therapies for conditions such as IBD, neurological, and metabolic disorders, and effectively overcome the limitations of probiotics. These are stable; fewer toxic compounds can be taken as supplements. But personalized treatment based on specific metabolite signatures remains challenging and long-term studies are limited [212].

2.40 Fecal Microbiota Transplantation (FMT)

FMT is used to restore microbial balance in patients with gut dysbiosis. FMT involves transferring stool from healthy donors to patients. This helps restore a balanced gut microbiome. It has proven highly effective in treating recurrent *Clostridioides difficile* infections. Microbiome-based therapies for this condition have also been approved by the US Food and Drug Administration. FMT is also used for other gut diseases, infections, and disorders linked to the gut-liver axis and GBA. Careful donor screening is needed before FMT to avoid side effects [273].

2.41 Traditional Chinese Medicine (TCM)

Traditional Chinese medicine (TCM) formulation affects metabolic and inflammatory processes. These formulations mainly regulate intestinal flora, SCFA production and it also regulates gene/protein. Ginsenosides, juniper extract, and polysaccharides improve energy use and reduce fat. They target SCFAs, IL-6, TNF-α, PYY, and GLP-1. These substances act through GPR41/43 receptors. They also affect PGC-1α, UCP1, and PPAR-γ to boost health. These interventions suggest that obesity, metabolic syndrome, and inflammatory diseases may be treated by improving the balance of gut microbiota and metabolic homeostasis [274].

Chinese herbal medicines are very effective in controlling intestinal flora in various diseases. Medicines like Shenlian Decoction and Qisheng Pills help control bacteria such as Prevotellaceae, *Verrucomicrobia*, and Bacteroidetes. This improves gut health in diabetes and Alzheimer's patients. Magnolia and chicory extracts fight obesity and liver fibrosis. They work by targeting key gut microorganisms. Active ingredients such as cythiazine A and Mori alkaloids produced positive changes in the liver and kidney microbiota. Overall, these treatments increased beneficial bacteria like *Bacteroides* and *Akkermansia*. They also reduced harmful or inflammation-related bacteria [275].

2.42 Ketogenic Diet

The KD is a diet with low carbohydrates and moderate protein. This diet is designed to mimic a fasting state. The KD has the ability to induce ketosis and produce ketone bodies from fat. The KD has the potential to improve cognitive function and regulate mood. Research on its safety and effectiveness focuses primarily on its anti-inflammatory properties and its effects on neurological health and the GBA [276].

KD has therapeutic effects on gut microbiota in neurological diseases and obesity. According to current knowledge, KD plays an important role in improving disease symptoms, mainly by increasing the ratio of Bacteroidetes to Firmicutes and in some cases by reducing Proteobacteria [277].

2.43 Conclusion

Microbiome dysbiosis plays an important role in many chronic and systemic diseases. These include gut, heart, brain, and reproductive disorders. When the balance of good and bad microbes is disturbed, it weakens immunity, damages the gut barrier, and affects metabolism. This can lead to disease. New research has helped develop ways to fix this imbalance. These include changes in diet, special probiotic treatments, and engineering of helpful microbes. Fecal transplants, certain probiotics, and diets rich in fiber and polyphenols are showing good results. Personalized treatment based on a person's unique gut microbes may improve outcomes. More research is needed to understand how microbes and the body interact. This will help create better treatments and improve overall health.

Conflicts of interests: The authors do not declare any conflicts of interest.

References

[1]   Akçelik M, Akçelik N, Şanlıbaba P, Uymaz Tezel B. Genetic modification and sequence analysis of probiotic microorganisms. Advances in Probiotics. 2021;101–112. https://doi.org/10.1016/b978-0-12-822909-5.00006-x.

[2]   Berg G, Rybakova D, Fischer D, et al. Microbiome definition re-visited: Old concepts and new challenges. Microbiome. 2020;8(1). https://doi.org/10.1186/s40168-020-00875-0.

[3]   Sender R, Fuchs S, Milo R. Revised estimates for the number of human and bacteria cells in the body. PLOS Biology. 2016;14(8):e1002533. https://doi.org/10.1371/journal.pbio.1002533.

[4]   Afridi OK, Ali J, Chang JH. Fecal microbiome and resistome profiling of healthy and diseased pakistani individuals using next-generation sequencing. Microorganisms. 2021;9(3):616. https://doi.org/10.3390/microorganisms9030616.

[5]   Safarchi A, Al-Qadami G, Tran CD, Conlon M. Understanding dysbiosis and resilience in the human gut microbiome: Biomarkers, interventions, and challenges. Frontiers in Microbiology. 2025;16. https://doi.org/10.3389/fmicb.2025.1559521.

[6]   Ali AY, Zahran SA, Eissa M, Kashef MT, Ali AE. Gut microbiota dysbiosis and associated immune response in systemic lupus erythematosus: Impact of disease and treatment. Gut Pathogens. 2025;17(1). https://doi.org/10.1186/s13099-025-00683-7.

[7]   Zhao Y, Liu Y, Jia L. Gut microbial dysbiosis and inflammation: Impact on periodontal health. Journal of Periodontal Research. Published online July 11, 2024. https://doi.org/10.1111/jre.13324.

[8]   Deo PN, Deshmukh R. Oral microbiome: Unveiling the fundamentals. Journal of Oral and Maxillofacial Pathology. 2019;23(1):122–128. https://doi.org/10.4103/jomfp.JOMFP_304_18.

[9]   Tao R, Li R, Wang R. Skin microbiome alterations in seborrheic dermatitis and dandruff: A systematic review. Experimental Dermatology. 2021;30(10):1546–1553. https://doi.org/10.1111/exd.14450.

[10] Schmid B, Künstner A, Fähnrich A, et al. Dysbiosis of skin microbiota with increased fungal diversity is associated with severity of disease in atopic dermatitis. Journal of the European Academy of Dermatology and Venereology. 2022;36(10):1811–1819. https://doi.org/10.1111/jdv.18347.

[11] Findley K, Oh J, Yang J, et al. human skin fungal diversity. Nature. 2013;498(7454):367–370. https://doi.org/10.1038/nature12171.

[12] Stankovic MM. Lung microbiota: From healthy lungs to development of chronic obstructive pulmonary disease. International Journal of Molecular Sciences. 2025;26(4):1403. https://doi.org/10.3390/ijms26041403.

[13] Saint-Criq V, Lugo-Villarino G, Thomas M. Dysbiosis, malnutrition and enhanced gut-lung axis contribute to age-related respiratory diseases. Ageing Research Reviews. 2021;66:101235. https://doi.org/10.1016/j.arr.2020.101235.

[14] Jia Q, Wang H, Wang Y, et al. Investigation of the mechanism of silica-induced pulmonary fibrosis: The role of lung microbiota dysbiosis and the LPS/TLR4 signaling pathway. Science of the Total Environment. 2024;912:168948. https://doi.org/10.1016/j.scitotenv.2023.168948.

[15] Chung KF. Airway microbial dysbiosis in asthmatic patients: A target for prevention and treatment?. Journal of Allergy and Clinical Immunology. 2017;139(4):1071–1081. https://doi.org/10.1016/j.jaci.2017.02.004.

[16] Valeriano VD, Lahtinen E, Hwang IC, Zhang Y, Du J, Schuppe-Koistinen I. Vaginal dysbiosis and the potential of vaginal microbiome-directed therapeutics. Frontiers in Microbiomes. 2024;3. https://doi.org/10.3389/frmbi.2024.1363089.

[17] Machado A, Foschi C, Marangoni A. Editorial: Vaginal dysbiosis and biofilms, volume II. Frontiers in Cellular and Infection Microbiology. 2025;15. https://doi.org/10.3389/fcimb.2025.1588434.

[18] Zhao M, Chu J, Feng S, et al. Immunological mechanisms of inflammatory diseases caused by gut microbiota dysbiosis: A review. Biomedicine & Pharmacotherapy. 2023;164(1):114985. https://doi.org/10.1016/j.biopha.2023.114985.

[19] Maciel-Fiuza MF, Muller GC, Campos DMS, et al. Role of gut microbiota in infectious and inflammatory diseases. Frontiers in Microbiology. 2023;14:1098386. https://doi.org/10.3389/fmicb.2023.1098386.

[20] Rolhion N, Danne C, Creusot L, et al. P0085 *Faecalibacterium prausnitzii* induces an anti-inflammatory response and a metabolic reprogramming in human monocytes. Journal of Crohn's and Colitis. 2025;19(Supplement_1):i450–i450. https://doi.org/10.1093/ecco-jcc/jjae190.0259.

[21] Zheng J, Sun Q, Zhang J, Ng SC. The role of gut microbiome in inflammatory bowel disease diagnosis and prognosis. United European Gastroenterology Journal. 2022;10(10):1091–1102. https://doi.org/10.1002/ueg2.12338.

[22] Acevedo-Román A, Pagán-Zayas N, Velázquez-Rivera LI, Torres-Ventura AC, Godoy-Vitorino F. Insights into gut dysbiosis: Inflammatory diseases, obesity, and restoration approaches. International Journal of Molecular Sciences. 2024;25(17):9715. https://doi.org/10.3390/ijms25179715.

[23] Morris G, Gamage E, Travica N, et al. Polyphenols as adjunctive treatments in psychiatric and neurodegenerative disorders: Efficacy, mechanisms of action, and factors influencing inter-individual response. Free Radical Biology and Medicine. 2021;172:101–122. https://doi.org/10.1016/j.freeradbiomed.2021.05.036.

[24] Flores-Dorantes MT, Díaz-López YE, Gutiérrez-Aguilar R. Environment and gene association with obesity and their impact on neurodegenerative and neurodevelopmental diseases. Frontiers in Neuroscience. 2020;14(863):863. https://doi.org/10.3389/fnins.2020.00863.

[25] Neto A, Fernandes A, Barateiro A. The complex relationship between obesity and neurodegenerative diseases: An updated review. Frontiers in Cellular Neuroscience. 2023;17. https://doi.org/10.3389/fncel.2023.1294420.

[26] Kuneš J, Hojná S, Mráziková L, Montezano AC, Touyz RM, Maletínská L. Obesity, cardiovascular and neurodegenerative diseases: Potential common mechanisms. Physiological Research. Published online July 31 2023;S73–S90. https://doi.org/10.33549/physiolres.935109.

[27] Rondanelli M, Borromeo S, Cavioni A, et al. Therapeutic strategies to modulate gut microbial health: Approaches for chronic metabolic disorder management. Metabolites. 2025;15(2):127–127. https://doi.org/10.3390/metabo15020127.

[28] Qin L, Fan B, Zhou Y, et al. Targeted gut microbiome therapy: Applications and prospects of probiotics, fecal microbiota transplantation and natural products in the management of type 2 diabetes. Pharmacological Research. 2025;213:107625. https://doi.org/10.1016/j.phrs.2025.107625.

[29] Wang Y, Yan H, Zheng Q, Sun X. The crucial function of gut microbiota on gut-liver repair. hLife. Published online January 2025. https://doi.org/10.1016/j.hlife.2025.01.001.

[30] Abbasi AF, Marinkovic A, Prakash S, Sanyaolu A, Smith S. COVID-19 and the human gut microbiome: An under-recognized association. Chonnam Medical Journal. 2022;58(3):96. https://doi.org/10.4068/cmj.2022.58.3.96.

[31] Zuo T, Zhang F, Lui GCY, et al. Alterations in gut microbiota of patients with COVID-19 during time of hospitalization. Gastroenterology. 2020;159(3). https://doi.org/10.1053/j.gastro.2020.05.048.

[32] Yeoh YK, Zuo T, Lui GCY, et al. Gut microbiota composition reflects disease severity and dysfunctional immune responses in patients with COVID-19. Gut. 2021;70(4). https://doi.org/10.1136/gutjnl-2020-323020.

[33] Winter S, Bäumler AJ. Gut dysbiosis: Ecological causes and causative effects on human disease. Proceedings of the National Academy of Sciences of the United States of America. 2023;120(50). https://doi.org/10.1073/pnas.2316579120.

[34] Mazumder MHH, Hussain S. Air-pollution-mediated microbial dysbiosis in health and disease: Lung–gut axis and beyond. Journal of Xenobiotics. 2024;14(4):1595–1612. https://doi.org/10.3390/jox14040086.

[35] Sierra-Vargas MP, Montero-Vargas JM, Debray-García Y, Vizuet-de-rueda JC, Loaeza-Román A, Terán LM. Oxidative stress and air pollution: Its impact on chronic respiratory diseases. International Journal of Molecular Sciences. 2023;24(1):853. https://doi.org/10.3390/ijms24010853.

[36] Mao Q, Jiang F, Yin R, et al. Interplay between the lung microbiome and lung cancer. Cancer Letters. 2018;415:40–48. https://doi.org/10.1016/j.canlet.2017.11.036.

[37] Zhou P, Zhou Y, Liu B, et al. Perinatal antibiotic exposure affects the transmission between maternal and neonatal microbiota and is associated with early-onset sepsis. Bradford PA, ed. mSphere. 2020;5(1). https://doi.org/10.1128/msphere.00984-19.

[38] Li S, Liu J, Zhang X, et al. The potential impact of antibiotic exposure on the microbiome and human health. Microorganisms. 2025;13(3):602. https://doi.org/10.3390/microorganisms13030602.

[39] Patangia DV, Anthony Ryan C, Dempsey E, Paul Ross R, Stanton C. Impact of antibiotics on the human microbiome and consequences for host health. MicrobiologyOpen. 2022;11(1). https://doi.org/10.1002/mbo3.1260.

[40] Suárez-Martínez C, Santaella-Pascual M, Yagüe-Guirao G, Martínez-Graciá C. Infant gut microbiota colonization: Influence of prenatal and postnatal factors, focusing on diet. Frontiers in Microbiology. 2023;14:1236254. https://doi.org/10.3389/fmicb.2023.1236254.

[41] Huang H, Jiang J, Wang X, Jiang K, Cao H. Exposure to prescribed medication in early life and impacts on gut microbiota and disease development. eClinicalMedicine. 2024;68:102428. https://doi.org/10.1016/j.eclinm.2024.102428.

[42] Zhang X, Mushajiang S, Luo B, Tian F, Ni Y, Yan W. The composition and concordance of lactobacillus populations of infant gut and the corresponding breast-milk and maternal gut. Frontiers in Microbiology. 2020;11. https://doi.org/10.3389/fmicb.2020.597911.

[43] Chen YY, Zhao X, Moeder W, et al. Impact of maternal intrapartum antibiotics, and caesarean section with and without labour on *Bifidobacterium* and other infant gut microbiota. Microorganisms. 2021;9(9):1847-1847. https://doi.org/10.3390/microorganisms9091847.

[44] Gostner JM, Becker K, Kofler H, Strasser B, Fuchs D. Tryptophan metabolism in allergic disorders. International Archives of Allergy and Immunology. 2016;169(4):203–215. https://doi.org/10.1159/000445500.

[45] O'Riordan KJ, Collins MK, Moloney GM, et al. Short chain fatty acids: Microbial metabolites for gut-brain axis signalling. Molecular and Cellular Endocrinology. 2022;546:111572. https://doi.org/10.1016/j.mce.2022.111572.

[46] Bermúdez-Sánchez S, Bahl MI, Hansen EB, Licht TR, Laursen MF. Oral amoxicillin treatment disrupts the gut microbiome and metabolome without interfering with luminal redox potential in the intestine of Wistar Han rats. FEMS Microbiology Ecology. Published online January 8 2025. https://doi.org/10.1093/femsec/fiaf003.

[47] Lekang K, Shekhar S, Berild D, Petersen FC, Winther-Larsen HC. Effects of different amoxicillin treatment durations on microbiome diversity and composition in the gut. Loor JJ, ed. PLOS ONE. 2022;17(10):e0275737. https://doi.org/10.1371/journal.pone.0275737.

[48] Chopyk J, Georgina A, Ramirez-Sanchez C, et al. Common antibiotics, azithromycin and amoxicillin, affect gut metagenomics within a household. BMC Microbiology. 2023;23(1). https://doi.org/10.1186/s12866-023-02949-z.

[49] Siegle RJ, Fekety R, Sarbone PD, Finch RN, Deery HG, Voorhees JJ. Effects of topical clindamycin on intestinal microflora in patients with acne. Journal of the American Academy of Dermatology. 1986;15(2):180–185. https://doi.org/10.1016/s0190-9622(86)70153-9.

[50] Olías-Molero AI, Botías P, Cuquerella M, et al. Effect of clindamycin on intestinal microbiome and miltefosine pharmacology in hamsters infected with *Leishmania infantum*. Antibiotics. 2023;12(2):362–362. https://doi.org/10.3390/antibiotics12020362.

[51] Hertz FB, Budding AE, van SPH, Løbner-Olesen A, Frimodt-Møller N. Effects of antibiotics on the intestinal microbiota of mice. Antibiotics. 2020;9(4):191–191. https://doi.org/10.3390/antibiotics9040191.

[52] Kumari R, Yadav Y, Misra R, et al. Emerging frontiers of antibiotics use and their impacts on the human gut microbiome. Microbiological Research. 2022;263:127127. https://doi.org/10.1016/j.micres.2022.127127.

[53] Janas-Naze A, Torbicka G, Chybicki D, Lipczyńska-Lewandowska M, Zhang W. Comparative efficacy of different oral doses of clindamycin in preventing post-operative sequelae of lower third molar surgery – a randomized, triple-blind study. Medicina-lithuania. 2022;58(5):668–668. https://doi.org/10.3390/medicina58050668.

[54] Isaac S, Scher JU, Djukovic A, et al. Short- and long-term effects of oral vancomycin on the human intestinal microbiota. Journal of Antimicrobial Chemotherapy. 2017;72(1):128–136. https://doi.org/10.1093/jac/dkw383.

[55] Liu L, Wang Q, Wu X, et al. Vancomycin exposure caused opportunistic pathogens bloom in intestinal microbiome by simulator of the human intestinal microbial ecosystem (SHIME). Environmental Pollution. 2020;265:114399-114399. https://doi.org/10.1016/j.envpol.2020.114399.

[56] Rodriguez-Ruiz JP, Lin Q, Van Heirstraeten L, et al. Long-term effects of ciprofloxacin treatment on the gastrointestinal and oropharyngeal microbiome are more pronounced after longer antibiotic courses. International Journal of Antimicrobial Agents. 2024;64(3):107259. https://doi.org/10.1016/j.ijantimicag.2024.107259.

[57] Rahman H, Anggadiredja K, Sasongko L. Mechanisms of oral ciprofloxacin-induced depressive-like behavior and the potential benefit of lactulose: A correlation analysis. Toxicology Reports. 2025;14:101920-101920. https://doi.org/10.1016/j.toxrep.2025.101920.

[58] Liu Y, Wang Y, Wang H. Effects of ciprofloxacin and levofloxacin on initial colonization of intestinal microbiota in *Bufo gargarizans* at embryonic stages. Chemosphere. 2024;361:142587. https://doi.org/10.1016/j.chemosphere.2024.142587.

[59] Tang H, Zhang X, Yang F, et al. Effect of ciprofloxacin on the composition of intestinal microbiota in *Sarcophaga peregrina* (Diptera: Sarcophagidae). Microorganisms. 2023;11(12):2867. https://doi.org/10.3390/microorganisms11122867.

[60] Wei S, Mortensen MS, Stokholm J, et al. Short- and long-term impacts of azithromycin treatment on the gut microbiota in children: A double-blind, randomized, placebo-controlled trial. EBioMedicine. 2018;38:265–272. https://doi.org/10.1016/j.ebiom.2018.11.035.

[61] Du S, Shang L, Zou X, et al. Azithromycin exposure induces transient microbial composition shifts and decreases the airway microbiota resilience from outdoor $PM_{2.5}$ stress in healthy adults: A randomized, double-blind, placebo-controlled trial. Microbiology Spectrum. 2023;11(3). https://doi.org/10.1128/spectrum.02066-22.

[62] Shayista H, Prasad MNN, Raj SN, et al. Impact of macrolide antibiotics on gut microbiota diversity with age-specific implications and scientific insights. Medicine in Microecology. Published online February 1 2025:100122-100122. https://doi.org/10.1016/j.medmic.2025.100122.

[63]   Raymond F, Ouameur AA, Déraspe M, et al. The initial state of the human gut microbiome determines its reshaping by antibiotics. The ISME Journal. 2016;10(3):707–720. https://doi.org/10.1038/ismej.2015.148.

[64]   Yang L, Bajinka O, Jarju PO, Tan Y, Taal AM, Ozdemir G. The varying effects of antibiotics on gut microbiota. AMB Express. 2021;11(1). https://doi.org/10.1186/s13568-021-01274-w.

[65]   Jarmusch AK, Vrbanac A, Momper JD, et al. Enhanced characterization of drug metabolism and the influence of the intestinal microbiome: A pharmacokinetic, microbiome, and untargeted metabolomics study. Clinical and Translational Science. 2020;13(5):972–984. https://doi.org/10.1111/cts.12785.

[66]   Kadry AA, El-Antrawy MA, El-Ganiny AM. Impact of Short Chain Fatty Acids (Scfas) on Antimicrobial Activity of New β-lactam/β-lactamase Inhibitor Combinations and on Virulence of Escherichia Coli Isolates. Published online February 1 2023, https://doi.org/10.1038/s41429-023-00595-1.

[67]   Shayista H, Prasad MNN, Raj SN, et al. Complexity of antibiotic resistance and its impact on gut microbiota dynamics. Engineering Microbiology. Published online December 2024:100187. https://doi.org/10.1016/j.engmic.2024.100187.

[68]   Chakraborty R, Lam V, Kommineni S, et al. Ceftriaxone administration disrupts intestinal homeostasis, mediating noninflammatory proliferation and dissemination of commensal enterococci. Raffatellu M, ed. Infection and Immunity. 2018;86(12). https://doi.org/10.1128/iai.00674-18.

[69]   Zou W, Liu Y, Zhang W, et al. Short-term use of ceftriaxone sodium leads to intestinal barrier disruption and ultrastructural changes of kidney in SD rats. Renal Failure. 2023;45(1):2230322. https://doi.org/10.1080/0886022X.2023.2230322.

[70]   Florian D, Wu L, Wilkinson CS, Kabbaj M, Knackstedt LA. Ceftriaxone alters the gut microbiome composition and reduces alcohol intake in male and female Sprague–Dawley rats. Alcohol. 2024;120:169–178. https://doi.org/10.1016/j.alcohol.2024.01.006.

[71]   Andrei C, Zanfirescu A, Ormeneanu VP, Negreş S. Evaluating the efficacy of secondary metabolites in antibiotic-induced dysbiosis: A narrative review of preclinical studies. Antibiotics. 2025;14(2):138. https://doi.org/10.3390/antibiotics14020138.

[72]   Khan KN, Fujishita A, Muto H, et al. Levofloxacin or gonadotropin releasing hormone agonist treatment decreases intrauterine microbial colonization in human endometriosis. European Journal of Obstetrics, Gynecology, and Reproductive Biology. 2021;264:103–116. https://doi.org/10.1016/j.ejogrb.2021.07.014.

[73]   Juhong J, Kongwattananon W, Surawatsatien N. Nathapon Treewipanon, Tanittha Chatsuwan. Effects of preoperative topical levofloxacin on conjunctival microbiome in patients undergoing intravitreal injections. PLoS ONE. 2025;20(3):e0320785–e0320785. https://doi.org/10.1371/journal.pone.0320785.

[74]   Prasad R, Rehman A, Rehman L, et al. Antibiotic-induced loss of gut microbiome metabolic output correlates with clinical responses to CAR T-cell therapy. Blood. 2025;145(8):823–839. https://doi.org/10.1182/blood.2024025366.

[75]   Kang GG, Trevaskis NL, Murphy AJ, Febbraio MA. Diet-induced gut dysbiosis and inflammation: Key drivers of obesity-driven NASH. iScience. 2023;26(1):105905. https://doi.org/10.1016/j.isci.2022.105905.

[76]   Fallani M, Young D, Scott J, et al. Intestinal microbiota of 6-week-old infants across Europe: Geographic influence beyond delivery mode, breast-feeding, and antibiotics. Journal of Pediatric Gastroenterology and Nutrition. 2010;51(1):77–84. https://doi.org/10.1097/MPG.0b013e3181d1b11e.

[77]   Santacruz A, Marcos A, Wärnberg J, et al. Interplay between weight loss and gut microbiota composition in overweight adolescents. Obesity. 2009;17(10):1906–1915. https://doi.org/10.1038/oby.2009.112.

[78] Brown K, DeCoffe D, Molcan E, Gibson DL. Diet-induced dysbiosis of the intestinal microbiota and the effects on immunity and disease. Nutrients. 2012;4(8):1095–1119. https://doi.org/10.3390/nu4081095.

[79] Moreno-Indias I, Cardona F, Tinahones FJ, Queipo-ortuã±o MI. Impact of the gut microbiota on the development of obesity and type 2 diabetes mellitus. Frontiers in Microbiology. 2014;5. https://doi.org/10.3389/fmicb.2014.00190.

[80] Martínez A, Velázquez L, Díaz R, et al. Impact of novel foods on the human gut microbiome: current status. Microorganisms. 2024;12(9):1750-1750. https://doi.org/10.3390/microorganisms12091750.

[81] Davani-Davari D, Negahdaripour M, Karimzadeh I, et al. Prebiotics: Definition, types, sources, mechanisms, and clinical applications. Foods. 2019;8(3):92. https://doi.org/10.3390/foods8030092.

[82] Perna S, Alalwan TA, Alaali Z, et al. The role of glutamine in the complex interaction between gut microbiota and health: A narrative review. International Journal of Molecular Sciences. 2019;20(20):5232. https://doi.org/10.3390/ijms20205232.

[83] Zambom de Souza AZ, Zambom AZ, Abboud KY, et al. Oral supplementation with l-glutamine alters gut microbiota of obese and overweight adults: A pilot study. Nutrition. 2015;31(6):884–889. https://doi.org/10.1016/j.nut.2015.01.004.

[84] Bailén M, Bressa C, Martínez-López S, et al. Microbiota features associated with a high-fat/low-fiber diet in healthy adults. Frontiers in Nutrition. 2020;7. https://doi.org/10.3389/fnut.2020.583608.

[85] Basak S, Banerjee A, Pathak S, Duttaroy AK. Dietary Fats and the Gut Microbiota: Their impacts on lipid-induced metabolic syndrome. Journal of Functional Foods. 2022;91:105026. https://doi.org/10.1016/j.jff.2022.105026.

[86] Zhang D, Jian YP, Zhang Y, et al. Short-chain fatty acids in diseases. Cell Communication and Signaling. 2023;21(1). https://doi.org/10.1186/s12964-023-01219-9.

[87] Deng J, Yun J, Gu Y, Yan B, Yin B, Huang C. Evaluating the in vitro and in vivo prebiotic effects of different xylo-oligosaccharides obtained from bamboo shoots by hydrothermal pretreatment combined with endo-xylanase hydrolysis. International Journal of Molecular Sciences. 2023;24(17):13422. https://doi.org/10.3390/ijms241713422.

[88] Charoensiddhi S, Conlon M, Methacanon P, Thayanukul P, Hongsprabhas P, Zhang W. Gut microbiome modulation and gastrointestinal digestibility in vitro of polysaccharide-enriched extracts and seaweeds from *Ulva rigida* and *Gracilaria fisheri*. Journal of Functional Foods. 2022;96:105204. https://doi.org/10.1016/j.jff.2022.105204.

[89] Jia L, Peng X, Deng Z, Zhang B, Li H. The structural characterization of polysaccharides from three cultivars of *Moringa oleifera Lam*. root and their effects on human intestinal microflora. Food Bioscience. 2023;52:102482-102482. https://doi.org/10.1016/j.fbio.2023.102482.

[90] Kleigrewe K, Haack M, Baudin M, et al. Dietary modulation of the human gut microbiota and metabolome with flaxseed preparations. International Journal of Molecular Sciences. 2022;23(18):10473. https://doi.org/10.3390/ijms231810473.

[91] Siddiqui NZ, Rehman AU, Yousuf W, et al. Effect of crude polysaccharide from seaweed, *Dictyopteris divaricata* (CDDP) on gut microbiota restoration and anti-diabetic activity in streptozotocin (STZ)-induced T1DM mice. Gut Pathogens. 2022;14(1). https://doi.org/10.1186/s13099-022-00512-1.

[92] Jacky D, Bibi C, Meng LMC, et al. Effects of osomefood clean label plant-based meals on the gut microbiome. BMC Microbiology. 2023;23(1). https://doi.org/10.1186/s12866-023-02822-z.

[93] Stewart CJ, Auchtung TA, Ajami NJ, et al. Effects of tobacco smoke and electronic cigarette vapor exposure on the oral and gut microbiota in humans: A pilot study. PeerJ. 2018;6:e4693. https://doi.org/10.7717/peerj.4693.

[94] Nolan-Kenney R, Wu F, Hu J, et al. The association between smoking and gut microbiome in Bangladesh. Nicotine and Tobacco Research. 2020;22(8):1339. https://doi.org/10.1093/ntr/ntz220.

[95]   Hajek P, Phillips-Waller A, Przulj D, et al. A randomized trial of e-cigarettes versus nicotine-replacement therapy. The New England Journal of Medicine. 2019;380(7):629–637. https://doi.org/10.1056/NEJMoa1808779.

[96]   Lim MY, Yoon HS, Rho M, et al. Analysis of the association between host genetics, smoking, and sputum microbiota in healthy humans. Scientific Reports. 2016;6(1). https://doi.org/10.1038/srep23745.

[97]   Biedermann L, Zeitz J, Mwinyi J, et al. Smoking cessation induces profound changes in the composition of the intestinal microbiota in humans. Heimesaat MM, ed. PLoS ONE. 2013;8(3):e59260. https://doi.org/10.1371/journal.pone.0059260.

[98]   Martinez JE, Kahana DD, Ghuman S, et al. Unhealthy lifestyle and gut dysbiosis: A better understanding of the effects of poor diet and nicotine on the intestinal microbiome. Frontiers in Endocrinology. 2021;12(12). https://doi.org/10.3389/fendo.2021.667066.

[99]   Leiten EO, Nielsen R, Wiker HG, et al. The airway microbiota and exacerbations of COPD. ERJ Open Research. 2020;6(3):00168–2020. https://doi.org/10.1183/23120541.00168-2020.

[100] Yang D, Xing Y, Song X, Qian Y. The impact of lung microbiota dysbiosis on inflammation. Immunology. 2020;159(2):156–166. https://doi.org/10.1111/imm.13139.

[101] Natalini JG, Singh S, Segal NN. The dynamic lung microbiome in health and disease. Nature Reviews Microbiology. 2022;21. https://doi.org/10.1038/s41579-022-00821-x.

[102] Li K, Chen Z, Huang Y, et al. Dysbiosis of lower respiratory tract microbiome are associated with inflammation and microbial function variety. Respiratory Research. 2019;20(1). https://doi.org/10.1186/s12931-019-1246-0.

[103] Russo R, Karadja E, De Seta F. Evidence-based mixture containing *Lactobacillus* strains and lactoferrin to prevent recurrent bacterial vaginosis: A double blind, placebo controlled, randomised clinical trial. Beneficial Microbes. 2019;10(1):19–26. https://doi.org/10.3920/BM2018.0075.

[104] Chen X, Lu Y, Chen T, Li R. The Female vaginal microbiome in health and bacterial vaginosis. Frontiers in Cellular and Infection Microbiology. 2021;11(631972).

[105] Ughade PA, Shrivastava D, Chaudhari K. Navigating the microbial landscape: Understanding dysbiosis in human genital tracts and its impact on fertility. Cureus. Published online August 16 2024. https://doi.org/10.7759/cureus.67040.

[106] Elahi Z, Mokhtaryan M, Mahmoodi S, et al. All properties of infertility microbiome in a review article. Journal of Clinical Laboratory Analysis. Published online March 9 2025. https://doi.org/10.1002/jcla.25158.

[107] Hamory BH, Parisi JT. Staphylococcus epidermidis: A significant nosocomial pathogen. American Journal of Infection Control. 1987;15(2):59–74. https://doi.org/10.1016/0196-6553(87)90003-4.

[108] Otto M. Staphylococcus epidermidis – The "accidental" pathogen. Nature Reviews Microbiology. 2009;7(8):555–567. https://doi.org/10.1038/nrmicro2182.

[109] Sabaté Brescó M, Harris LG, Thompson K, et al. Pathogenic mechanisms and host interactions in *Staphylococcus epidermidis* device-related infection. Frontiers in Microbiology. 2017;8(8):1401. https://doi.org/10.3389/fmicb.2017.01401.

[110] Doan T, Akileswaran L, Andersen D, et al. Paucibacterial microbiome and resident DNA virome of the healthy conjunctiva. Investigative Ophthalmology and Visual Science. 2016;57(13):5116–5126. https://doi.org/10.1167/iovs.16-19803.

[111] Wang Y, Li X, Gu S, Fu J. Characterization of dysbiosis of the conjunctival microbiome and nasal microbiome associated with allergic rhinoconjunctivitis and allergic rhinitis. Frontiers in Immunology. 2023;14. https://doi.org/10.3389/fimmu.2023.1079154.

[112] Glatthardt T, Lima RD, Monteiro R, Barreto R. Microbe interactions within the skin microbiome. Antibiotics. 2024;13(1):49–49. https://doi.org/10.3390/antibiotics13010049.

[113] Smythe P, Wilkinson HN. The skin microbiome: Current landscape and future opportunities. International Journal of Molecular Sciences. 2023;24(4):3950-3950. https://doi.org/10.3390/ijms24043950.

[114] Sutaria AH, Schlessinger J. Acne vulgaris. National Library of Medicine. Published August 17, 2023. https://www.ncbi.nlm.nih.gov/books/NBK459173/

[115] Borrego-Ruiz A, Borrego JJ. Microbial dysbiosis in the skin microbiome and its psychological consequences. Microorganisms. 2024;12(9):1908-1908. https://doi.org/10.3390/microorganisms12091908.

[116] AlEdani EM, Ashour Y. All about acne vulgaris and adolescent acne plus its mechanism. Updates in Clinical Dermatology. Published online 2025;1–30. https://doi.org/10.1007/978-3-031-83677-0_1.

[117] Kim JE, Kim HS. Microbiome of the Skin and gut in atopic dermatitis (AD): Understanding the pathophysiology and finding novel management strategies. Journal of Clinical Medicine. 2019;8(4):444. https://doi.org/10.3390/jcm8040444.

[118] Mustari AP, Agrawal I, Das A, Vinay K. Role of cutaneous microbiome in dermatology. PubMed. 2023;68(3):303–312. https://doi.org/10.4103/ijd.ijd_560_22.

[119] Bjerre RD, Holm JB, Palleja A, Sølberg J, Skov L, Johansen JD. Skin dysbiosis in the microbiome in atopic dermatitis is site-specific and involves bacteria, fungus and virus. BMC Microbiology. 2021;21(1). https://doi.org/10.1186/s12866-021-02302-2.

[120] Liang Q, Li J, Zou Y, et al. Metagenomic analysis reveals the heterogeneity of conjunctival microbiota dysbiosis in dry eye disease. Frontiers in Cell and Developmental Biology. 2021;9. https://doi.org/10.3389/fcell.2021.731867.

[121] Afzaal M, Saeed F, Shah YA, et al. Human gut microbiota in health and disease: Unveiling the relationship. Frontiers in Microbiology. 2022;13(13). https://doi.org/10.3389/fmicb.2022.999001.

[122] Madhogaria B, Bhowmik P, Kundu A. Correlation between human gut microbiome and diseases. Infectious Medicine. 2022;1(3). https://doi.org/10.1016/j.imj.2022.08.004.

[123] Simonyté Sjödin K, Hammarström M, Rydén L, et al. Temporal and long-term gut microbiota variation in allergic disease: A prospective study from infancy to school age. Allergy. 2018;74(1):176–185. https://doi.org/10.1111/all.13485.

[124] Alcazar CGM, Paes VM, Shao Y, et al. The association between early-life gut microbiota and childhood respiratory diseases: A systematic review. The Lancet Microbe. 2022;3(11):e867–e880. https://doi.org/10.1016/s2666-5247(22)00184-7.

[125] Hickman B, Salonen A, Ponsero AJ, et al. Gut microbiota wellbeing index predicts overall health in a cohort of 1000 infants. Nature Communications. 2024;15(1). https://doi.org/10.1038/s41467-024-52561-6.

[126] Quesada-Vázquez S, Bone C, Saha S, et al. Microbiota dysbiosis and gut barrier dysfunction associated with non-alcoholic fatty liver disease are modulated by a specific metabolic cofactors. Combination. 2022;23(22):13675-13675. https://doi.org/10.3390/ijms232213675.

[127] Yan D, Sun Y, Zhou X, et al. Regulatory effect of gut microbes on blood pressure. 2022;5(6):513–531. https://doi.org/10.1002/ame2.12233.

[128] Tokarek J, Budny E, Saar M, et al. Does the composition of gut microbiota affect hypertension? molecular mechanisms involved in increasing blood pressure. International Journal of Molecular Sciences. 2023;24(2):1377. https://doi.org/10.3390/ijms24021377.

[129] Yang F, Chen H, Gao Y, et al. Gut microbiota-derived short-chain fatty acids and hypertension: Mechanism and treatment. Biomedicine & Pharmacotherapy. 2020;130:110503. https://doi.org/10.1016/j.biopha.2020.110503.

[130] Tsafack PB, Li C, Tsopmo A. Food peptides, gut microbiota modulation, and antihypertensive effects. Molecules. 2022;27(24):8806-8806. https://doi.org/10.3390/molecules27248806.

[131] Mardanparvar H. Aggravation of hypertension by gut microbiota dysbiosis; a short-review on new concepts. Journal of Renal Endocrinology. 2025;11:e25185. https://doi.org/10.34172/jre.2025.25185.

[132] Bier A, Braun T, Khasbab R, et al. A high salt diet modulates the gut microbiota and short chain fatty acids production in a salt-sensitive hypertension rat model. Nutrients. 2018;10(9):1154. https://doi.org/10.3390/nu10091154.

[133] Wilck N, Matus MG, Kearney SM, et al. Salt-responsive gut commensal modulates TH17 axis and disease. Nature. 2017;551(7682):585–589. https://doi.org/10.1038/nature24628.

[134] Sun D, Xiang H, Yan J, He L. Intestinal microbiota: A promising therapeutic target for hypertension. Frontiers in Cardiovascular Medicine. 2022;9:970036. https://doi.org/10.3389/fcvm.2022.970036.

[135] Singh P, Meenatchi R, Ahmed ZHT, et al. Implications of the gut microbiome in cardiovascular diseases: Association of gut microbiome with cardiovascular diseases, therapeutic interventions and multi-omics approach for precision medicine. Medicine in Microecology. 2024;19:100096. https://doi.org/10.1016/j.medmic.2023.100096.

[136] Motiani KK, Collado MC, Eskelinen JJ, et al. Exercise training modulates gut microbiota profile and improves endotoxemia. Medicine and Science in Sports and Exercise. Published online August 2019;1. https://doi.org/10.1249/mss.0000000000002112.

[137] Xu H, Wang X, Feng W, et al. The gut microbiota and its interactions with cardiovascular disease. Microbial Biotechnology. 2020;13(3):637–656. https://doi.org/10.1111/1751-7915.13524.

[138] Caradonna E, Abate F, Schiano E, et al. Trimethylamine-N-Oxide (TMAO) as a rising-star metabolite: Implications for human health. Metabolites. 2025;15(4):220–220. https://doi.org/10.3390/metabo15040220.

[139] Dolkar P, Deyang T, Anand N, et al. Trimethylamine-N-oxide and cerebral stroke risk: A review. Neurobiology of Disease. 2024;192:106423-106423. https://doi.org/10.1016/j.nbd.2024.106423.

[140] Shanmugham M, Bellanger S, Leo CH. Gut-derived metabolite, trimethylamine-N-oxide (TMAO) in cardio-metabolic diseases: Detection, mechanism, and potential therapeutics. Pharmaceuticals. 2023;16(4):504. https://doi.org/10.3390/ph16040504.

[141] Zuo K, Li J, Li K, et al. Disordered gut microbiota and alterations in metabolic patterns are associated with atrial fibrillation. GigaScience. 2019;8(6). https://doi.org/10.1093/gigascience/giz058.

[142] Abrignani V, Salvo A, Pacinella G, Tuttolomondo A. The Mediterranean diet, its microbiome connections, and cardiovascular health: A narrative review. International Journal of Molecular Sciences. 2024;25(9):4942. https://doi.org/10.3390/ijms25094942.

[143] Shayista H, Prasad MNN, Raj SN, Ranjini HK, Manju K, Baker S. Mechanistic overview of gut microbiota and mucosal pathogens with respect to cardiovascular diseases. The Microbe. 2024;5:100160. https://doi.org/10.1016/j.microb.2024.100160.

[144] Lupu VV, Adam Raileanu A, Mihai CM, et al. The implication of the gut microbiome in heart failure. Cells. 2023;12(8):1158. https://doi.org/10.3390/cells12081158.

[145] Huang J, Lin Y, Ding X, et al. Alteration of the gut microbiome in patients with heart failure: A systematic review and meta-analysis. Microbial Pathogenesis. 2024;192:106647. https://doi.org/10.1016/j.micpath.2024.106647.

[146] Li Z, Wu Z, Yan J, et al. Gut microbe-derived metabolite trimethylamine N-oxide induces cardiac hypertrophy and fibrosis. Laboratory Investigation. 2018;99(3):346–357. https://doi.org/10.1038/s41374-018-0091-y.

[147] Jarmukhanov Z, Mukhanbetzhanov N, Kozhakhmetov S, et al. The association between the gut microbiota metabolite trimethylamine N-oxide and heart failure. Frontiers in Microbiology. 2024;15. https://doi.org/10.3389/fmicb.2024.1440241.

[148] Kim B, Choi HN, Yim JE. Effect of diet on the gut microbiota associated with obesity. Journal of Obesity and Metabolic Syndrome. 2019;28(4):216–224. https://doi.org/10.7570/jomes.2019.28.4.216.

[149] Clemente-Suárez VJ, Beltrán-Velasco AI, Redondo-Flórez L, Martín-Rodríguez A, Tornero-Aguilera JF. Global impacts of western diet and its effects on metabolism and health: A narrative review. Nutrients. 2023;15(12):2749-2749. https://doi.org/10.3390/nu15122749.

[150] Chong S, Lin M, Chong D, Jensen S, Lau NS. A systematic review on gut microbiota in type 2 diabetes mellitus. Frontiers in Endocrinology. 2025;15:1486793. https://doi.org/10.3389/fendo.2024.1486793.

[151] Jandhyala SM. Role of the normal gut microbiota. World Journal of Gastroenterology. 2015;21 (29):8787. https://doi.org/10.3748/wjg.v21.i29.8787.

[152] Gurung M, Li Z, You H, et al. Role of gut microbiota in type 2 diabetes pathophysiology. EBioMedicine. 2020;51(102590). https://doi.org/10.1016/j.ebiom.2019.11.051.

[153] Slouha E, Rezazadah A, Farahbod K, Gerts A, Clunes LA, Kollias TF. Type-2 diabetes mellitus and the gut microbiota: Systematic review. Cureus. Published online November 30 2023. https://doi.org/10.7759/cureus.49740.

[154] Olša Fliegerová K, Mahayri TM, Sechovcová H, et al. Diabetes and gut microbiome. Frontiers in Microbiology. 2025;15. https://doi.org/10.3389/fmicb.2024.1451054.

[155] Arora T, Tremaroli V. Therapeutic potential of butyrate for treatment of type 2 diabetes. Frontiers in Endocrinology. 2021;12. https://doi.org/10.3389/fendo.2021.761834.

[156] Engelsberger V, Gerhard M, Mejías-Luque R. Effects of Helicobacter pylori infection on intestinal microbiota, immunity and colorectal cancer risk. Frontiers in Cellular and Infection Microbiology. 2024;14. https://doi.org/10.3389/fcimb.2024.1339750.

[157] Wroblewski LE, Peek RM, Wilson KT. Helicobacter pylori and gastric cancer: Factors that modulate disease risk. Clinical Microbiology Reviews. 2010;23(4):713–739. https://doi.org/10.1128/cmr.00011-10.

[158] Mishra Y, Ranjan A, Mishra V, et al. The role of the gut microbiome in gastrointestinal cancers. Cellular Signalling. 2024;115:111013-111013. https://doi.org/10.1016/j.cellsig.2023.111013.

[159] LaCourse KD, Johnston CD, Bullman S. The relationship between gastrointestinal cancers and the microbiota. The Lancet Gastroenterology & Hepatology. 2021;6(6):498–509. https://doi.org/10.1016/s2468-1253(20)30362-9.

[160] Dahmus JD, Kotler DL, Kastenberg DM, Kistler CA. The gut microbiome and colorectal cancer: A review of bacterial pathogenesis. Journal of Gastrointestinal Oncology. 2018;9(4):769–777. https://doi.org/10.21037/jgo.2018.04.07.

[161] Rajasekaran JJ, Krishnamurthy HK, Bosco J, et al. Oral microbiome: A review of its impact on oral and systemic health. Microorganisms. 2024;12(9):1797. https://doi.org/10.3390/microorganisms12091797.

[162] Peng X, Cheng L, You Y, et al. Oral microbiota in human systematic diseases. International Journal of Oral Science. 2022;14(1):1–11. https://doi.org/10.1038/s41368-022-00163-7.

[163] Weinberg F, Dickson RP, Nagrath D, Ramnath N. The lung microbiome: A central mediator of host inflammation and metabolism in lung cancer patients?. Cancers. 2020;13(1):13. https://doi.org/10.3390/cancers13010013.

[164] Belaid A, Roméo B, Rignol G, et al. Impact of the lung microbiota on development and progression of lung cancer. Cancers. 2024;16(19):3342-3342. https://doi.org/10.3390/cancers16193342.

[165] Voigt AY, Emiola A, Johnson JS, et al. Skin Microbiome Variation With Cancer Progression In Human Cutaneous Squamous Cell Carcinoma. The Journal of Investigative Dermatology. 2022;142(10):2773–2782.e16. https://doi.org/10.1016/j.jid.2022.03.017.

[166] Tian Z, Zhang X, Yao G, et al. Intestinal flora and pregnancy complications: Current insights and future prospects. iMeta. Published online January 22 2024. https://doi.org/10.1002/imt2.167.

[167] Agus A, Clément K, Sokol H. Gut microbiota-derived metabolites as central regulators in metabolic disorders. Gut. 2020;70(6):gutjnl-2020-323071. https://doi.org/10.1136/gutjnl-2020-323071.

[168] Zhang D, Huang Y, Ye D. Intestinal dysbiosis: An emerging cause of pregnancy complications?. Medical Hypotheses. 2015;84(3):223–226. https://doi.org/10.1016/j.mehy.2014.12.029.

[169] Beckers KF, Flanagan JP, Sones JL. Microbiome and pregnancy: Focus on microbial dysbiosis coupled with maternal obesity. International Journal of Obesity. Published online December 25 2023:1–10. https://doi.org/10.1038/s41366-023-01438-7.

[1730] Saadaoui M, Singh P, Al Khodor S. Oral microbiome and pregnancy: A bidirectional relationship. Journal of Reproductive Immunology. 2021;145:103293. https://doi.org/10.1016/j.jri.2021.103293.

[171] Jang H, Patoine A, Wu TT, Castillo DA, Xiao J. Oral microflora and pregnancy: A systematic review and meta-analysis. Scientific Reports. 2021;11(1):16870. https://doi.org/10.1038/s41598-021-96495-1.

[172] Giannella L, Grelloni C, Quintili D, et al. Microbiome changes in pregnancy disorders. Antioxidants. 2023;12(2):463. https://doi.org/10.3390/antiox12020463.

[173] Liu S, Gao J, Zhu M, Liu K, Zhang HL. Gut microbiota and dysbiosis in Alzheimer's disease: Implications for pathogenesis and treatment. Molecular Neurobiology. 2020;57(12):5026–5043. https://doi.org/10.1007/s12035-020-02073-3.

[174] Turjeman S, Collado MC, Koren O. The gut microbiome in pregnancy and pregnancy complications. Current Opinion in Endocrine and Metabolic Research. Published online March 2021. https://doi.org/10.1016/j.coemr.2021.03.004.

[175] Komiya S, Naito Y, Okada H, et al. Characterizing the gut microbiota in females with infertility and preliminary results of a water-soluble dietary fiber intervention study. Journal of Clinical Biochemistry and Nutrition. 2020;67(1):105–111. https://doi.org/10.3164/jcbn.20-53.

[176] d'Afflitto M, Upadhyaya A, Green A, Peiris M. Association between sex hormone levels and gut microbiota composition and diversity – a systematic review. Journal of Clinical Gastroenterology. 2022;56(5):384–392. https://doi.org/10.1097/mcg.0000000000001676.

[177] Yoon K, Kim N. Roles of sex hormones and gender in the gut microbiota. Journal of Neurogastroenterology and Motility. 2021;27(3):314–325. https://doi.org/10.5056/jnm20208.

[178] Santos-Marcos JA, Mora-Ortiz M, Tena-Sempere M, López-Miranda J, Camargo A. Interaction between gut microbiota and sex hormones and their relation to sexual dimorphism in metabolic diseases. Biology of Sex Differences. 2023;14(1). https://doi.org/10.1186/s13293-023-00490-2.

[179] Han H, Zhou Y, Xiong B, et al. Alginate oligosaccharides increase boar semen quality by affecting gut microbiota and metabolites in blood and sperm. Frontiers in Microbiology. 2022;13. https://doi.org/10.3389/fmicb.2022.982152.

[180] Alahmar A. Role of oxidative stress in male infertility: An updated review. Journal of Human Reproductive Sciences. 2019;12(1):4. https://doi.org/10.4103/jhrs.jhrs_150_18.

[181] Sengupta P, Pinggera G, Calogero AE, Agarwal A. Oxidative stress affects sperm health and fertility – Time to apply facts learned at the bench to help the patient: Lessons for busy clinicians. Reproductive Medicine and Biology. 2024;23(1). https://doi.org/10.1002/rmb2.12598.

[182] Kaltsas A, Zachariou A, Markou E, Dimitriadis F, Sofikitis N, Pournaras S. Microbial dysbiosis and male infertility: Understanding the impact and exploring therapeutic interventions. Journal of Personalized Medicine. 2023;13(10):1491-1491. https://doi.org/10.3390/jpm13101491.

[183] Dutta S, Sengupta P, Izuka E, Menuba I, Jegasothy R, Nwagha U. Staphylococcal infections and infertility: Mechanisms and management. Molecular and Cellular Biochemistry. Published online July 20 2020. https://doi.org/10.1007/s11010-020-03833-4.

[184] Qi X, Yun C, Pang Y, Qiao J. The impact of the gut microbiota on the reproductive and metabolic endocrine system. Gut Microbes. 2021;13(1):1894070. https://doi.org/10.1080/19490976.2021.1894070.

[185] Ashonibare VJ, Akorede BA, Ashonibare PJ, Akhigbe TM, Akhigbe RE. Gut microbiota-gonadal axis: The impact of gut microbiota on reproductive functions. Frontiers in Immunology. 2024;15. https://doi.org/10.3389/fimmu.2024.1346035.

[186] Hu S, Ding Q, Zhang W, Kang MI, Ma J, Zhao L. Gut microbial beta-glucuronidase: A vital regulator in female estrogen metabolism. Gut Microbes. 2023;15(1). https://doi.org/10.1080/19490976.2023.2236749.

[187] Beni FA, Saffarfar H, Elhami A, Kazemi M. Gut microbiota dysbiosis: A neglected risk factor for male and female fertility. Cellular Microbiology. 2024;2024(1). https://doi.org/10.1155/cmi/7808354.

[188] Ludgate ME, Masetti G, Soares P. The relationship between the gut microbiota and thyroid disorders. Nature Reviews Endocrinology. 2024;20(9):511–525. https://doi.org/10.1038/s41574-024-01003-w.

[189] Lathakumari RH, Vajravelu LK, Satheesan A, Ravi S, Thulukanam J. Antibiotics and the gut microbiome: Understanding the impact on human health. Medicine in Microecology. 2024;20(100106):100106-100106. https://doi.org/10.1016/j.medmic.2024.100106.

[190] Shah T, Baloch Z, Shah Z, Cui X, Xia X. The intestinal microbiota: Impacts of antibiotics therapy, colonization resistance, and diseases. International Journal of Molecular Sciences. 2021;22(12):6597. https://doi.org/10.3390/ijms22126597.

[191] Mougiou D, Gioula G, Skoura L, Anastassopoulou C, Kachrimanidou M. Insights into the interaction between *Clostridioides difficile* and the gut microbiome. Journal of Personalized Medicine. 2025;15(3):94. https://doi.org/10.3390/jpm15030094.

[192] Bu F, Chen K, Chen S, Jiang Y. Gut microbiota and intestinal immunity interaction in ulcerative colitis and its application in treatment. Frontiers in Cellular and Infection Microbiology. 2025;15. https://doi.org/10.3389/fcimb.2025.1565082.

[193] Kushkevych I, Dvořáková M, Dordevic D, et al. Advances in gut microbiota functions in inflammatory bowel disease: Dysbiosis, management, cytotoxicity assessment, and therapeutic perspectives. Computational and Structural Biotechnology Journal. 2025;27:851–868. https://doi.org/10.1016/j.csbj.2025.02.026.

[194] Silva YP, Bernardi A, Frozza RL. The role of short-chain fatty acids from gut microbiota in gut-brain communication. Frontiers in Endocrinology. 2020;11(25). https://doi.org/10.3389/fendo.2020.00025.

[195] Akhtar M, Chen Y, Ma Z, et al. Gut microbiota-derived short chain fatty acids are potential mediators in gut inflammation. Animal Nutrition. 2022;8:350–360. https://doi.org/10.1016/j.aninu.2021.11.005.

[196] Park J, Kim M, Kang SG, et al. Short-chain fatty acids induce both effector and regulatory T cells by suppression of histone deacetylases and regulation of the mTOR–S6K pathway. Mucosal Immunology. 2014;8(1):80–93. https://doi.org/10.1038/mi.2014.44.

[197] Du Y, He C, An Y, et al. The role of short chain fatty acids in inflammation and body health. International Journal of Molecular Sciences. 2024;25(13):7379-7379. https://doi.org/10.3390/ijms25137379.

[198] Strober W, Fuss IJ. Proinflammatory Cytokines in the pathogenesis of inflammatory bowel diseases. Gastroenterology. 2011;140(6):1756–1767. https://doi.org/10.1053/j.gastro.2011.02.016.

[199] Sanchez-Muñoz F, Dominguez-Lopez A, Yamamoto-Furusho JK. Role of cytokines in inflammatory bowel disease. World Journal of Gastroenterology. 2008;14(27):4280–4288. https://doi.org/10.3748/wjg.14.4280.

[200] Lin L, Zhang K, Xiong Q, et al. Gut microbiota in pre-clinical rheumatoid arthritis: From pathogenesis to preventing progression. Published online March 1, 2023:103001-103001. https://doi.org/10.1016/j.jaut.2023.103001.

[201] Lu J, Wang Y, Wu J, Duan Y, Zhang H, Du H. Linking microbial communities to rheumatoid arthritis: Focus on gut, oral microbiome and their extracellular vesicles. Frontiers in Immunology. 2025;16. https://doi.org/10.3389/fimmu.2025.1503474.

[202] Zheng D, Liwinski T, Elinav E. Interaction between microbiota and immunity in health and disease. Cell Research. 2020;30(6):492–506. https://doi.org/10.1038/s41422-020-0332-7.

[203] Zheng P, Li Z, Zhou Z. Gut microbiome in type 1 diabetes: A comprehensive review. Diabetes/Metabolism Research and Reviews. 2018;34(7):e3043. https://doi.org/10.1002/dmrr.3043.

[204] Mokhtari P, Metos J, Anandh Babu PV. Impact of type 1 diabetes on the composition and functional potential of gut microbiome in children and adolescents: Possible mechanisms, current knowledge, and challenges. Gut Microbes. 2021;13(1):1926841. https://doi.org/10.1080/19490976.2021.1926841.

[205] Yoo J, Groer M, Dutra S, Sarkar A, McSkimming D. Gut microbiota and immune system interactions. Microorganisms. 2020;8(10):1587. https://doi.org/10.3390/microorganisms8101587.

[206] Candelli M, Franza L, Pignataro G, et al. Interaction between lipopolysaccharide and gut microbiota in inflammatory bowel diseases. International Journal of Molecular Sciences. 2021;22(12):6242. https://doi.org/10.3390/ijms22126242.

[207] Cox LM, Maghzi AH, Liu S, et al. Gut microbiome in progressive multiple sclerosis. Annals of Neurology. 2021;89(6):1195–1211. https://doi.org/10.1002/ana.26084.

[208] Campagnoli LIM, Marchesi N, Varesi A, et al. New therapeutic avenues in multiple sclerosis: Is there a place for gut microbiota-based treatments?. Pharmacological Research. 2024;209:107456. https://doi.org/10.1016/j.phrs.2024.107456.

[209] Tang W, Zhu H, Feng Y, Guo R, Wan D. The impact of gut microbiota disorders on the blood–brain barrier. Infection and Drug Resistance. 2020;13:3351–3363. https://doi.org/10.2147/IDR.S254403.

[210] Suganya K, Koo BS. Gut–brain axis: Role of gut microbiota on neurological disorders and how probiotics/prebiotics beneficially modulate microbial and immune pathways to improve brain functions. International Journal of Molecular Sciences. 2020;21(20):7551. https://doi.org/10.3390/ijms21207551.

[211] Li J, Zhao F, Wang Y, et al. Gut microbiota dysbiosis contributes to the development of hypertension. Microbiome. 2017;5(1). https://doi.org/10.1186/s40168-016-0222-x.

[212] Overby HB, Ferguson JF. Gut microbiota-derived short-chain fatty acids facilitate microbiota:host cross talk and modulate obesity and hypertension. Current Hypertension Reports. 2021;23(2). https://doi.org/10.1007/s11906-020-01125-2.

[213] Malicevic U, Rai V, Skrbic R, Agrawal DK. NLRP3 Inflammasome and gut dysbiosis linking diabetes mellitus and inflammatory bowel disease. Archives of Internal Medicine Research. 2024;7(3). https://doi.org/10.26502/aimr.0178.

[214] Ruiz-Sánchez C, Escudero-López B, Fernández-Pachón MS. Evaluation of the efficacy of probiotics as treatment in irritable bowel syndrome. Endocrinología, Diabetes Y Nutrición (English Ed). 2024;71(1):19–30. https://doi.org/10.1016/j.endien.2024.01.003.

[215] Vatanen T, Franzosa EA, Schwager R, et al. The human gut microbiome in early-onset type 1 diabetes from the TEDDY study. Nature. 2018;562(7728):589–594. https://doi.org/10.1038/s41586-018-0620-2.

[216] Gradisteanu Pircalabioru G, Corcionivoschi N, Gundogdu O, et al. Dysbiosis in the development of type i diabetes and associated complications: From mechanisms to targeted gut microbes manipulation therapies. International Journal of Molecular Sciences. 2021;22(5):2763. https://doi.org/10.3390/ijms22052763.

[217] Loverdos K, Bellos G, Kokolatou L, et al. Lung microbiome in asthma: Current perspectives. Journal of Clinical Medicine. 2019;8(11). https://doi.org/10.3390/jcm8111967.

[218] Pathak JL, Yan Y, Zhang Q, Wang L, Ge L. The role of oral microbiome in respiratory health and diseases. Respiratory Medicine. 2021;185:106475. https://doi.org/10.1016/j.rmed.2021.106475.

[219] Mu Q, Tavella VJ, Luo XM. Role of *Lactobacillus reuteri* in human health and diseases. Frontiers in Microbiology. 2018;9(757). https://doi.org/10.3389/fmicb.2018.00757.

[220] Peng Y, Ma Y, Luo Z, Jiang Y, Xu Z, Yu R. *Lactobacillus reuteri* in digestive system diseases: Focus on clinical trials and mechanisms. Frontiers in Cellular and Infection Microbiology. 2023;13:1254198. https://doi.org/10.3389/fcimb.2023.1254198.

[221] Iddrisu I, Monteagudo-Mera A, Poveda C, et al. Malnutrition and gut microbiota in children. Nutrients. 2021;13(8):2727. https://doi.org/10.3390/nu13082727.

[222] Radjabzadeh D, Bosch JA, Uitterlinden AG, et al. Gut microbiome-wide association study of depressive symptoms. Nature Communications. 2022;13(1):7128. https://doi.org/10.1038/s41467-022-34502-3.

[223] Patel RA, Panche AN, Harke SN. Gut microbiome-gut brain axis-depression: Interconnection. The World Journal of Biological Psychiatry. 2024;26(1):1–36. https://doi.org/10.1080/15622975.2024.2436854.

[224] Tian P, Chen Y, Zhu H, et al. *Bifidobacterium breve* CCFM1025 attenuates major depression disorder via regulating gut microbiome and tryptophan metabolism: A randomized clinical trial. Brain, Behavior, and Immunity. 2022;100:233–241. https://doi.org/10.1016/j.bbi.2021.11.023.

[225] Zhang X, Chen S, Zhang M, et al. Effects of fermented milk containing lacticaseibacillus paracasei strain shirota on constipation in patients with depression: A randomized, double-blind, placebo-controlled trial. Nutrients. 2021;13(7):2238. https://doi.org/10.3390/nu13072238.

[226] Kim SY, Woo SY, Raza S, et al. Association between gut microbiota and anxiety symptoms: A large population-based study examining sex differences. Journal of Affective Disorders. Published online April 2023. https://doi.org/10.1016/j.jad.2023.04.003.

[227] Boustany A, Feuerstadt P, Tillotson G. The 3 Ds: Depression, dysbiosis, and clostridiodes difficile. Advances in Therapy. 2024;41(11):3982–3995. https://doi.org/10.1007/s12325-024-02972-0.

[228] Cheng Y, Wang Y, Zhang W, Yin J, Dong J, Liu J. Relationship between intestinal flora, inflammation, BDNF gene polymorphism and generalized anxiety disorder: A clinical investigation. Medicine. 2022;101(29):e28910. https://doi.org/10.1097/md.0000000000028910.

[229] Zhu R, Fang Y, Li H, et al. Psychobiotic Lactobacillus plantarum JYLP-326 relieves anxiety, depression, and insomnia symptoms in test anxious college via modulating the gut microbiota and its metabolism. Frontiers in Immunology. 2023;14. https://doi.org/10.3389/fimmu.2023.1158137.

[230] Lewis-Mikhael AM, Davoodvandi A, Jafarnejad S. Effect of Lactobacillusplantarum containing probiotics on blood pressure: A systematic review and meta-analysis. Pharmacological Research. 2020;153:104663. https://doi.org/10.1016/j.phrs.2020.104663.

[231] Kim ED, Lee HS, Kim KT, Paik HD. Antioxidant and angiotensin-converting enzyme (ACE) inhibitory activities of yogurt supplemented with *Lactiplantibacillus plantarum* NK181 and *Lactobacillus delbrueckii* KU200171 and sensory evaluation. Foods. 2021;10(10):2324. https://doi.org/10.3390/foods10102324.

[232] Yuan L, Li Y, Chen M, et al. Antihypertensive activity of milk fermented by *Lactiplantibacillus plantarum* SR37-3 and SR61-2 in L-NAME-induced hypertensive rats. In: Foods (*Basel, Switzerland*). 2022;vol. 11(15):2332. https://doi.org/10.3390/foods11152332.

[233] Solanki D, Sakure A, Prakash S, Hati S. Characterization of Angiotensin I-Converting Enzyme (ACE) inhibitory peptides produced in fermented camel milk (Indian breed) by Lactobacillus acidophilus NCDC-15. Journal of Food Science and Technology. Published online January 14 2022. https://doi.org/10.1007/s13197-022-05357-9.

[234] Abdulrahim AO, Sai N, Salman N, et al. The gut–heart axis: A review of gut microbiota, dysbiosis, and cardiovascular disease development. Annals of Medicine and Surgery. 2024;87(1):177–191. https://doi.org/10.1097/ms9.0000000000002789.

[235] Wang X, Zhou S, Hu X, et al. Candida albicans accelerates atherosclerosis by activating intestinal hypoxia-inducible factor2α signaling. Cell Host & Microbe. 2024;32(6):964–979.e7. https://doi.org/10.1016/j.chom.2024.04.017.

[236] Adi A, Lebrun S, Kondo M, Alvarez Villela M, Fontes JD. Culture-negative subacute lactobacillus endocarditis diagnosed by microbial cell-free DNA sequencing. JACC: Case Reports. 2025;30 (12):103505. https://doi.org/10.1016/j.jaccas.2025.103505.

[237] Ioannou P, Ziogou A, Giannakodimos I, Giannakodimos A, Baliou S, Samonis G. Infective endocarditis by lactobacillus species – A narrative review. Antibiotics. 2024;13(1):53–53. https://doi.org/10.3390/antibiotics13010053.

[238] KavianFar A, Taherkhani H, Lanjian H, et al. Keystone bacteria dynamics in chronic obstructive pulmonary disease (COPD): Towards differential diagnosis and probiotic candidates. Heliyon. 2025;11(4):e42719. https://doi.org/10.1016/j.heliyon.2025.e42719.

[239] Cruz Mosquera FE, Perlaza CL, Naranjo Rojas A, et al. Effectiveness of probiotics, prebiotics, and symbiotic supplementation in cystic fibrosis patients: A systematic review and meta-analysis of clinical trials. Medicina (Kaunas, Lithuania). 2025;61(3):489. https://doi.org/10.3390/medicina61030489.

[240] De Freitas MB, Moreira EAM, Tomio C, et al. Altered intestinal microbiota composition, antibiotic therapy and intestinal inflammation in children and adolescents with cystic fibrosis. Palaniyar N, ed. PLOS ONE. 2018;13(6):e0198457. https://doi.org/10.1371/journal.pone.0198457.

[241] Chen TA, Chuang YT, Pai SC, Zheng JF. The potential of probiotics in reducing ventilator-associated pneumonia: A literature-based analysis. Microorganisms. 2025;13(4):856–856. https://doi.org/10.3390/microorganisms13040856.

[242] Fernández-Barat L, López-Aladid R, Torres A. Reconsidering ventilator-associated pneumonia from a new dimension of the lung microbiome. EBioMedicine. 2020;60:102995. https://doi.org/10.1016/j.ebiom.2020.102995.

[243] Cheng J, Wang H. Lung microbiome alterations correlate with immune imbalance in non-small cell lung cancer. Frontiers in Immunology. 2025;16. https://doi.org/10.3389/fimmu.2025.1589843.

[244] Yiminniyaze R, Zhang Y, Zhu N, et al. Characterizations of lung cancer microbiome and exploration of potential microbial risk factors for lung cancer. Scientific Reports. 2025;15(1). https://doi.org/10.1038/s41598-025-98424-y.

[245] Panebianco C, Pisati F, Ulaszewska MM, et al. Tuning gut microbiota through a probiotic blend in gemcitabine-treated pancreatic cancer xenografted mice. 2021;11(11). https://doi.org/10.1002/ctm2.580.

[246] Alagiakrishnan K, Morgadinho J, Halverson T. Approach to the diagnosis and management of dysbiosis. Frontiers in Nutrition. 2024;11. https://doi.org/10.3389/fnut.2024.1330903.

[247] Abdulaal R, Tlaiss Y, Jammal F, et al. The role of microbiome dysbiosis in cardiovascular disease: Mechanisms and therapeutic implications. Global Cardiology Science and Practice. 2025;2025(1). https://doi.org/10.21542/gcsp.2025.3.

[248] Mutalub YB, Abdulwahab M, Mohammed A, et al. Gut microbiota modulation as a novel therapeutic strategy in cardiometabolic diseases. Foods. 2022;11(17):2575. https://doi.org/10.3390/foods11172575.

[249] Gong X, Liu S, Xia B, et al. Oral delivery of therapeutic proteins by engineered bacterial type zero secretion system. Nature Communications. 2025;16(1). https://doi.org/10.1038/s41467-025-57153-6.

[250] Yu M, Hu S, Tang B, Yang H, Sun D. Engineering *Escherichia coli Nissle* 1917 as a microbial chassis for therapeutic and industrial applications. Biotechnology Advances. 2023;67:108202-108202. https://doi.org/10.1016/j.biotechadv.2023.108202.

[251] Kan A, Gelfat I, Emani S, Praveschotinunt P, Joshi NS. Plasmid vectors for in vivo selection-free use with the Probiotic *E. coli Nissle* 1917. ACS Synthetic Biology. 2020;10(1):94–106. https://doi.org/10.1021/acssynbio.0c00466.

[252] Zou ZP, Du Y, Fang TT, Zhou Y, Ye B. Biomarker-responsive engineered probiotic diagnoses, records, and ameliorates inflammatory bowel disease in mice. Cell Host & Microbe. 2023;31(2):199–212. e5. https://doi.org/10.1016/j.chom.2022.12.004.

[253] Li S, Jiang W, Zheng C, et al. Oral delivery of bacteria: Basic principles and biomedical applications. Journal of Controlled Release. 2020;327:801–833. https://doi.org/10.1016/j.jconrel.2020.09.011.

[254] Xu C, Ban Q, Wang W, Hou J, Jiang Z. Novel nano-encapsulated probiotic agents: Encapsulate materials, delivery, and encapsulation systems. Journal of Controlled Release. 2022;349:184–205. https://doi.org/10.1016/j.jconrel.2022.06.061.

[255] Han K, Xu J, Xie F, Crowther J, Moon JJ. Engineering strategies to modulate the gut microbiome and immune system. Journal of Immunology. 2024;212(2):208–215. https://doi.org/10.4049/jimmunol.2300480.

[256] Maftei NM, Raileanu CR, Balta AA, et al. The potential impact of probiotics on human health: an update on their health-promoting properties. Microorganisms. 2024;12(2):234. https://doi.org/10.3390/microorganisms12020234.

[257] El-Saadony MT, Alagawany M, Patra AK, et al. The functionality of probiotics in aquaculture: An overview. Fish and Shellfish Immunology. 2021;117:36–52. https://doi.org/10.1016/j.fsi.2021.07.007.

[258] Liber A, Szajewska H. Effect of oligofructose supplementation on body weight in overweight and obese children: A randomised, double-blind, placebo-controlled trial. British Journal of Nutrition. 2014;112(12):2068–2074. https://doi.org/10.1017/s0007114514003110.

[259] Bedu-Ferrari C, Biscarrat P, Langella P, Cherbuy C. Prebiotics and the human gut microbiota: from breakdown mechanisms to the impact on metabolic health. Nutrients. 2022;14(10):2096. https://doi.org/10.3390/nu14102096.

[260] Roy S, Dhaneshwar S. Role of prebiotics, probiotics, and synbiotics in management of inflammatory bowel disease: Current perspectives. World Journal of Gastroenterology. 2023;29(14):2078–2100. https://doi.org/10.3748/wjg.v29.i14.2078.

[261] Nicolucci AC, Hume MP, Martínez I, Mayengbam S, Walter J, Reimer RA. Prebiotics reduce body fat and alter intestinal microbiota in children who are overweight or with obesity. Gastroenterology. 2017;153(3):711–722. https://doi.org/10.1053/j.gastro.2017.05.055.

[262] Ouwehand AC, Tiihonen K, Mäkivuokko H, Rautonen N. Synbiotics: Combining the benefits of pre- and probiotics. Woodhead Publishing. 2014;195–213. https://doi.org/10.1533/9781845693107.2.195.

[263] Ganatsios V, Nigam P, Plessas S, Terpou A. Kefir as a functional beverage gaining momentum towards its health promoting attributes. Beverages. 2021;7(3):48. https://doi.org/10.3390/beverages7030048.

[264] Moncada E, Bulut N, Li S, Johnson T, Hamaker B, Reddivari L. Dietary fiber's physicochemical properties and gut bacterial dysbiosis determine fiber metabolism in the gut. Nutrients. 2024;16(15):2446. https://doi.org/10.3390/nu16152446.

[265] Núñez-Gómez V, Jesús Periago M, Luis Ordóñez-Díaz J, Pereira-Caro G, Moreno-Rojas JM, González-Barrio R. Dietary fibre fractions rich in (poly)phenols from orange by-products and their metabolisation by in vitro digestion and colonic fermentation. Food Research International. Published online November 1 2023:113718-113718. https://doi.org/10.1016/j.foodres.2023.113718.

[266] TabibzadehTehrani P, Nazari M, Rastgoo P, et al. Bacterial-based drug delivery systems: A new way to combat infectious disease. Medicine in Drug Discovery. Published online March 1 2025:100205-100205. https://doi.org/10.1016/j.medidd.2025.100205.

[267] Zhou L, Liu Q, Fang Z, Li QL, Wong HM. Targeted antimicrobial self-assembly peptide hydrogel with in situ bio-mimic remineralization for caries management. Bioactive Materials. 2025;44:428–446. https://doi.org/10.1016/j.bioactmat.2024.10.022.

[268] Ma X, Yang N, Mao R, et al. Self-assembly antimicrobial peptide for treatment of multidrug-resistant bacterial infection. Journal of Nanobiotechnology. 2024;22(1). https://doi.org/10.1186/s12951-024-02896-5.

[269] Behrouzi A, Ashrafian F, Mazaheri H, et al. The importance of interaction between MicroRNAs and gut microbiota in several pathways. Microbial Pathogenesis. 2020;144:104200. https://doi.org/10.1016/j.micpath.2020.104200.

[270] Liu S, Rezende RM, Garcias Moreira T, et al. Oral administration of miR-30d from feces of MS patients suppresses MS-like symptoms in mice by expanding *Akkermansia Muciniphila*. Cell Host & Microbe. 2019;26(6):779–794.e8. https://doi.org/10.1016/j.chom.2019.10.008.

[271] Cano R, Bermúdez V, Galban N, et al. Dietary Polyphenols and gut microbiota cross-talk: Molecular and therapeutic perspectives for cardiometabolic disease: A narrative review. International Journal of Molecular Sciences. 2024;25(16):9118-9118. https://doi.org/10.3390/ijms25169118.

[272] Perrone P, D'Angelo S. Gut microbiota modulation through Mediterranean diet foods: Implications for human health. Nutrients. 2025;17(6):948. https://doi.org/10.3390/nu17060948.

[273] Cha RR, Sonu I. Fecal microbiota transplantation: Present and future. Clinical Endoscopy. Published online March 25 2025. https://doi.org/10.5946/ce.2024.270.

[274] Song Z, Bu S, Sang S, et al. The Active components of traditional Chinese medicines regulate the multi-target signaling pathways of metabolic dysfunction-associated fatty liver disease. Drug Design, Development and Therapy. 2025;19:2693–2715. https://doi.org/10.2147/dddt.s514498.

[275] Zhang R, Gao X, Bai H, Ning K. Traditional Chinese medicine and gut microbiome: Their respective and concert effects on healthcare. Frontiers in Pharmacology. 2020;11. https://doi.org/10.3389/fphar.2020.00538.

[276] Liu Y, Li X, Chen Y, et al. Fecal microbiota transplantation: Application scenarios, efficacy prediction, and factors impacting donor-recipient interplay. Frontiers in Microbiology. 2025;16. https://doi.org/10.3389/fmicb.2025.1556827.

[277] Liu Y, Li X, Chen Y, et al. Fecal microbiota transplantation: Application scenarios, efficacy prediction, and factors impacting donor-recipient interplay. Frontiers in Microbiology. 2025;16. https://doi.org/10.3389/fmicb.2025.1556827.

Harshada Joshi, Jyoti, Vidhi Jain, Tamanna Ajmera, Moomal Acharya,
Namita Ashish Singh*

Chapter 3
Introduction to Probiotics, Prebiotics, Synbiotics, and Postbiotics: Potent Functional Foods

Abstract: Functional foods have drawn a lot of scientific interest throughout the years, particularly in the fields of technology and better dietary health. Prebiotics, probiotics, synbiotics, and postbiotics are included in functional foods. Probiotics are live microorganisms that provide health benefits when consumed in appropriate amounts and prebiotics are indigestible fibers that primarily encourage the proliferation and action of good bacteria. Prebiotics have demonstrated immense potential in altering the makeup as well as function of the microbiota in the gut. While postbiotics function as biologically active compounds with potential antimicrobial and anti-inflammatory benefits, synbiotics combine probiotics and prebiotics to enhance gut well-being and immunity. These foods can be used into a preventative healthcare and wellness strategy. The chapter highlights the role and types of probiotics, prebiotics, synbiotics, and postbiotics along with the classes, functions, and mechanisms of these elements in addressing many facets of good health and fostering general well-being. The health benefits of functional foods in maintaining the equilibrium of gut flora, increasing immunity, improving metabolic parameters, and lowering the risk of certain diseases are also highlighted. By combining the functional foods into our meals, we can proactively establish a harmonic, balanced environment within ourselves, which will contribute to a healthy future.

Keywords: functional food, probiotic, prebiotic, synbiotic, postbiotics

3.1 Introduction

An individual's daily diet influences their health and demands, so the inclusion of foods that offer greater advantages than merely providing energy, minerals, trace ele-

---
*Corresponding author: **Namita Ashish Singh**, Department of Microbiology, Mohanlal Sukhadia University, Udaipur 313001, Rajasthan, India, e-mail:namita.singh@mlsu.ac.in
Harshada Joshi, Tamanna Ajmera, Moomal Acharya, Department of Biotechnology, Mohanlal Sukhadia University, Udaipur 313001, Rajasthan, India
Jyoti, Vidhi Jain, Department of Microbiology, Mohanlal Sukhadia University, Udaipur 313001, Rajasthan, India

https://doi.org/10.1515/9783112205150-003

ments, and vitamins are required such as functional foods [1, 2]. The idea of endorsing functional foods originated in Japan in 1984 due to research showing the connection between nutrition, taste, stimulation of physiological systems (like immunity), and food enhancement. Functional foods may be natural or processed dietary staples that deliver essential nutrition and also offer potential beneficial effects on the host's health, such as decreasing disease incidence by enhancing the immune system's ability to prevent and manage infections from pathogens and disorders that lead to functional changes in the host [3]. The type of food and the necessary consumption must serve distinct functions; because functional food cannot exist as pills or capsules. These are foods that satisfy the body's nutritional requirements while also positively impacting health, provided they are eaten in appropriate amounts and in line with proper nutrition guidelines [4]. Aside from conventional fermented foods, key types of functional foods consist of probiotics, prebiotics, synbiotics (a combination of probiotics and prebiotics), and postbiotics (PPSP). Functional foods have been noted to alter physiological processes in the gastrointestinal tract (GIT) by enhancing biochemical markers and boosting neuronal functions [5].

The market for functional foods, particularly those with probiotic bacteria that make up 60% of all functional foods, is continually growing. Research in science has thoroughly documented numerous health benefits of these functional food components. These advantages include boosting the immune system, alleviating symptoms of irritable bowel disease, preventing and treating diarrhea, reducing the severity of allergies, and showing anti-inflammatory and antimicrobial properties [6]. Additionally, recent findings suggest that PPSPs can address metabolic issues by affecting the composition of the gut microbiome, enhancing the intestinal barrier's integrity, regulating the output of metabolites from the intestinal microbiota, and boosting overall immune responses [7].

Probiotics, which are live beneficial bacteria, can be found in fermented products such as yogurt, kimchi, and kefir. They promote several health benefits including healthy gut microbiome, aiding in immunity, digestion, and potentially decreasing inflammation [8]. Prebiotics are indigestible fibers present in food that particularly promote the growth and development of helpful gut microbes [9], providing them with the vital nutrients required for their thriving [10]. Foods such as banana, garlic, and onion contain compounds that support beneficial bacteria in the gut and enhance the growth as well as function of probiotics. Synbiotics, made up of prebiotics and probiotics [11], work together to support the survival and growth of beneficial gut bacteria, leading to improved gastrointestinal (GI) health and overall immunity [12]. In recent years, new terms have emerged, including paraprobiotics (inactive probiotic cells) and postbiotics (probiotic metabolites); as studies have shown that dead cells, whether whole or broken, can significantly impact human health. The health impacts of intestinal microbiomes rely on the microbiome's viability and on nonviable microbiome products, while postbiotics, which originate from probiotic cells, provide health advantages to the host when consumed in sufficient amounts [13]. Thus, postbiotics can be described as metabolites or elements created by microbiota that influence human health. The International Scientific Association for Probiotic and Prebiotics (ISAPP) describes postbiotics as "a preparation of non-living microorganisms and/or their components that provides a health advantage to the host."

3.2 Probiotics

Probiotics, available as supplements or incorporated into food products, have risen to prominence as key ingredients in the field of functional foods and have long been marketed for their potential to offer health benefits [14]. In 1907, Elie Metchnikoff observed that the frequent intake of fermented dairy products was linked to the enhanced health and increased lifespan of Bulgarian peasants [15]. The Food and Agriculture Organization and the World Health Organization have defined them as "live microorganisms that, when consumed in sufficient amounts, offer health advantages to the host" [16]. Probiotics are widely used in the food industry, especially in fermented dairy products. Fermented milk products containing probiotics become a valuable part of the diet since they improve the functional properties of food and have a positive effect on the health status of the consumer. They can be a preventive and therapeutic tool in many diseases due to the modulation of the composition of intestinal microflora [17]. They are nonpathogenic and have been engineered for compatibility with industrial processes. They produce antimicrobial agents (AAs) and are capable of surviving exposure to stomach acid and bile [18]. In addition to these, the bacteria produce short-chain fatty acids (SCFAs), hydrogen peroxide (H_2O_2), and diacetyl, which influence the intestinal microflora, leading to advantageous health outcomes [19].

A range of bacteria from the genera *Lactobacillus, Bifidobacterium, Pediococcus, Lactococcus, Enterococcus, Streptococcus, Propionibacterium*, and *Bacillus* are identified as potential candidates for achieving probiotic status [20]. They are involved in detoxifying xenobiotics (metals, polycyclic aromatic hydrocarbons, etc.), environmental pollutants (pesticides, drugs, etc.), transform mycotoxins present in foods, produce vitamin K, riboflavin, and folate, and break down undigested fiber in the colon [21–23]. Probiotics play a role in fortifying the epithelial barrier and facilitating the attachment of beneficial microbes to the intestinal mucosa, which helps to block harmful pathogens from adhering [24]. The benefits of probiotics include the regulation of gut microbiota, the reduction of nutritional intolerances such as lactose intolerance, the improvement of macro- and micronutrient bioavailability, and the reduction of allergic reactions in individuals who are prone to them [25]. Probiotics may be ingested through their incorporation into food products or beverages, whether they are dairy or nondairy, or by taking them as dietary supplements [26].

Any microbial strain can be utilized as a probiotic, if it meets specific safety and functionality standards. These standards encompass genetic stability, resistance to acid and bile, the capacity to adhere to the intestinal lining, anti-genotoxic character-istics, a nonpathogenic nature, lactic acid production, resilience to rigorous processing conditions, and a shorter generation time [27]. Considering the effective dosage, pro-biotics are being added to a variety of foods such as beverages, ice cream, yogurt, bread, and more by the food industry. The primary challenge with probiotics in the food sector is their vulnerability to processing conditions and sensitivity to GI stresses. Nevertheless, due to their health advantages, consumers consistently show a strong interest in probiotic products [28].

3.2.1 Mechanism of Action

Significant progress has been achieved in the realm of probiotics, yet a crucial break-through in understanding their mechanisms of action remain elusive. Probiotics may benefit the human body through several primary mechanisms: competitive exclusion of harmful microorganisms, enhancement of intestinal barrier functions, modulation of the host's immune system, and neurotransmitter production (Figure 3.1). By com-peting with pathogens for nutrients and receptor sites, probiotics hinder their sur-vival in the gut [29].

Additionally, probiotics function as AAs by producing substances such as SCFAs, organic acids, H_2O_2 [30], and bacteriocins [31], thereby reducing pathogenic bacteria in the gut. Furthermore, probiotics enhance intestinal barrier function by promoting mucin protein production [32], regulating the expression of tight junction proteins like occludin and claudin 1, and modulating the immune response in the gut [33]. Pro-biotics exert influence on both innate and adaptive immune responses by modulating dendritic cells (DCs), macrophages, and B and T lymphocytes. They enhance the pro-duction of anti-inflammatory cytokines through interactions with intestinal epithelial cells, and the recruitment of macrophages and mononuclear cells [34]. Furthermore, probiotics are capable of producing neurotransmitters in the gut via the gut-brain axis. Specific probiotic strains can modify levels of serotonin, gamma-aminobutyric acid (GABA), and dopamine, thereby affecting mood, behavior, gut motility, and stress-related pathways [35].

3.2.2 Types of Probiotics

Currently, a diverse array of microorganisms is employed as probiotics [19]. For ex-ample, *Lactobacillus* genus produces lactic acid in the GIT, *Bifidobacterium* produces acetic and lactic acids as metabolic by-products, and *Bacillus coagulans* is effective in inhibiting pathogen colonization and restoring normal intestinal microbiota [36]. Var-ious prevalent probiotic bacteria found in yogurt, fermented foods, and dietary sup-plements are mentioned below [37].

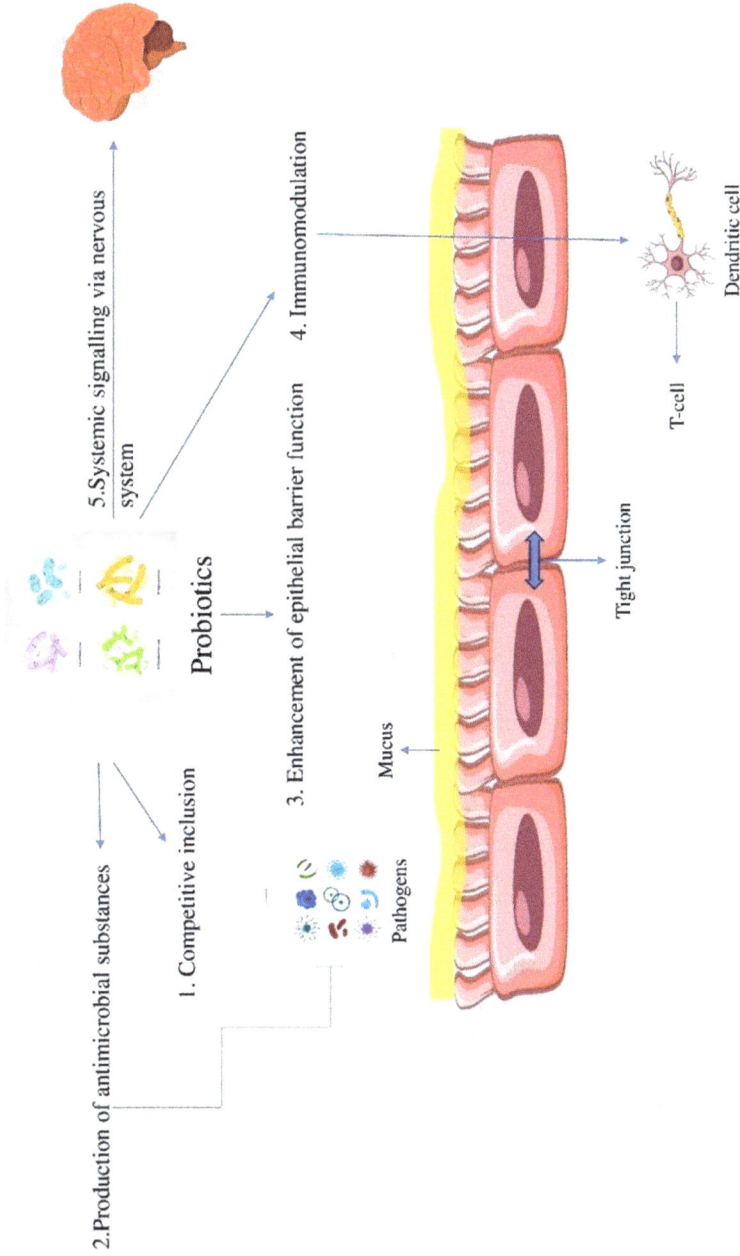

Figure 3.1: Mechanism of action of probiotics.

3.2.2.1 *Lactobacillus acidophilus*

Lactobacillus acidophilus is a type of probiotic bacteria that is extensively used for its potential health benefits. This strain is capable of adhering to various intestinal cells, can withstand bile, and is resistant to acid, which are crucial characteristics for a probiotic. However, laboratory studies have shown that some strains of *L. acidophilus* can reduce the level of cholesterol by more than 50%. This highlights their potential importance in promoting cardiovascular health, especially when combined with other probiotic strains. Additionally, it has proven effective in preventing GI diseases in adults and alleviating symptoms of the common cold in children. The commercially available strains of *L. acidophilus* include LA-1, LA-5, NCFM, DDS-1, and SBT-2026 [37].

3.2.2.2 *Lactobacillus rhamnosus*

The *Lactobacillus rhamnosus* strain has evolved unique traits that allow it to thrive in both acidic and alkaline conditions found within the human body. Its ability to adhere to and colonize the intestinal lining offers potential long-term benefits. As a result, it is commonly incorporated into yogurts, cheeses, milk, and other dairy products to enhance probiotic content. Furthermore, *L. rhamnosus* is crucial in the cheese-ripening process, contributing to improved flavor. Additionally, certain strains of *L. rhamnosus* have shown beneficial effects for both adults and children, particularly in managing irritable bowel syndrome (IBS), eczema, allergies, and supporting the immune system [38].

3.2.2.3 *Bifidobacterium breve*

Bifidobacterium breve, a species within the *Bifidobacterium* genus, is well-known for its probiotic properties. This bacterium, which lives symbiotically in the human gut, has been used to treat various conditions, including constipation, diarrhea, IBS, and even the common cold and flu. Scientific studies have confirmed several uses, highlighting the potential health benefits of *B. breve*. *B. breve* strains are widely used in pediatrics and are the dominant species in the GIT of breastfed infants. Additionally, they have been isolated from human milk, underscoring their natural presence in infants' digestive systems [39].

3.2.2.4 *Bacillus coagulans*

B. coagulans is a probiotic bacterium known for its ability to form spores, which grants it remarkable resistance to challenging environments and a wide range of probiotic benefits. This resilience allows it to remain dormant in tough conditions, such as high stomach acidity. Its natural robustness makes it highly effective in alleviating GI issues and other health problems. Moreover, it can regulate the host's beneficial microbiota and inhibit the growth of harmful bacteria, thereby enhancing overall gut health and supporting both the digestive and immune systems. *B. coagulans* is present in natural food sources, including fermented foods like sauerkraut, kimchi, and yogurt. Additionally, it is used in various probiotic food additives, highlighting its applicability for industrial use in the food industry [40].

3.2.2.5 *Streptococcus thermophilus*

Streptococcus thermophilus is a bacterium commonly used in the production of various dairy products, including cheese and yogurt. It plays a role in breaking down lactose in milk, contributing to yogurt's unique taste and texture. Additionally, this bacterium is recognized for its ability to reduce fat content in certain cheeses, like Swiss cheese, by producing natural polymer extracts. This specific probiotic strain is associated with numerous health benefits, such as enhancing the immune system and reducing inflammation in the GI and urogenital tracts. Moreover, it has shown promise in fighting viral, fungal, and parasitic infections. The presence of both *Bifidobacterium bifidum* and *S. thermophilus* in infants has been linked to a lower occurrence of rotavirus diarrhea. This combination may also help lessen inflammation-related damage caused by sepsis, highlighting its potential as a dietary supplement [41].

3.2.2.6 *Saccharomyces cerevisiae*

Widely recognized for its probiotic properties, *Saccharomyces cerevisiae*, particularly the *Saccharomyces boulardii* variant, has been thoroughly studied for its beneficial effects on GI health in both humans and animals. It is often used as a supplementary treatment for GI disorders, such as inflammatory bowel disease, and for managing various types of diarrhea. This organism's protective actions include binding to and neutralizing intestinal pathogens or their toxins, reducing inflammation, and promoting immunoglobulin A (IgA) secretion. These strains have the potential to be used in functional food products and can protect DNA from damage. The possible probiotic characteristics of *S. cerevisiae* yeast include its ability to auto-aggregate, co-aggregate with pathogens, exhibit hydrophobicity, survive in simulated GI environments, and adhere to Caco-2 cells. These traits make them strong candidates for therapeutic applications [42].

3.2.2.7 *Escherichia coli* Nissle 1917

This particular *Escherichia coli* strain serves as a probiotic, primarily utilized for its efficacy in addressing specific GI issues. It is beneficial in managing ulcerative colitis and preventing its recurrence. Additionally, it promotes gut health by supporting the intestinal barrier and influencing the immune system [43].

3.2.2.8 Other Probiotic Species

Leuconostoc, frequently utilized in the fermentation of vegetables, generates lactic acid and plays a role in promoting gut health. *Pediococcus*, often present in fermented foods, aids in preserving and improving food quality. *Propionibacterium*, used in making Swiss cheese, can enhance gut health by producing beneficial substances such as propionic acid [44, 45]. Table 3.1 delineates various probiotics strains, sources with their health benefits to mankind.

Table 3.1: Probiotic strains containing sources with their health benefits.

Probiotics	Sources	Health benefits	References
Lactobacillus acidophilus and *Lactiplantibacillus plantarum*	Yogurt, milk, kefir, sauerkraut, kimchi, some cheeses, and sourdough	Cholesterol lowering activities	[46, 47]
Lactobacillus acidophilus, *Bifidobacterium infantis*, and *Saccharomyces boulardii*	Yogurt, kefir, some cheeses, tempeh, apples, dates, lentils, blueberries, and broccoli	Prevent many diseases and disorder like inflammatory bowel disease and syndrome	[10, 48–51]
Lactobacillus rhamnosus	Yogurt, kefir, sauerkraut, kimchi, some cheeses, and sourdough	Beneficial in treatment of urinary tract infections	[52]
Bacillus subtilis, Saccharomyces boulardii, Bifidobacterium longum, Bifidobacterium bifidum, Lacticaseibacillus casei, and *Lactobacillus acidophilus*	Yogurt, kefir, some cheeses, tempeh, sauerkraut, and kimchi	Plays vital role in the treatment of diarrhea	[53–57]
Streptococcus thermophilus and *Lactobacillus acidophilus*	Fermented dairy products like yogurt, cheese, and kefir, as well as in some fermented foods like miso and tempeh	Anticancer and antitumor activities	[58–60]

Table 3.1 (continued)

Probiotics	Sources	Health benefits	References
Lactiplantibacillus plantarum and *Lactobacillus rhamnosus*	Yogurt, kefir, sauerkraut, kimchi, some cheeses, and sourdough	Antimicrobial activity	[61, 62]
Lactobacillus gasseri	Fermented foods like yogurt, kefir, sauerkraut, kimchi, and tempeh, as well as in some probiotic supplements	Helps in belly weight loss	[63]

3.2.3 Classes of Probiotic Products

Currently, the food industry is experiencing substantial growth in the development of functional foods, primarily through the incorporation of probiotic bacteria into product formulations and as food additives [64]. Probiotics have been classified into two main sources: dairy-based and nondairy-based products, each with fermented and non-fermented varieties [65].

3.2.3.1 Dairy-Based Probiotic Products

Numerous lactic acid bacteria (LAB) have been isolated from fermented dairy products and has demonstrated beneficial effects. Consequently, a variety of foods exhibiting probiotic activity have been developed from these fermented dairy products, including yogurt, kefir, and dairy serum containing lactobacilli and *Bifidobacterium* [66].

3.2.3.1.1 Fresh and Fermented Milks

Scientists have explored the probiotic attributes, prebiotic fermentability, and GABA production potential of microorganisms extracted from Mexican milk kefir grains obtained from the Guadalajara area in Jalisco, Mexico [67]. It was discovered that a diet including dairy items, such as yogurt, fermented milk, and cheese, enhanced with potential probiotic strains such as *L. acidophilus*, *Lacticaseibacillus casei*, and *Bifidobacterium*, along with prebiotics or synbiotics (a mix of probiotics and prebiotics), improved type 2 diabetes mellitus symptoms by lowering oxidative stress, reducing pro-inflammatory markers, and alleviating intestinal dysbiosis [68].

3.2.3.1.2 Yogurt

In the food sector, there is an ever-increasing demand for yogurt production that incorporates probiotics. Mani-López et al. [69] focused on the viability and storage stability of probiotics in various yogurt and fermented milk products, which were made using different combinations of LAB, primarily from the *Lactobacillus delbrueckii* subsp. *bulgaricus* and *S. thermophilus* genera.

3.2.3.1.3 Cheese

Cheese is a dairy product derived from milk curd. Often, the flavor, taste, consistency, texture, and appearance of cheese are influenced by fermentation processes involving microorganisms like LAB. In vitro and in situ probiotic potential of *Lacticaseibacillus paracasei,* isolated from Tenate cheese was analyzed [70] and it was suggested that this strain is a potential probiotic with strong antagonistic properties against pathogens in food technology.

3.2.3.1.4 Other Dairy Products

Probiotic bacteria have been isolated from various dairy products, including butter, where lactobacilli strains have demonstrated antagonistic properties against pathogens such as *Salmonella* and *Listeria* [71]. The incorporation of *Lacticaseibacillus casei* subsp. *casei* AB 16–65 and *Lactobacillus maltaramicus* AC 3–16 strains into butter has been shown to reduce cholesterol levels and saturated fatty acids, thereby mitigating the cardiovascular risks associated with regular butter consumption [72]. Functional dairy foods have also been derived from cream, offering notable health benefits. Scientists have incorporated *B. bifidum, L. acidophilus, S. thermophilus,* and *L. delbrueckii* subsp. *bulgaricus* into cream enriched with sunflower, hazelnut, and soybean oils, facilitating fermentation to produce high levels of capric, butyric, and caproic acids, which are recognized for their substantial health benefits [73]. Ice cream has also emerged as an innovative source for probiotic isolation and supplementation. An example is the addition of *L. acidophilus* strain ATCC 4356 to ice cream to assess the strain's survival conditions and its beneficial impact on reducing fat content from high-fat ice cream ingredients, thereby proposing a healthier low-fat ice cream alternative [74].

3.2.3.2 Plant-Based Probiotic Products

Research indicates that the consumption of vegetables, which are rich in cellulose due to their cell walls, can provide nutrients for LAB, serving a symbiotic role and enhancing their probiotic potential within the gut environment. An example of this is kimchi, a traditional Korean dish made by fermenting vegetables with probiotic bacteria such as *Latilactobacillus sakei, Leuconostoc citreum, Leuconostoc gasicomitatum, Leuconostoc gelidum,* and *Levilactobacillus brevis,* which perform lactic acid fermentation. The dynamic nutritional profile and interaction of bacterial communities such as *Weissella, Sphingomonas, Leuconostoc, Pediococcus,* and *Psychrobacter* strains, and their role in metabolite production in *Raphanus sativus* (a type of radish rich in proteins, glucosinolates, flavonoids, β-carotene, and minerals), when treated as a fermented pickle and supplemented with wheat bran, enable its use as a synbiotic in food preservation and enhance its nutritional value [75].

3.2.3.3 Fruit-Based Probiotic Products

Fruit juices with 100% fruit content, nectars containing 25–99% fruit, and juice drinks with up to 25% fruit are widely consumed beverages, and there have been proposals to enhance them with probiotic bacteria. The incorporation of probiotics has been demonstrated to produce bioactive compounds through the fermentation of sucrose and sugars present in the fruit, thereby classifying these beverages as functional foods. This classification is due to their ability to enhance the antioxidant properties of both the fruit and the consumer's intestinal tract [76]. Strains such as *L. delbrueckii* subsp. *bulgaricus*, *S. thermophilus*, *S. cerevisiae*, and *S. boulardii* have been predominantly utilized. Their inclusion in fruit juices has been shown to facilitate the intestinal absorption of calcium, iron, and magnesium, as well as contribute to the supply of ascorbic acid (vitamin C) and offer a protective effect.

3.2.3.4 Cereal-Based Probiotic Products

One area where LAB is applied is in cereal fermentation. This process primarily yields traditional beverages and other cereal-based fermented foods [77] as well as bakery and pastry products known as sourdoughs. These consist of flour and water mixtures, containing both yeast and LAB. Sourdoughs are utilized in creating various products such as sourdough bread, buns, and panettone [78]. The LAB involved in these fermentations is mostly derived from raw materials, such as the surface microbiota of cereal seeds. They can also be introduced through the surrounding air and various environments encountered during cereal processing. The LAB most frequently found in traditional sourdoughs and fermented drinks include lactobacilli, *Leuconostoc, Weissella, and Pediococcus*, with less common genera such as *Streptococcus, Enterococcus*, and *Lactococcus* also present.

3.2.3.5 Meat-Based Probiotic Products

The meat industry is persistently investigating innovative strategies for livestock nutrition. To enhance the nutritional quality of meat, particularly in cattle, a combination of ionophores, yeasts, and probiotics, including specific strains of lactobacilli and *S. cerevisiae*, is employed. This methodology alters the microbial composition in ruminants by promoting gut colonization, thereby improving nutrient absorption and facilitating muscle development in cattle. The inclusion of probiotics in feed has been shown to improve meat quality by diminishing the prevalence of harmful bacteria such as *Prevotella ruminicola, Selenomonas ruminantium,* and the commensal *Streptococcus bovis*, which can hinder nutrient uptake [79].

3.2.3.6 Chocolate-Based Probiotic Products

They are gaining attention as chocolate is a globally favored product, appreciated for its pleasant taste and aroma, high nutritional energy, rapid metabolism, and easy digestibility. As a functional food, chocolate is abundant in polyphenolic antioxidants and flavonoids, which are associated with its health benefits. Currently, manufacturers are able to augment functional foods by incorporating probiotic bacteria into chocolate, thereby offering health benefits previously unattainable [80]. However, developing a probiotic chocolate that is both affordable and nutritious for a broader audience remains a challenge. Efforts to create such probiotic chocolate products have been limited. Additionally, a chocolate product was assessed as a potential protective carrier for the oral delivery of a microencapsulated blend of *Lactobacillus helveticus* CNCM I-1722 and *Bifidobacterium longum* CNCMI-3470 [81]. The study's findings suggested that coating probiotics in chocolate is an effective method to protect them from environmental stress and ensure optimal delivery.

3.3 Prebiotics

The term *prebiotic* is formally used to define nondigestible food ingredients that are utilized by gut microbiome to enhance their metabolism and growth. Prebiotic is a substrate that is selectively utilized by host microorganisms conferring a healthy benefit. The idea of prebiotics has drawn a lot of interest from both the scientific and the business community. Nevertheless, a lot of food ingredients, particularly a lot of food oligosaccharides and polysaccharides, e.g., dietary fiber, have been asserted to have prebiotic properties.

Not all dietary carbohydrates are prebiotics, but specific criteria need to be established for categorizing a food item as a prebiotic [82]. The criteria include resistance to GI absorption, hydrolysis by mammalian enzymes, and gastric acidity; fermentation by intestinal microflora; and selective encouragement of the growth and/or activity of those intestinal bacteria that support health and well-being. According to these conditions, resistance does not always indicate that the prebiotic is completely indigestible; rather, it should ensure that significant amount of the substance is present in the colon, particularly the large bowel, to act as a substrate for fermentation.

Demonstrating a selective stimulation of growth and/or activity of these intestinal bacteria that contribute to health and well-being requires anaerobic sampling of feces followed by reliable and quantitative microbiological analysis of a wide variety of bacterial genera, e.g., total aerobes/anaerobes, *Bacteroides*, bifidobacteria, clostridia, Enterobacteria, Eubacteria, and lactobacilli. Molecular-based microbiological methodologies have been developed and should make prebiotic demonstration easier. Among the foods that naturally contain prebiotics are peas, beans, seaweeds, microalgae, asparagus, sugar beet, garlic, chicory, onion, Jerusalem artichoke, wheat, honey, banana, barley, tomato, rye, soybean, human and cow's milk, etc. [83]. Majority of them are oligosaccharide carbohydrates, a particular kind of carbohydrates. The structure of carbohydrates affects the colonic ecology and the fermentation of gut microbes. Varied prebiotics have varied impacts on gut flora. The type of enzymes produced by a microorganism will strongly influence the carbohydrate structures that it can degrade to provide nutrition and energy, although this is extremely difficult to characterize in a mixed culture environment [84].

3.3.1 Types of Prebiotics

3.3.1.1 Galactooligosaccharides (GOSs)

Galactooligosaccharides (GOS) are alternatively known as trans-GOS. GOS, acknowledged globally as a prebiotic, which is an organic by-product of lactose extension that comes from human milk [83, 85]. GOS are formed by the elongation of lactose, a disaccharide sugar present in milk and other foodstuffs [86]. GOSs can greatly stimulate bifidobacteria and lactobacilli. Another isomer of lactose, lactulose can also be a source of GOS. Lactulose is an artificial disaccharide with a molecule of galactose and fructose linked together by a bond that is not digestible by lactase. Lactulose resists digestion in the upper gut and it is not absorbed by the small intestine; however, it undergoes fermentation to yield SCFAs, carbon dioxide, and hydrogen, which consequentially results in reducing the fecal pH [87].

3.3.1.2 Fructans

Fructans are made up of linearly organized fructose chains linked by $\beta(2, 1)$ bonds. Oligofructose and fructooligosaccharide (FOS) or inulins are combined in this subset [83]. Several plant species that are commonly consumed as vegetables like chicory roots, Jerusalem artichokes, garlic, leek, onion, and asparagus are high in fructans [88]. Specific LAB strains exhibit strain-dependent utilization of FOS and inulin, highlighting the importance of prebiotic-substrate compatibility [89]. There have been some studies in recent years revealing the chain length of fructans as a significant condition in determining which bacteria are capable of fermenting them [90].

3.3.1.3 Starch and Glucose-Derived Oligosaccharides

Two well-known examples of this group are polydextrose and resistant starch, which is found in foods like rice, whole grains, legumes, green bananas, raw potatoes, cooked-and-cooled potatoes, rice, pasta, and industrially modified starches used in processed foods [83, 91–93]. Prebiotics derived from starch include dextrins and resistant starch. Foods like grains and legumes include resistant starch, which resists digestion in the upper GIT and makes it to the colon undigested. There, it is fermented by gut bacteria like *Faecalibacterium prausnitzii* and bifidobacteria. SCFAs, such as butyrate, are produced during this fermentation and have anti-inflammatory and gut-healthy effects [94]. The structural complexity of resistant starch and derived oligosaccharide exhibit effective prebiotic potential by enabling selective fermentation by fostering microbial diversity and SCFA production especially butyrate which supports gut health. The crystalline structure and specific glycosidic linkage also prevent enzymatic hydrolysis in upper GIT. Their intricate molecular arrangement ensures they reach the colon intact for microbial utilization.

3.3.1.4 Pectin Oligosaccharides

Pectin, a polysaccharide presents in fruits including citrus, apples, and berries, is hydrolyzed to produce pectin oligosaccharides (POS), a promising family of prebiotics. These oligosaccharides, which contain constituents such as galacturonic acid, rhamnogalacturonan, and xylogalacturonan, are distinguished by their low molecular weight and structural variety. POS has shown great promise in inhibiting harmful strains like *E. coli* and selectively promoting the growth of good gut bacteria like *Bifidobacterium* and *Lactobacillus* [95]. Agro-industrial wastes such as sugar beet pulp and orange peel can be used to make POS, which makes its manufacturing economic and environment-friendly [96]. Through species-specific stimulation, enzymatically generated POS from various sources have been demonstrated to improve gut health and increase the growth of LAB [97]. The structural complexity of POS makes them effective prebiotic because they not only selectively ferment beneficial bacteria, produce SCFA, but also have valuable component in functional foods and therapeutic interventions [98].

3.3.1.5 Other Types

Prebiotics that are not members of the carbohydrate family but yet show selective fermentation by good gut bacteria are known as noncarbohydrate oligosaccharides. By interacting with the gut microbiota, these oligosaccharides promote the growth of good bacteria like *Lactobacillus* and *Bifidobacterium* while suppressing harmful ones. They offer prebiotic advantages over conventional carbohydrate-based prebiotics like FOS or GOS because of their distinct chemical structures. Noncarbohydrate oligosaccharide commonly includes cocoa-derived flavanols and polyphenols. Cocoa-derived flavonoids have been documented to modulate microbial diversity, either by promoting the growth of beneficial bacteria or by inhibiting pathogenic bacteria, hence providing a prebiotic effect in the gut [99]. They also demonstrated great promise as a prebiotic as most of these substances reach to the colon, where they interact with gut flora, because they are not well absorbed in the small intestine. High dosages of cocoa flavanols have been shown in studies to reduce dangerous bacteria like *Clostridium* while selectively increasing populations of *Lactobacillus* and *Bifidobacterium*. In one study, participants who consumed 494 mg of cocoa flavanols daily for 4 weeks showed a decrease in inflammatory markers like C-reactive protein [100] and an increase in bifidobacteria and *Lactobacillus* populations. Dietary polyphenols have been documented to initiate interaction with gut microbiota to selectively promote or inhibit microbial growth or proliferation [101]. These polyphenols rely on the diverse and specific nature of these gut microbiota to synthesize the secondary bioactive metabolites that are used in human biochemical pathways.

3.3.2 Role of Prebiotics

Through a number of interrelated mechanisms, prebiotics improve the development and function of probiotic microorganisms. As indigestible compounds, prebiotics resist digestion in the upper GIT and reach the colon intact, where they serve as a carbon source for probiotics such as *Bifidobacterium* and *Lactobacillus*. SCFAs, such as acetate, propionate, and butyrate, are produced when probiotics ferment prebiotics. These SCFAs create an acidic environment in the gut that inhibits harmful bacteria and promotes the growth of probiotics. SCFA acts as energy source for colocytes (cells that are present on the lining of colon), which helps maintain the healthy intact gut lining. A robust epithelial barrier supports the colonization of probiotics. It also supports mucosal lining, which indirectly lead to probiotic growth [102].

Additionally, by giving particular probiotic strains a competitive edge over pathogenic bacteria, prebiotics selectively promote their growth. For instance, it has been demonstrated that polysaccharide-based prebiotics, such as inulin and FOS, increase microbial stability and diversity by encouraging the growth of *Lactobacillus* and *Bifidobacterium*. Moreover, probiotics can release antimicrobial peptides that inhibit pathogenic bacteria while prebiotics can competitively bind to epithelial cell receptors that prevent pathogens from adhering to the gut lining [11]. The efficiency of prebiotics is significantly influenced by their degree of polymerization (DP). Because of their exposed active groups, lower DP prebiotics are easier for bacteria to use, which promotes increased SCFA synthesis and faster fermentation rates. For example, lower DP concentrations of β-glucan and inulin have been shown to have greater effects on SCFA generation and microbial composition [17]. Together, these processes produce an environment that is conducive to the growth of probiotic bacteria, enhancing gut health, and general well-being. Table 3.2 outlines different prebiotics, their sources, and their health benefits.

Table 3.2: Different prebiotics, their sources, and their health benefits.

Prebiotics	Sources	Heath benefits	References
Inulin	Chicory root, Jerusalem artichoke, and onions	Improved digestion, enhanced mineral absorption, and immune support	[103–105]
Fructooligosaccharides (FOS)	Bananas, garlic, and onion	Increased satiety, reduced blood glucose levels, and improved gut health	[51, 106]
Galactooligosaccharides (GOS)	Legumes and human breast milk	Enhanced immune function, improved calcium absorption, and gut health	[51, 107]
Resistant starch	Green bananas and cooked-and-cooled potatoes	Improved insulin sensitivity, reduced appetite, and enhanced gut health	[108, 109]
Pectin	Fruits and vegetables	Lowered cholesterol levels and improved gut barrier function	[110]
Lactulose	Synthetic disaccharide	Treatment of constipation and improved gut health	[111, 112]
Xylooligosaccharides (XOS)	Hardwood, oat bran, and wheat bran	Improved bowel regularity and enhanced immune response	[113, 114]

3.4 Synbiotics and Postbiotics

According to the definition given by ISAPP, a synbiotic is a mixture that includes live microorganisms and substrates used selectively by host microorganisms that confers a health benefit to the host [115]. In synbiotic food products, probiotic bacteria selectively utilize the prebiotics as substrate for their growth [116]. Synbiotics are classified into two groups: complementary and synergistic synbiotics. Complementary synbiotics consist of probiotics and prebiotics that synergistically offer health benefits without needing any interdependent actions. A synergistic synbiotics, on the contrary, provides a substrate that is specifically utilized by the coadministered living microorganisms [115]. Synbiotic approaches target metabolic disorders by fostering a conducive environment for gut microbiota stability and enhancing host health and nutrition positively [117]. The application of synbiotics has notably reduces various cardiovascular risk factors, the occurrence of metabolic syndrome, and markers of insulin resistance in older patients [118].

Postbiotics refer to the structural and metabolic products generated by microbes, encompassing cell-free supernatants, bacterial lysates, fragments of cell walls, teichoic acid, SCFAs, vitamins, enzymes, exopolysaccharides, various peptides, amino acids, and by-products of fermentation [119]. Postbiotics are bioactive substances generated by probiotic microbes during fermentation, which can be classified into two types: low molecular weight (such as H_2O_2, carbon dioxide, and di-acetylene) and high molecular weight (including bacteriocins and bacteriocin-like materials) [120, 121]. These substances, such as SCFAs, peptides, enzymes, vitamins, and organic acids provide health advantages by influencing the immune system, decreasing inflammation, and engaging with the gut microbiome. Postbiotics are gaining recognition as possible therapeutic agents for addressing diverse health issues owing to their stability and specific bioactivity. They are endotoxins of gram-negative bacteria or elements of cell walls or other cellular parts of both gram-negative and gram-positive bacteria, which may offer beneficial biological effects for the host [122]. Effective postbiotics should include inactivated microbial cells or their components, along with or without metabolites that provide observable health advantages, including immunomodulatory, antimicrobial, and anticancer effects. It leads to decreased inflammation, oxidative stress, proliferation, lower blood pressure, cholesterol levels, and body weight. Postbiotics are well tolerated in healthy people [123, 124]. Nonetheless, specific groups of individuals should refrain from raising their intake of postbiotics via consumption of probiotic-rich foods, such as those who have undergone recent surgery, suffer from structural heart conditions, have digestive tract issues, are pregnant, or are children. These groups often possess weakened immune systems, which may put them at a higher risk for an adverse reaction. Some foods can enhance postbiotics in the gut, such as buttermilk, cottage cheese, fermented pickles, yogurt, and high-fiber options like oats, flaxseed, and garlic. Various researchers have reported that both internal and external elements could influence the efficacy of postbiotics. The interaction between active metabolites of postbiotic and internal factors like the resident microbiota, enzymes, and various food components can inhibit metabolites functions [125, 126].

Since postbiotics cannot reproduce, they are considered safer than probiotics [123]. A key factor in postbiotic research is their natural stability, which is important for both industrial processes and storage. Postbiotic supplements are not widely available yet, but postbiotics surpass probiotics in purity, preparation simplicity, shelf stability, mass production potential, specific action, and more targeted responses through particular ligand-receptor interactions [127]. Postbiotics are characterized by the microorganisms used as the starting point for its production and also the inactivation procedure or technique used for its production, as each procedure affects the quality and quantity of the final postbiotic product and results in a different postbiotic with varied effects. Postbiotics are defined by the microorganisms utilized as the foundation for their creation and the inactivation method or technique applied during production, as each method influences the quality and quantity of the end postbiotic, leading to different postbiotics with diverse effects. The consensus statement from the ISAPP unveiled the enigma of postbiotic effects through five mechanisms: adjusting the resident microbiota, improving epithelial barrier functions, altering local and systemic immune responses, influencing systemic metabolic reactions, and systemic signaling through the nervous system [124].

3.4.1 Postbiotic Production Process

Postbiotics occur naturally in numerous fermented foods such as yogurt, sauerkraut, and pickled vegetables, and they are generated by different bacterial and fungal species, primarily including *Lactobacillus*, *Bifidobacterium*, *Streptococcus*, *Eubacterium*, *Faecalibacterium*, and *Saccharomyces* strains [128–130]. The quantity of postbiotics created during natural fermentation cannot be regulated and might be low or inadequate to elicit a physiological response [131]. Consequently, researchers have investigated production methods to generate postbiotics in a controlled and efficient way, facilitating their study and application in food, pharmaceutical, and nutraceutical sectors. Postbiotics can originate from compounds produced by microbes as well as from microbial processes that involve the creation of metabolites and products resulting from microbial activity within a nutrient matrix. Postbiotic production technologies typically encompass cell disruption methods such as heat, high pressure, enzymatic processes, co-culturing, solvent extraction, chemical treatments (e.g., formalin), and sonication [132, 133]. Extractions and purification processes, including centrifugation, dialysis, lyophilization/freeze-drying, spray drying, and column purification, are also employed to facilitate the procedures. The selection of a method for postbiotic production relies on several factors, such as the specific postbiotic compound sought, the microbial strain employed, the planned application, and considerations for scalability. The primary source of postbiotics in the food industry is fermentation. The producer strains that can be used for extracting postbiotics in situ mainly include *Lactobacillus* and *Bifidobacterium* strains, but they might also encompass *Streptococcus*, *Akkermansia muciniphila*, *Eubacterium hallii*, *Faecalibacterium*, and *S. boulardii* [134–136].

3.4.2 Applications of Postbiotics and Synbiotics

Because of their different ways of working, postbiotics and synbiotics are innovative fields in nutritional science that offer substantial health benefits. Probiotic bacterial fermentation produces postbiotics, which have been associated with better gut health, immunological modulation, possible illness prevention, and food safety. Probiotics and prebiotics are examples of synbiotics, which combine to increase the growth and survival of beneficial gut bacteria, improving GI health, and immunity in general. This demonstrates a thorough approach to using food supplements to improve health outcomes by affecting the gut flora.

3.4.2.1 Antimicrobial Properties

Both postbiotics and synbiotics have shown beneficial role in lowering the quantity of bacteria that cause disease [137]. *Lactobacillus* spp. and its nutraceutical compounds reduce intestinal inflammation in mice by lowering the infectivity of *E. coli* [138]. The mechanism of immunomodulation provoked by probiotic *Lactobacillus* spp. in the existence of *E. coli* was justified by the suppression effect of *Lactobacillus* spp. against the expression of nuclear factor-kappa B in the case of *E. coli* [139]. *Lactobacillus* strains attribute the bactericidal activity due to the synergistic action of lactic acid and secrets bacteriocins [140]. By altering the gut microbiota's composition and the liver's metabolic processes, postbiotics made from *L. paracasei* CCFM1224 successfully treat nonalcoholic fatty liver disease (NAFLD) [141].

3.4.2.2 Antioxidant Properties

It has been demonstrated that postbiotics and synbiotics can enhance antioxidant qualities [142, 143]. Dietary postbiotics containing *L. plantarum* strain used for lambs improved antioxidant activity in serum and rumen barrier function in the postbiotic group [144]. *Lactobacillus plantarum*'s significant scavenging ability against hydroxyl, superoxide, and 2,2-diphenyl-1-picrylhydrazyl (DPPH)-free radicals, as well as its resistance to H_2O_2 are the primary factors contributing to its efficacy as an antioxidant. Lactic acid buildup in the intestinal epithelium is thought to decrease the intestinal absorption of glucose [145].

Intervention of *L. acidophilus*, *L. casei*, and *B. bifidum* in combination with inulin (i.e., a synbiotics capsule) considerably improved health metrics for 60 diabetic patients [146]. It was observed that synbiotic supplementation significantly lowers fasting plasma glucose, insulin levels, and insulin resistance while raising insulin sensitivity. Consuming probiotics and synbiotics may improve the activity of antioxidant enzymes, decrease lipid peroxidation, and inhibit the production of pro-inflammatory cytokines [147].

3.4.2.3 Anti-inflammatory Properties

The potential health advantages of postbiotics and synbiotics as anti-inflammatory medicines have been the subject of numerous investigations in recent decades [148]. Strains of *Lactobacillus bulgaricus* and *S. thermophilus* slow down the course of the inflammatory illness colitis in mice as the postbiotics reduces the disease progression [149]. Prebiotic inulin and the synbiotic produced when it is combined with yeast *S. cerevisiae* had a good effect on developing several morphological features in the rumen and gut of calves. Synbiotics positively influence the atherogenic index of plasma, inflammatory markers, and oxidative stress markers, which are all recognized risk factors for cardiovascular diseases [150]. Several studies are recommended to confirm the beneficial effects of synbiotics and postbiotics, which offer anti-inflammatory properties that enhance GI health. However, to confirm these results and prevent chronic diseases, more clinical research and optimization are required.

3.4.2.4 Perspectives for Nonclinical Application

3.4.2.4.1 Biopreservation

In biopreservation, particular microorganisms and their by-products are utilized as postbiotics to inhibit the microbial degradation of food and to improve their longevity. Postbiotics demonstrate antimicrobial effects on both pathogenic and spoilage microbes through various mechanisms such as forming pores in cell membrane, influencing cell wall proteins, and lowering the pH of bacterial cytoplasm, thereby showing considerable importance in the food industry [151]. Nonvegetarian food can be preserved by directly applying the postbiotic coating (such as fish fillets and meat slices) or by spraying (like ground fish and meat), depending on the characteristics of the postbiotic and the meat type, to safeguard their nutritional properties and sensory changes. Postbiotics with flavonoids and phenolics derived from *Pediococcus acidilactici* and *Latilactobacillus sakei/Staphylococcus xylosus* diminish the count of *Salmonella typhimurium* in chicken drumsticks [152]. The addition of *S. cerevisiae* fermentation products to the diet could serve as a possible method to manage *Salmonella enterica* in poultry food [153]. Preservatives containing postbiotics were shown to be as effective as standard commercial preservatives in maintaining the quality of vacuum-packaged cooked sausages, offering natural preservation methods [154]. Postbiotics may also serve as sanitizers within the food industry [155].

3.4.2.4.2 Food Packaging

The longevity of food is influenced by a mix of factors, which includes the food item, the type of packaging utilized, and different environmental conditions. An active packaging system safeguards food from microbial decay during transit and storage by incorporating AAs derived from plant, animal, and microbial sources or their metabolites, as well as antimicrobial nanoparticles, among others, into the packaging [156]. To enhance stability, the postbiotic can be utilized in different methods including thin coatings, covalent bonding for immobilization, incorporation into the packaging matrix, or lamination onto the polymer [157]. Different forms of postbiotic packaging utilize organic acids, peptides, and bacteriocins. Organic acids are important contenders for postbiotics and serve as effective AAs. Lactic acid, citric acid, and acetic acid inhibit pathogen growth by lowering cytoplasmic pH and affecting membrane function [157].

3.4.2.4.3 Sensory Acceptability of Postbiotics

The existence of postbiotics can alter the sensory properties of foods, which highlights the importance of assessing various foods that contain postbiotics for their sensory appeal. Sensory testing of functional yogurt enriched with cape gooseberry (*Physalis peruviana* L.) that has postbiotics from *E. coli* showed that consumer approval was notably greater for various sensory factors such as appearance, smoothness, sourness, mouthfeel, and general acceptance [158]. The sensory acceptability of lamb meat slices treated with an edible coating that includes postbiotics from *L. paracasei* ATCC 55544 (PLP) is superior to that of the uncoated samples, and there were no notable alterations in color, appearance, or overall acceptance during the storage duration [159]. The use of postbiotics through spraying decreases the bacterial count on the surfaces of cold beef carcasses while not influencing the sensory qualities of the meat. Postbiotics notably decreases the fungal count in cheese while preserving sensory acceptance [160].

3.4.2.4.4 Removal of Biofilms

Biofilms consist of a matrix composed of microbial polysaccharides or proteins. *Listeria monocytogenes*, *Yersinia enterocolitica*, *Campylobacter jejuni*, *Staphylococcus aureus*, and *Bacillus cereus* are microorganisms that create biofilms within the food industry. Biofilms are challenging to eliminate using standard cleaning techniques and disinfection [161, 162]. Postbiotics possess antibacterial qualities and are capable of eliminating biofilms created by bacteria. *L. acidophilus* LA5, *Lactobacillus casei* 431, and *Lactobacillus salivarius* exhibit antimicrobial activity against a biofilm developed by *L. monocytogenes* on polystyrene surfaces. The decrease of biofilm in *L. monocytogenes* is achieved through bacteriocin and organic acid-derived postbiotics. Consequently, postbiotics are employed as a means to diminish biofilm development by pathogens within the food industry [151].

3.5 Health Benefits of Functional Foods

3.5.1 Gut Health and Digestive Support

Microbial ecology of the human GIT is a fascinating and complex area of study. The GIT is home to a wide variety of microorganisms that are vital to human well-being and disease. Studies have shown that the gut microbiota plays a key role in mediating health benefits. Understanding the distribution and dynamics of these microorganisms throughout the gut is necessary to ascertain their function in maintaining homeostasis and preventing multiple diseases. Different gut domains support a diverse microbial population that is influenced by host genetic makeup, nutrition and oral hygiene.

Enhancing gut health is one of the main health advantages of functional foods. In particular, *Lactobacillus* and *Bifidobacterium* species are important LAB that help keep the gut microbiota in balance. They limit the colonization of hazardous pathogens such as *Salmonella enteritidis*, *E. coli*, and *L. monocytogenes* while fostering the growth of beneficial bacteria [163]. This preventive action lessens the risk of diarrhea, prevents GI infections, and eases the symptoms of inflammatory bowel illnesses, including Crohn's disease and ulcerative colitis [164, 165]. They can help alleviate symptoms by controlling gut bacteria and lowering inflammation. The intestinal barrier acts as a defense against pathogens, and research has indicated that probiotics enhance this barrier along with the immune system functioning. They also encourage the release of anti-inflammatory substances that prevent the growth of pathogenic bacteria in the gut [166]. Additionally, probiotics improve nutrient absorption and digestion. Because of inadequate levels of lactase enzyme activity in humans, about 60% of the population have a decreased capacity to digest lactose sugar. In addition to improving lactose digestion, probiotics lessens symptoms such as gas, bloating, and diarrhea [167]. They help minimize the amount of toxins and infections that are absorbed into the bloodstream by fortifying the gut lining. This supports a healthier digestive tract and overall well-being. Probiotics could be particularly beneficial for those who have malabsorption issues as they improve intestinal flora, which facilitates the absorption of nutrients such as vitamins, minerals, and fatty acids. Prebiotic fermentation increases the absorption of fat from the jejunum, improves insulin sensitivity and glucose tolerance, modifies lipid metabolism in the liver, regulates cholesterol and bile acids, and ultimately lowers the risk of atherosclerosis and kidney damage. Through a variety of mechanisms, including the production of SCFAs, the balancing of gut microbiota, the enhancement of gut barrier function, the improvement of digestive disorders, the enhancement of nutrient absorption, and immune system modulation, they significantly contribute to the improvement of digestive health. Together, these actions provide protection against food allergies and lower the risk of systemic disorders and a variety of GI pathologies. Taking probiotic supplements or eating foods high in probiotics, such as yogurt, kefir, sauerkraut, kimchi, and kombucha, can help maintain a healthy balance of gut bacteria [168].

3.5.2 Remodeling of the Immune System

The immune system's response is influenced by gut health. A healthy gut microbiota promotes and enhances the immunological system in organs such as the skin, lungs, and other membranous surfaces. It has been demonstrated that functional foods alter the immune system, strengthening the body's defenses against inflammation and infections.

Probiotics have an impact on Peyer's patches and mesenteric lymph nodes, two immune cell types present in the gut-associated lymphoid tissue (GALT) [169]. Through their interactions with immune cells in GALT, probiotics help regulate and improve immune responses. These cells boost the synthesis of immunoglobulins, such as IgA, which are important for mucosal immunity and may help prevent infections and autoimmune illnesses [170]. By assisting the body in developing immunological tolerance, probiotics prevent the immune system from overreacting to harmless stimuli. Additionally, they aid in the maturation of immune cells critical to mounting a successful immune response, including macrophages and DCs [171]. Probiotics can boost the synthesis of anti-inflammatory cytokines (interleukin (IL)-10) and inhibit the synthesis of pro-inflammatory cytokines (tumor necrosis factor-alpha and IL-6) [172].

When taken with or after antibiotics, probiotics can aid in the quicker restoration of the gut microbiota, reducing the likelihood of diarrhea or other digestive issues linked to antibiotics [173]. Gupta et al.'s in vivo investigation shows that *Lactobacillus* species and their nutraceutical components decrease intestinal inflammation in mice by reducing *E. coli*'s infectivity [138]. It has been discovered that probiotic strains, especially those belonging to the *Lactobacillus* and *Bifidobacterium* genera, lessen the severity of respiratory infections, lower the frequency of allergic reactions, and lessen the inflammatory response in diseases including atopic dermatitis and eczema. They also help prevent chronic inflammatory illnesses by regulating the production of cytokines [87, 174]. After being taken orally, the probiotic bacteria will interact with the intestinal epithelial cells lining the intestinal microvilli. They will then attach to the immune cells linked to the lamina propria via Toll-like receptors, which will cause the production of various cytokines and chemokines [175]. This results in colonization and selective adherence of probiotic bacteria to the GIT that prevents the attachment of harmful bacteria to the tract. The pH of the microhabitat is reduced due to the production of lactic acid by prebiotic bacteria [166]. Probiotics compete with pathogenic bacteria for resources and available space in the gut. They produce bacteriocins and other antimicrobial compounds that prevent the spread of dangerous bacteria [176]. Probiotics promote tight junction proteins, which maintain the integrity of the gut barrier, improve the synthesis of antibodies, exclude competing bacteria and antimicrobial production, generation of SCFAs, passive immune system assistance, metabolism, and utilization of nutrients [177]. This barrier reduces the burden on the immune system by preventing harmful infections and contaminants from entering the bloodstream. SCFAs support immune cell modulation, inflammation lowering, and gut health upkeep [178]. Minerals (like zinc) and vitamins (like vitamin D) that are essential for immune function are broken down and assimilated more easily by probiotics [179].

3.5.3 Cardiovascular Well-Being

Probiotics may potentially improve cardiovascular health by affecting the metabolism of fats. Lipid molecules such as cholesterol are indispensable for the body's cellular makeup, production of hormones, and bile acid generation. Conversely, a higher risk of cardiovascular diseases is associated with elevated levels of low-density lipoprotein (LDL), frequently referred to as "bad" cholesterol [180]. Bile salt hydrolase enzymes found in some strains of *Lactobacillus* and *Bifidobacterium* can lower blood cholesterol levels by breaking down bile salts. This results in decreased reabsorption and increased use of cholesterol by the liver for bile synthesis, which lowers blood cholesterol levels [181]. Furthermore, probiotics produce bioactive compounds such as conjugated linoleic acid and *n*-3 fatty acids that are linked with decrease in blood pressure and a reduced likelihood of cardiovascular ailments [182, 183].

In many controlled clinical trials, the impact of specific probiotic strains on cholesterol levels has been studied [184]. For instance, a randomized controlled trial by Shimizu et al. showed that administering *L. acidophilus* supplements daily for 12 weeks significantly decreased the levels of LDL cholesterol in people with mild to moderate hypercholesterolemia [185]. An additional investigation into the cholesterol-lowering properties of *L. rhamnosus* BFE5264, which was isolated from fermented milk, was carried out by Park et al. [186], where the treatment outcomes showed significantly lowered cholesterol levels. Following a comprehensive examination, Pan et al. reported that postbiotics generated from *L. paracasei* CCFM1224 successfully prevent NAFLD by changing the gut microbiota's composition and the liver's metabolic activities. The scientists looked into how *L. paracasei*-derived postbiotics might prevent NAFLD [141].

3.5.4 Weight Control and Metabolic Health

Functional foods have a major impact on the makeup and function of the gut microbiota, which in turn affects metabolic health. By regulating gut microbiota, reducing inflammation, improving digestion and food absorption, producing SCFAs, and regulating hunger and satiety, probiotic use has been linked to weight management and control [187]. Probiotics have an impact on fat storage, energy expenditure, and appetite control. A balanced gut flora is linked to better metabolic health and a decreased risk of obesity [188]. Better nutritional and energy absorption from food is made possible by probiotics' assistance in the breakdown of complex carbs, fibers, and fats that the body may not be able to digest on its own. When certain bacteria ferment dietary fibers, they produce SCFAs such as acetate, propionate, and butyrate. In addition to being used by the body as an energy source, SCFAs also affect hormones related to satiety and fat storage, which aid in regulating metabolism [189].

Some probiotics may inhibit intestinal absorption of fat from diet, hence reducing total caloric intake [190]. Certain probiotics have been shown to improve insulin sensitivity, which helps the body use insulin more effectively. This is important for the treatment of metabolic diseases like type 2 diabetes and obesity [191]. This may help regulate weight and reduce the risk of significant blood sugar increases [192]. Probiotics can lower chronic low-grade inflammation, which is a major contributing factor to the development of metabolic syndrome and associated disorders, by altering the gut microbiota [193]. Some probiotic strains have been linked to decreased waist circumference and body fat, suggesting that they could be used as an adjuvant treatment for obesity [167]. By altering these hormones, probiotics may be able to regulate appetite and fullness, which could help with weight management and food consumption. Through their impact on the creation of particular proteins and enzymes involved in fat metabolism, probiotics may have an impact on the body's energy storage and utilization. This might affect the balance of energy [194].

3.5.5 Cancer Prevention

Cancer is one of the most common causes of illness and death. It is a fatal neoplasm that arises from the uncontrollably repeated growth of abnormal cells. As of right now, there is no precise or conclusive treatment. Standard chemotherapeutic medications and synthetic chemicals used to treat cancer are extremely harmful to the human body in terms of stability and safety. They result in after effects that harm healthy cells and organs throughout the body [195]. These medication regimens have an impact on life quality or fuel the emergence of drug resistance.

Probiotics that are metabolically active or inert, along with their metabolites (SCFAs, ferrichrome, polysaccharides, protein inhibitory substances, and nucleic acid) are a safe and efficient treatment [166]. By altering the gut microbiota and strengthening the immune system, probiotics may help prevent cancer. According to certain research, probiotics can improve the body's detoxification processes, lower the generation of toxic metabolites, and stop the growth of bacteria that cause cancer [167]. The presence of alternative biomolecules and biotherapeutics against cancer, which release chemicals with antioxidant capacity that hinder the multiplication of cancer cells, has led to the proposal of this treatment as an alternative cancer prevention and treatment [196]. Numerous mechanisms have demonstrated the antitumor properties of probiotics. Probiotics change the composition and diversity of the gut microbiota, which reduces the conditions that support tumor formation. By promoting the immune response, probiotics increase immune cell activity that contributes to tumor monitoring and eradication [197]. SCFAs and bacteriocins are two metabolites produced by probiotics that have antitumor effects. By reducing intestinal inflammation, probiotics indirectly block pathways that promote tumor growth [198]. Probiotics may also stop tumor cell growth, especially in colorectal cancer, and cause cancer cells to undergo apoptosis, or programmed cell death [199].

The metabolic activity of the bacteria that are present in the GI system can either produce or consume the carcinogens. Through detoxification, the production of metabolic products that trigger apoptosis, immune response stimulation, and intestinal environmental changes that inhibit tumor cells, probiotics may reduce exposure to carcinogens [200]. Additionally, probiotics prevent bacteria from growing and turning pro-carcinogens into carcinogens. The main microbial enzymes that cause colon cancer to be induced significantly decreased when bile-resistant *L. acidophilus* of human origin was added to the diet [116]. Preclinical studies using mice and other animal models have demonstrated the inhibitory effects of probiotics on the formation and progression of tumors. Although human clinical studies have also produced promising outcomes, further research is required to identify the optimal strains, doses, and treatment durations [56].

3.5.6 Mental Health and the Gut-Brain Axis

Recent studies have brought attention to the function that probiotics play in mental health, specifically in relation to the gut-brain axis. Via the vagus nerve, immune system, and microbial compounds including SCFAs and neurotransmitters, the gut microbiota interacts with the brain. Probiotics have an impact on this axis by altering the composition and activity of the gut microbiota, which may have an impact on mood, stress response, and cognitive function [201, 202].

Numerous studies have demonstrated that taking probiotic supplements help lessen the symptoms of stress-related illnesses, anxiety, and depression. For example, strains of *Bifidobacterium* and *Lactobacillus* have been implicated in the synthesis of GABA, a neurotransmitter that regulates the neurological system. In order to promote general mental health, this implies that probiotics may have therapeutic potential in treating disorders like anxiety and depression [163].

Probiotics may have an effect on the hypothalamic-pituitary-adrenal axis, which controls the stress response. The physiological effects of stress on mental health may be mitigated by this modification [203]. SCFAs influence the central nervous system, which may affect mood and cognition, in addition to promoting gut health [204]. Antioxidant properties in some probiotic strains reduce oxidative stress in the body, which may have a beneficial effect on mental health issues [205]. Certain probiotics have been shown to create compounds with neuroprotective properties, which may help protect neurons from harm and promote brain health in general [206].

Probiotics help produce neuroactive chemicals and vitamins like B vitamins, which might affect mood and cognitive performance [207]. Serotonin is mostly produced in the GI system and is widely recognized for its role in mood modulation [208]. Probiotics aid in the production of neurotransmitters such as serotonin in the stomach. Through their interactions with GALT, probiotics modify immune responses. Cytokine production is impacted by this connection, and cytokines impact mood and behavior. Probiotics, such as *L. plantarum* help maintain the integrity of the gut barrier, which reduces intestinal permeability. By decreasing the transfer of toxic chemicals into the bloodstream [209], this action reduces systemic inflammation [210], which has been connected to mental health issues. Probiotics improve the microbiota-gut-brain axis, reduce inflammation in the brain, produce vitamins and neuroactive compounds, modulate the immune system, produce neurotransmitters, have anti-inflammatory effects, regulate stress hormones, have antioxidant and neuroprotective qualities, and improve the microbiota-gut-brain axis [211]. This shows probiotics may be therapeutically useful in treating mental health issues including sadness and anxiety.

3.6 Conclusion

In conclusion, there is encouraging evidence that PPSP can improve human health and support sustainable food systems. People are become more conscious of food quality and the health advantages of various foods as an outcome of the popularity of functional foods. As a result, there is now a much greater demand for healthy food products and a greater interest in consuming healthier foods. The intricate relationships between these elements provide a wholesome gut environment, which is essential for immunity, digestion, and even mental health. As our knowledge of these elements increases, so does the awareness of their potential to ameliorate a range of health issues. By incorporating these functional foods into our meals, we can proactively care for our bodies and establish a harmonic, balanced environment within ourselves, which will contribute to a healthy future. Including these functional foods in a holistic lifestyle emphasizes how they can support us in our quest for better health. There are still many unanswered questions about the mechanisms that regulate the activity of bacterial probiotics and the reasons behind their ability to alleviate so many ailments, despite the fact that clinical investigations have been the main source of useful knowledge. Given that foods associated with bacterial probiotics may help lower medicine usage and prevent some diseases when combined with a balanced diet and lifestyle, further study involving clinical studies is therefore required in the future. This study may result in new approaches to improving food production and cutting down on food waste. As a result, the food industry and research in the fields of nutrition, medicine, and food science stand to gain a great deal. To fully grasp this functional food components' potential to improve human health and well-being, more research is needed.

Conflicts of interest: The authors do not declare any conflicts of interest.

References

[1] Kasote D, Tiozon RN, Sartagoda KJD, Itagi H, Roy P, Kohli A, Regina A, Sreenivasulu N. Food processing technologies to develop functional foods with enriched bioactive phenolic compounds in cereals. Frontiers in Plant Science. 2021;12,771276.

[2] Topolska K, Florkiewicz A, Filipiak-Florkiewicz A. Functional food-consumer motivations and expectations. International Journal of Environmental Research and Public Health. 2021;18:5327.

[3] Granato D, Barba FJ, Kovâceviíc DB, Lorenzo JM, Cruz AG, Putnik P. Functional foods: Product development, technological trends, efficacy testing, and safety. Annual Review of Food Science and Technology. 2020;11:93–118.

[4] European Commission. European Research Area Studies and Reports. Functional Foods. 2010. Available online: http://publications.europa.eu/resource/cellar/238407ee-0301-4309-9fac-e180e33a3f89.0001.02/DOC_1 (accessed on 10 January 2023).

[5] Di Cerbo A, Morales-Medina JC, Palmieri B, Pezzuto F, Cocco R, Flores G, Iannitti T. Functional foods in pet nutrition: Focus on dogs and cats. Research in Veterinary Science. 2017;112:161–166.

[6] Watson RR, Preedy VR. Probiotics, Prebiotics, and Synbiotics: Bioactive Foods in Health Promotion. 2015, Cambridge, MA, USA: Academic Press.

[7] Zommiti M, Feuilloley MGJ, Connil N. Update of probiotics in human world: A nonstop source of benefactions till the end of time. Microorganisms. 2020;8:1907.

[8] Deepali S, Mahesh G, Ganesh G. Probiotics, their health benefits and applications for development of human health: A review. Journal of Drug Delivery and Therapeutics. 2019;9:631–640.

[9] Rastall RA, Diez-Municio M, Forssten SD, Hamaker B, Meynier A, Moreno FJ, Respondek F, Stahl B, Venema K, Wiese M. Structure and function of non-digestible carbohydrates in the gut microbiome. Benef Microbes. 2022;13:95–168.

[10] Li HY, Zhou DD, Gan RY, Huang SY, Zhao CN, Shang AO, Xu XY, Li HB. Effects and mechanisms of probiotics, prebiotics, synbiotics, and postbiotics on metabolic diseases targeting gut microbiota: A narrative review. Nutrition. 2021;13:3211.

[11] Pandey K, Naik S, Vakil B. Probiotics, prebiotics and synbiotics – A review. Journal of Food Science and Technology. 2015;52:7577–7587.

[12] Thorakkattu P, Khanashyam AC, Shah K, Babu KS, Mundanat AS, Deliephan A, Deokar GS, Santivarangkna C, Nirmal NP. Postbiotics: Current trends in food and pharmaceutical industry. Foods. 2022;11:3094.

[13] Sabahi S, Homayouni Rad A, Aghebati-Maleki L, Sangtarash N, Ozma MA, Karimi A, Hosseini H, Abbasi A. Postbiotics as the new frontier in food and pharmaceutical research. Critical Reviews in Food Science and nutrition. 2023;63:8375–8402.

[14] Sanz Y, Portune K, Del Pulgar EG, Benítez-Páez A. Targeting the microbiota: Considerations for developing probiotics as functional foods. In: Hyland N, Stanton C. (eds.) The Gut-brain Axis. 2016, pp. 17–30, Academic Press.

[15] Martín R, Langella P, Chatel JM. Microbiote et vieillissement. Innovations in Agronomy. 2018;65:55–66.
[16] Munir A, Javed GA, Javed S, Arshad N. Levilactobacillus brevis from carnivores can ameliorate hypercholesterolemia: In vitro and in vivo mechanistic evidence. Journal of Applied Microbiology. 2022;133:1725–1742.
[17] Markowiak P, Śliżewska K. Effects of probiotics, prebiotics, and synbiotics on human health. Journal of nutritional. 2017,;9:1021.
[18] Tamang JP, Watanabe K, Holzapfel WH. Diversity of microorganisms in global fermented foods and beverages. Frontiers in Microbiology. 2016;7:377.
[19] Hawrelak J, Nat B. Probiotics. In: Pizzorno JE, Murray MT. (eds.) Textbook of Natural Medicine. 2013, (4th ed , pp. 979–994). Churchill Livingstone Elsevier: Elsevier.
[20] Le Morvan de Sequeira C, Hengstberger C, Enck P, Mack I. Effect of probiotics on psychiatric symptoms and central nervous system functions in human health and disease: A systematic review and meta-analysis. Journal of Nutritional. 2022;14(3):621.
[21] Reid G. The growth potential for dairy probiotics. International Dairy Journal. 2015;49:16–22.
[22] Hamad G, Ombarak RA, Eskander M, Mehany T, Anees FR, Elfayoumy RA, Omar SA, Lorenzo JM, Abou-Alella SA. Detection and inhibition of Clostridium botulinum in some Egyptian fish products by probiotics cell-free supernatants as bio-preservation agents. LWT. 2022;163:113603.
[23] Warman DJ, Jia H, Kato H. The potential roles of probiotics, resistant starch, and resistant proteins in ameliorating inflammation during aging (inflammaging). Journal of Nutritional. 2022;14(4):747.
[24] Bermudez-Brito M, Plaza-Díaz J, Muñoz-Quezada S, Gómez-Llorente C, Gil A. Probiotic mechanisms of action. Annals of Nutrition and Metabolism. 2012;61:160–174.
[25] Roobab U, Batool Z, Manzoor MF, Shabbir MA, Khan MR, Aadil RM. Sources, formulations, advanced delivery and health benefits of probiotics. Current Opinion in Food Science. 2020;32:17–28.
[26] Fenster K, Freeburg B, Hollard C, Wong C, Rønhave Laursen R, Ouwehand AC. The production and delivery of probiotics: A review of a practical approach. Microorganisms. 2019;7:83.
[27] de Melo Pereira GV, de Oliveira Coelho B, Júnior AI, Thomaz-Soccol V, Soccol CR. How to select a probiotic? A review and update of methods and criteria. Biotechnology Advances. 2018;36:2060–2076.
[28] Konuray G, Erginkaya Z. Potential use of *Bacillus coagulans* in the food industry. Foods. 2018;7:92.
[29] Plaza-Diaz J, Ruiz-Ojeda FJ, Gil-Campos M, Gil A. Mechanisms of action of probiotics. Advances in Nutritional. 2019;10:49–66.
[30] Ahire JJ, Jakkamsetty C, Kashikar MS, Lakshmi SG, Madempudi RS. In vitro evaluation of probiotic properties of Lactobacillus plantarum UBLP40 isolated from traditional indigenous fermented food. Probiotics Antimicrob Proteins. 2021;5:1413–1424.
[31] Fantinato V, Camargo HR, Sousa AL. Probiotics study with Streptococcus salivarius and its ability to produce bacteriocins and adherence to KB cells. Revista de Odontologia da UNESP. 2019;48:20190029.
[32] Chang YH, Jeong CH, Cheng WN, Choi Y, Shin DM, Lee S, Han SG. Quality characteristics of yogurts fermented with short-chain fatty acid-producing probiotics and their effects on mucin production and probiotic adhesion onto human colon epithelial cells. Journal of Dairy Science. 2021;104:7415–7425.
[33] Bu Y, Liu Y, Liu Y, Wang S, Liu Q, Hao H, Yi H. Screening and probiotic potential evaluation of bacteriocin-producing *Lactiplantibacillus plantarum* in vitro. Foods. 2022;11:1575.
[34] Petruzziello C, Saviano A, Ojetti V. Probiotics, the immune response and acute appendicitis: A review. Vaccines. 2023;11:1170.
[35] Srivastav S, Neupane S, Bhurtel S, Katila N, Maharjan S, Choi H, Hong JT, Choi DY. Probiotics mixture increases butyrate, and subsequently rescues the nigral dopaminergic neurons from MPTP and rotenone-induced neurotoxicity. Journal of Nutritional Biochemistry. 2019;69:73–86.

[36] Cao J, Yu Z, Liu W, Zhao J, Zhang H, Zhai Q, Chen W. Probiotic characteristics of *Bacillus coagulans* and associated implications for human health and diseases. Journal of Functional Foods. 2020;64:103643.

[37] Elhossiny RM, Elshahawy HH, Mohamed HM, Abdelmageed RI. Assessment of probiotic strain *Lactobacillus acidophilus* LB supplementation as adjunctive management of attention-deficit hyperactivity disorder in children and adolescents: A randomized controlled clinical trial. BMC Psychiatry. 2023;23:823.

[38] Gao H, Li X, Chen X, Hai D, Wei C, Zhang L, Li P. The functional roles of *Lactobacillus acidophilus* in different physiological and pathological processes. JMB. 2022;32:1226.

[39] Zanjani SY, Eskandari MR, Kamali K, Mohseni M. The effect of probiotic bacteria (*Lactobacillus acidophilus* and *Bifidobacterium lactis*) on the accumulation of lead in rat brains. ESPR. 2017;24:1700–1705.

[40] Ma S, Cao J, Liliu R, Li N, Zhao J, Zhang H, Chen W, Zhai Q. Effects of *Bacillus coagulans* as an adjunct starter culture on yogurt quality and storage. Journal of Dairy Science. 2021;104:7466–7479.

[41] Qu R, Zhang Y, Ma Y, Zhou X, Sun L, Jiang C, Zhang Z, Fu W. Role of the gut microbiota and its metabolites in tumorigenesis or development of colorectal cancer. Advanced Science. 2023;10:2205563.

[42] Fernandez-Pacheco P, Arévalo-Villena M, Bevilacqua A, Corbo MR, Pérez AB. Probiotic characteristics in *Saccharomyces cerevisiae* strains: Properties for application in food industries. LWT. 2018;197:332–340.

[43] Roselli M, Finamore A. Use of synbiotics for ulcerative colitis treatment. Current Clinical Pharmacology. 2020;15:174–182.

[44] Traisaeng S, Batsukh A, Chuang TH, Herr DR, Huang YF, Chimeddorj B, Huang CM. Leuconostoc mesenteroides fermentation produces butyric acid and mediates Ffar2 to regulate blood glucose and insulin in type 1 diabetic mice. Scientific Report. 2020;10:7928.

[45] Greeshma K, Deokar CD, Raghuwanshi KS, Bhalerao VK. Probiotics as a biocontrol agent in management of post harvest diseases of mango. Current Journal of Applied Science and Technology. 2020;39:85–92.

[46] Oh JK, Kim YR, Lee B, Choi YM, Kim SH. Prevention of cholesterol gallstone formation by *Lactobacillus acidophilus* ATCC 43121 and Lactobacillus fermentum MF27 in lithogenic diet-induced mice. Food Science of Animal Resources. 2021;41:343.

[47] Tian L, Liu R, Zhou Z, Xu X, Feng S, Kushmaro A, Marks RS, Wang D, Sun Q. Probiotic characteristics of *Lactiplantibacillus plantarum* N-1 and its cholesterol-lowering effect in hypercholesterolemic rats. Probiotics Antimicrob Proteins. 2022;14:337–348.

[48] Al-Sadi R, Nighot P, Nighot M, Haque M, Rawat M, Ma TY. *Lactobacillus acidophilus* induces a strain-specific and toll-like receptor 2–dependent enhancement of intestinal epithelial tight junction barrier and protection against intestinal inflammation. American Journal of Pathology. 2021;191:872–884.

[49] Zhou LY, Xie Y, Li Y. *Bifidobacterium infantis* regulates the programmed cell death 1 pathway and immune response in mice with inflammatory bowel disease. WJG. 2022;28:3164.

[50] Li B, Zhang H, Shi L, Li R, Luo Y, Deng Y, Li S, Li R, Liu Z. *Saccharomyces boulardii* alleviates DSS-induced intestinal barrier dysfunction and inflammation in humanized mice. Food & Function. 2022;13:102–112.

[51] Li W, Wang K, Sun Y, Ye H, Hu B, Zeng X. Influences of structures of galactooligosaccharides and fructooligosaccharides on the fermentation in vitro by human intestinal microbiota. Journal of Functional Foods. 2015;13:158–168.

[52] Daniel M, Szymanik-Grzelak H, Turczyn A, Pańczyk-Tomaszewska M. *Lactobacillus rhamnosus* PL1 and Lactobacillus plantarum PM1 versus placebo as a prophylaxis for recurrence urinary tract infections in children: A study protocol for a randomised controlled trial. BMC Urology. 2020;20:168.

[53] Ghosh A, Sundaram B, Bhattacharya P, Mohanty N, Dheivamani N, Mane S, Acharyya B, Kamale V, Poddar S, Khobragade A, Thomas W. Effect of *Saccharomyces boulardii* cncm-I 3799 and Bacillus subtilis cu-1 on acute watery diarrhea: A randomized double-blind placebo-controlled study in Indian children. Pediatric Gastroenterology, Hepatology & Nutrition. 2021;24:423.

[54] Ivashkin V, Fomin V, Moiseev S, Brovko M, Maslennikov R, Ulyanin A, Sholomova V, Vasilyeva M, Trush E, Shifrin O, Poluektova E. Efficacy of a probiotic consisting of Lacticaseibacillus rhamnosus PDV 1705, *Bifidobacterium bifidum* PDV 0903, *Bifidobacterium longum* subsp. infantis PDV 1911, and *Bifidobacterium longum* subsp. longum PDV 2301 in the treatment of hospitalized patients with COVID-19: A randomized controlled trial. Probiotics Antimicrob Proteins. 2021;13:1–9.

[55] Bhuyan AA, Akbar Bhuiyan A, Memon AM, Zhang B, Alam J, He QG. The in vitro antiviral activity of *Lacticaseibacillus casei* MCJ protein-based metabolites on bovine viral diarrhea virus. Animal Biotechnology. 2023;34:340–349.

[56] Chen K, Xin J, Zhang G, Xie H, Luo L, Yuan S, Bu Y, Yang X, Ge Y, Liu C. A combination of three probiotic strains for treatment of acute diarrhoea in hospitalised children: An open label, randomised controlled trial. Benef Microbes. 2020;11:339–346.

[57] Chen SM, Chieng WW, Huang SW, Hsu LJ, Jan MS. The synergistic tumor growth-inhibitory effect of probiotic Lactobacillus on transgenic mouse model of pancreatic cancer treated with gemcitabine. Scientific Report. 2020;10:20319.

[58] Al-Nabulsi AA, Jaradat ZW, Al Qudsi FR, Elsalem L, Osaili TM, Olaimat AN, Esposito G, Liu SQ, Ayyash MM. Characterization and bioactive properties of exopolysaccharides produced by *Streptococcus thermophilus* and *Lactobacillus bulgaricus* isolated from labaneh. LWT. 2022;167:113817.

[59] Garbacz K. Anticancer activity of lactic acid bacteria. In: Seminars in Cancer Biology. 2022 Nov 1, (vol. 86, pp. 356–366), Academic Press.

[60] Isazadeh A, Hajazimian S, Shadman B, Safaei S, Bedoustani AB, Chavoshi R, Shanehbandi D, Mashayekhi M, Nahaei M, Baradaran B. Anti-cancer effects of probiotic *Lactobacillus acidophilus* for colorectal cancer cell line caco-2 through apoptosis induction. Journal of Pharmaceutical Research. 2020;27:262–267.

[61] Ruiz MJ, García MD, Canalejo LM, Krüger A, Padola NL, Etcheverría AI. Antimicrobial activity of *Lactiplantibacillus plantarum* against shiga toxin-producing *Escherichia coli*. Journal of Applied Microbiology. 2023;134:202.

[62] Rose Jørgensen M, Thestrup Rikvold P, Lichtenberg M, Østrup Jensen P, Kragelund C, Twetman S. *Lactobacillus rhamnosus* strains of oral and vaginal origin show strong antifungal activity in vitro. Journal of Oral Microbiology. 2020;12:1832832.

[63] Jung SP, Lee KM, Kang JH, Yun SI, Park HO, Moon Y, Kim JY. Effect of Lactobacillus gasseri BNR17 on overweight and obese adults: A randomized, double-blind clinical trial. Journal of Family Medicine. 2013;34:80.

[64] Min M, Bunt CR, Mason SL, Hussain MA. Non-dairy probiotic food products: An emerging group of functional foods. Critical Reviews in Food Science and Nutrition. 2019;59:2626–2641.

[65] Rastall RA, Gibson GR. Recent developments in prebiotics to selectively impact beneficial microbes and promote intestinal health. Cobiot. 2015;32:42–46.

[66] Nagaoka S. Yogurt production. In: Lactic Acid Bacteria: Methods and Protocols 2018 Dec 1, pp. 45–54, New York, NY: Springer New York.

[67] Hurtado-Romero A, Del Toro-Barbosa M, Gradilla-Hernández MS, Garcia-Amezquita LE, García-Cayuela T. Probiotic properties, prebiotic fermentability, and GABA-producing capacity of microorganisms isolated from Mexican milk kefir grains: A clustering evaluation for functional dairy food applications. Foods. 2021;10:2275.

[68] Zepeda-Hernández A, Garcia-Amezquita LE, Requena T, García-Cayuela T. Probiotics, prebiotics, and synbiotics added to dairy products: Uses and applications to manage type 2 diabetes. Food Research International. 2021;142:110208.

[69] Mani-López E, Palou E, López-Malo A. Probiotic viability and storage stability of yogurts and fermented milks prepared with several mixtures of lactic acid bacteria. Journal of Dairy Science. 2014;97:2578–2590.

[70] Falfán-Cortés RN, Mora-Peñaflor N, Gómez-Aldapa CA, Rangel-Vargas E, Acevedo-Sandoval OA, Franco-Fernández MJ, Castro-Rosas J. Characterization and evaluation of the probiotic potential in vitro and in situ of Lacticaseibacillus paracasei isolated from Tenate cheese. Journal of Food Protection. 2022;85:112–121.

[71] Klu YA, Chen J. Influence of probiotics, included in peanut butter, on the fate of selected Salmonella and Listeria strains under simulated gastrointestinal conditions. Journal of Applied Microbiology. 2016;120:1052–1060.

[72] Aloğlu H, Öner Z. Assimilation of cholesterol in broth, cream, and butter by probiotic bacteria. European Journal of Lipid Science and Technology. 2006;108:709–713.

[73] Ekinci FY, Okur OD, Ertekin B, Guzel-Seydim Z. Effects of probiotic bacteria and oils on fatty acid profiles of cultured cream. European Journal of Lipid Science and Technology. 2008;110:216–224.

[74] Arslan AA, Gocer EM, Demir M, Atamer Z, Hinrichs J, Kücükcetin A. Viability of *Lactobacillus acidophilus* ATCC 4356 incorporated into ice cream using three different methods. DST. 2016;96:477–487.

[75] Li X, Liu D. Nutritional content dynamics and correlation of bacterial communities and metabolites in fermented pickled radishes supplemented with wheat bran. Frontiers in Nutrition. 2022;9:840641.

[76] Žuntar I, Petric Z, Bursać Kovačević D, Putnik P. Safety of probiotics: Functional fruit beverages and nutraceuticals. Foods. 2020;9:947.

[77] Rivera-Espinoza Y, Gallardo-Navarro Y. Non-dairy probiotic products. International Journal of Food Microbiology. 2010;27:1–1.

[78] De Vuyst L, Vrancken G, Ravyts F, Rimaux T, Weckx S. Biodiversity, ecological determinants, and metabolic exploitation of sourdough microbiota. International Journal of Food Microbiology. 2009;26:666–675.

[79] Mombach MA, Da Silva Cabral L, Lima LR, Ferreira DC, Pedreira BC, Pereira DH. Association of ionophores, yeast, and bacterial probiotics alters the abundance of ruminal microbial species of pasture intensively finished beef cattle. JAHP. 2021;53:1–1.

[80] Kris-Etherton PM, Keen CL. Evidence that the antioxidant flavonoids in tea and cocoa are beneficial for cardiovascular health. Current Opinion in Lipidology. 2002;13:41–49.

[81] Possemiers S, Marzorati M, Verstraete W, Van de Wiele T. Bacteria and chocolate: A successful combination for probiotic delivery. International Journal of Food Microbiology. 2010;141:97–103.

[82] Gibson GR, Probert HM, Van Loo J, Rastall RA, Roberfroid MB. Dietary modulation of the human colonic microbiota: Updating the concept of prebiotics. Nutrition Research Reviews. 2004;17:259–275.

[83] Davani-Davari D, Negahdaripour M, Karimzadeh I, Seifan M, Mohkam M, Masoumi SJ, Berenjian A, Ghasemi Y. Prebiotics: Definition, types, sources, mechanisms, and clinical applications. Foods. 2019;8:92.

[84] Blatchford P, Ansell J, De Godoy MR, Fahey G, Garcia-Mazcorro JF, Gibson GR, Goh YJ, Hotchkiss AT, Hutkins R, LaCroix C, Rastall RA. Prebiotic mechanisms, functions and applications-A review. International Journal of Probiotics and Prebiotics. 2013;8.

[85] Niittynen L, Kajander K, Korpela R. Galacto-oligosaccharides and bowel function. Scandinavian Journal of Food and Nutrition. 2007;51:62–66.

[86] Arnold JW, Whittington HD, Dagher SF, Roach J, Azcarate-Peril MA, Bruno-Barcena JM. Safety and modulatory effects of humanized galacto-oligosaccharides on the gut microbiome. Frontiers in Nutrition. 2021;8:640100.

[87]   Huang YY, Liang YT, Wu JM, Wu WT, Liu XT, Ye TT, Chen XR, Zeng XA, Manzoor MF, Wang LH. Advances in the study of probiotics for immunomodulation and intervention in food allergy. Molecules. 2023;28:1242.

[88]   Franco-Robles E, López MG. Implication of fructans in health: Immunomodulatory and antioxidant mechanisms. The Scientific World Journal. 2015;289267.

[89]   Renye JA Jr, White AK, Hotchkiss AT Jr. Identification of Lactobacillus strains capable of fermenting fructo-oligosaccharides and inulin. Microorganisms. 2021;9:2020.

[90]   Scott KP, Martin JC, Duncan SH, Flint HJ. Prebiotic stimulation of human colonic butyrate-producing bacteria and bifidobacteria, in vitro. FEMS Microbiology Ecology. 2014;87:30–40.

[91]   Bamigbade GB, Subhash AJ, Kamal-Eldin A, Nyström L, Ayyash M. An updated review on prebiotics: Insights on potentials of food seeds waste as source of potential prebiotics. Molecules. 2022;27:5947.

[92]   Johnson CR, Combs GF Jr, Thavarajah P. Lentil (Lens culinaris L.): A prebiotic-rich whole food legume. Food Research International. 2013;51:107–113.

[93]   Fuentes-Zaragoza E, Sánchez-Zapata E, Sendra E, Sayas E, Navarro C, Fernández-López J, Pérez-Alvarez JA. Resistant starch as prebiotic: A review. Starch/Staerke. 2011;63:406–415.

[94]   Roberfroid M. Prebiotics: The concept revisited1. Journal of Nutritional. 2007;137:830–837.

[95]   de Oliveira DP, Todorov SD, Fabi JP. Exploring the prebiotic potentials of hydrolyzed pectins: Mechanisms of action and gut microbiota modulation. Journal of Nutritional. 2024;16:3689.

[96]   Gómez B, Gullón B, Remoroza C, Schols HA, Parajó JC, Alonso JL. Purification, characterization, and prebiotic properties of pectic oligosaccharides from orange peel wastes. Journal of Agricultural and Food Chemistry. 2014;62:9769–9782.

[97]   Manderson K, Pinart M, Tuohy KM, Grace WE, Hotchkiss AT, Widmer W, Yadhav MP, Gibson GR, Rastall RA. In vitro determination of prebiotic properties of oligosaccharides derived from an orange juice manufacturing by-product stream. AEM. 2005;71:8383–8389.

[98]   Freitas CM, Coimbra JS, Souza VG, Sousa RC. Structure and applications of pectin in food, biomedical, and pharmaceutical industry: A review. Coatings. 2021;11:922.

[99]   Blumberg JB, Ding EL, Dixon R, Pasinetti GM, Villarreal F. The science of cocoa flavanols: Bioavailability, emerging evidence, and proposed mechanisms. Advances in Nutritional. 2014;5:547–549.

[100] Tzounis X, Rodriguez-Mateos A, Vulevic J, Gibson GR, Kwik-Uribe C, Spencer JP. Prebiotic evaluation of cocoa-derived flavanols in healthy humans by using a randomized, controlled, double-blind, crossover intervention study. American Journal of Clinical Nutrition. 2011;93:62–72.

[101] Cardona F, Andrés-Lacueva C, Tulipani S, Tinahones FJ, Queipo-Ortuño MI. Benefits of polyphenols on gut microbiota and implications in human health. Nutrition Biochemistry. 2013;24:1415–1422.

[102] Fusco W, Lorenzo MB, Cintoni M, Porcari S, Rinninella E, Kaitsas F, Lener E, Mele MC, Gasbarrini A, Collado MC, Cammarota G. Short-chain fatty-acid-producing bacteria: Key components of the human gut microbiota. Journal of Nutritional. 2023;15, 2211.

[103] Jackson PP, Wijeyesekera A, Theis S, Van Harsselaar J, Rastall RA. Effects of food matrix on the prebiotic efficacy of inulin-type fructans: A randomised trial. Benef Microbes. 2023;14:317–334.

[104] Yang J, Martínez I, Walter J, Keshavarzian A, Rose DJ. In vitro characterization of the impact of selected dietary fibers on fecal microbiota composition and short chain fatty acid production. Anaerobe. 2013;23:74–81.

[105] Roberfroid M. Inulin-type Fructans: Functional Food Ingredients. 2004 Oct 28, CRC Press.

[106] Martyniak A, Medyńska-Przęczek A, Wędrychowicz A, Skoczeń S, Tomasik PJ. Prebiotics, probiotics, synbiotics, paraprobiotics and postbiotic compounds in IBD. Biomolecules. 2021;11:1903.

[107] Megur A, Daliri EB, Baltriukienė D, Burokas A. Prebiotics as a tool for the prevention and treatment of obesity and diabetes: Classification and ability to modulate the gut microbiota. International Journal of Molecular Sciences. 2022;23:6097.

[108] Walsh SK, Lucey A, Walter J, Zannini E, Arendt EK. Resistant starch – An accessible fiber ingredient acceptable to the Western palate. Comprehensive Reviews in Food Science and Food Safety. 2022;21:2930–2955.

[109] Plongbunjong V, Graidist P, Knudsen KE, Wichienchot S. Isomaltooligosaccharide synthesised from rice starch and its prebiotic properties in vitro. IJFST. 2017;52:2589–2595.

[110] Bang SJ, Kim G, Lim MY, Song EJ, Jung DH, Kum JS, Nam YD, Park CS, Seo DH. The influence of in vitro pectin fermentation on the human fecal microbiome. Amb Express. 2018;8:1–9.

[111] Karakan T, Tuohy KM, Janssen-van Solingen G. Low-dose lactulose as a prebiotic for improved gut health and enhanced mineral absorption. Frontiers in Nutrition. 2021;8:672925.

[112] Bai J, Wang B, Tan X, Huang L, Xiong S. Regulatory effect of lactulose on intestinal flora and serum metabolites in colitis mice: In vitro and in vivo evaluation. Food Chemistry: X. 2023;19:100821.

[113] Tuncil YE, Nakatsu CH, Kazem AE, Arioglu-Tuncil S, Reuhs B, Martens EC, Hamaker BR. Delayed utilization of some fast-fermenting soluble dietary fibers by human gut microbiota when presented in a mixture. Journal of Functional Foods. 2017;32:347.

[114] Bongiovanni T, Yin MO, Heaney LM. The athlete and gut microbiome: Short-chain fatty acids as potential ergogenic aids for exercise and training. International Journal of Sports Medicine. 2021;42:1143–1158.

[115] Swanson KS, Gibson GR, Hutkins R, Reimer RA, Reid G, Verbeke K, Scott KP, Holscher HD, Azad MB, Delzenne NM, Sanders ME. The International Scientific Association for Probiotics and Prebiotics (ISAPP) consensus statement on the definition and scope of synbiotics. Nat Rev Gastroenterol Hepatol. 2020;17:687–701.

[116] Sharma M, Shukla G. Metabiotics: One step ahead of probiotics; an insight into mechanisms involved in anticancerous effect in colorectal cancer. Frontiers in Microbiology. 2016;7:1940.

[117] Yadav M, Sehrawat N, Sharma AK, Kumar S, Singh R, Kumar A, Kumar A. Synbiotics as potent functional food: Recent updates on therapeutic potential and mechanistic insight. Journal of Food Science and Technology. 2024;61,1–5.

[118] Cicero AF, Fogacci F, Bove M, Giovannini M, Borghi C. Impact of a short-term synbiotic supplementation on metabolic syndrome and systemic inflammation in elderly patients: A randomized placebo-controlled clinical trial. European Journal of Nutrition. 2021;60:655–663.

[119] Wegh CA, Geerlings SY, Knol J, Roeselers G, Belzer C. Postbiotics and their potential applications in early life nutrition and beyond. International Journal of Molecular Sciences. 2019;20:4673.

[120] Šušković J, Kos B, Beganović J, Leboš Pavunc A, Habjanič K, Matošić S. Antimicrobial activity – The most important property of probiotic and starter lactic acid bacteria. Food Technology and Biotechnology. 2010;48:296–307.

[121] de Almada CN, Almada CN, Martinez RCR, Sant'Ana AS. Paraprobiotics: Evidences on their ability to modify biological responses, inactivation methods and perspectives on their application in foods. Trends Food and Science Technology. 2016;58:96–114.

[122] Vinderola G, Sanders ME, Cunningham M, Hill C. Frequently asked questions about the ISAPP postbiotic definition. Frontiers in Microbiology. 2024;1324565.

[123] Yelin I, Flett KB, Merakou C, Mehrotra P, Stam J, Snesrud E. Genomic and epidemiological evidence of bacterial transmission from probiotic capsule to blood in ICU patients. Nature Medicine. 2019;25:1728–1732.

[124] Salminen S, Collado MC, Endo A, Hill C, Lebeer S, Quigley EMM. The International Scientific Association of Probiotics and Prebiotics (ISAPP) consensus statement on the definition and scope of postbiotics. Nature Reviews Gastroenterology & Hepatology. 2021;18:649–667.

[125] Rad AH, Abbasi A, Kafil HS, Ganbarov K. Potential pharmaceutical and food applications of postbiotics: A review. Current Pharmaceutical Biotechnology. 2020;21:1576–1587.

[126] Humam AM, Loh TC, Foo HL, Izuddin WI, Zulkifli I, Samsudin AA. Supplementation of postbiotic RI11 improves antioxidant enzyme activity, upregulated gut barrier genes, and reduced cytokine, acute

phase protein, and heat shock protein 70 gene expression levels in heat-stressed broilers. Poultry Science. 2021;100:100908.

[127] Nataraj BH, Ali SA, Behare PV, Yadav H. Postbiotics-parabiotics: The new horizons in microbial biotherapy and functional foods. Microbial Cell Factories. 2020;19.

[128] Barros CP, Guimarães JT, Esmerino EA, Duarte MCKH, Silva MC, Silva R, Ferreira BM, Sant'Ana AS, Freitas MQ, Cruz AG. Paraprobiotics and postbiotics: Concepts and potential applications in dairy products. Current Opinion in Food Science. 2020;32:1–8.

[129] Aguilar-Toalá JE, Garcia-Varela R, Garcia HS, Mata-Haro V, González-Córdova AF, Vallejo-Cordoba B, Hernández-Mendoza A. Postbiotics: An evolving term within the functional foods field. Trends Food and Science Technology. 2018;75:105–114.

[130] Amores MCC, Vinderola CG, Salminen S. Postbiotics: Facts and open questions. A position paper on the need for a consensus definition. Benef Microbes. 2019;10:711–719.

[131] Chaluvadi S, Hotchkiss AT, Yam KL. Gut microbiota: Impact of probiotics, prebiotics, synbiotics, pharmabiotics, and postbiotics on human health. In: Probiotics, Prebiotics, and Synbiotics: Bioactive Foods in Health Promotion. 2015, pp. 515–523, Amsterdam, The Netherlands: Elsevier Inc.

[132] Shuwen Z. Antioxidative activity of lactic acid bacteria in yogurt. African Journal of Microbiology Research. 2011;5:5194–5201.

[133] Amaretti A, Di Nunzio M, Pompei A, Raimondi S, Rossi M, Bordoni A. Antioxidant properties of potentially probiotic bacteria: In vitro and in vivo activities. Applied Microbiology and Biotechnology. 2013;97:809–817.

[134] Gezginç Y, Karabekmez-Erdem T, Tatar HD, Ayman S, Ganiyusufoğlu E, Dayisoylu KS. Health promoting benefits of postbiotics produced by lactic acid bacteria: Exopolysaccharide. Biotechnology Studies. 2022;31:62–63.

[135] Hernández-Granados MJ, Franco-Robles E. Postbiotics in human health: Possible new functional ingredients?. Food Research International. 2020;137:109660.

[136] Żółkiewicz J, Marzec A, Ruszczyński M, Feleszko W. Postbiotics – A step beyond pre-and probiotics. Nutrients. 2020;12:1–17.

[137] Mehta JP, Ayakar S, Singhal RS. The potential of paraprobiotics and postbiotics to modulate the immune system: A review. Microbiological Research. 2023;275:127449.

[138] Gupta T, Kaur H, Kapila S, Kapila R. Potential probiotic Lacticaseibacillus rhamnosus MTCC-5897 attenuates *Escherichia coli* induced inflammatory response in intestinal cells. Archives of Microbiology. 2021;203:5703–5713.

[139] Ballan R, Saad SMI. Characteristics of the gut microbiota and potential effects of probiotic supplements in individuals with type 2 diabetes mellitus. Foods. 2021;10:2528.

[140] Cirat R, Capozzi V, Benmechernene Z, Spano G, Grieco F, Fragasso M. LAB antagonistic activities and their significance in food biotechnology: Molecular mechanisms, food targets, and other related traits of interest. Fermentation. 2024;10:222.

[141] Pan Z, Mao B, Zhang Q, Tang X, Yang B, Zhao J, Cui S, Zhang H. Postbiotics prepared using Lactobacillus paracasei ccfm1224 prevent non alcoholic fatty liver disease by modulating the gut microbiota and liver metabolism. International Journal of Molecular Sciences. 2022;23:13522.

[142] Ibrahim GA, Mabrouk AM, El-Ssayad MF, Mehaya FM, Sharaf OM, Ibrahim MI. Properties of postbiotics produced by probiotics: Antimicrobial, antioxidant activities and production of vitamins, organic acids. Research Square. 2024.

[143] Mounir M, Ibijbijen A, Farih K, Rabetafika HN, Razafindralambo HL. Synbiotics and their antioxidant properties, mechanisms, and benefits on human and animal health: A narrative review. Biomolecules. 2022;12:1443.

[144] Izuddin WI, Humam AM, Loh TC, Foo HL, Samsudin AA. Dietary postbiotic Lactobacillus plantarum improves serum and ruminal antioxidant activity and upregulates hepatic antioxidant enzymes and ruminal barrier function in post-weaning lambs. Antioxidants. 2020;9:250.

[145] Lee E, Jung SR, Lee SY, Lee NK, Paik HD, Lim SI. Lactobacillus plantarum strain ln4 attenuates diet-induced obesity, insulin resistance, and changes in hepatic mRNA levels associated with glucose and lipid metabolism. Nutrients. 2018;10:643.

[146] Soleimani A, Motamedzadeh A, Zarrati Mojarrad M, Bahmani F, Amirani E, Ostadmohammadi V, Tajabadi-Ebrahimi M, Asemi Z. The effects of synbiotic supplementation on metabolic status in diabetic patients undergoing hemodialysis: A randomized, double-blinded, placebo-controlled trial. Probiotics Antimicrob Proteins. 2019;11:1248–1256.

[147] Miraghajani M, Dehsoukhteh SS, Rafie N, Hamedani SG, Sabihi S, Ghiasvand R. Potential mechanisms linking probiotics to diabetes: A narrative review of the literature. São Paulo Medical Journal. 2017;135:169–178.

[148] Mori P, Chauhan M, Modasiya I, Kumar V. Dietary modulation of the nervous and immune system: Role of probiotics/prebiotics/synbiotics/postbiotics. In: Kothari V, Kumar P, Ray S. (eds.) Probiotics, Prebiotics, Synbiotics, and Postbiotics. 2023, pp. 307–328, Singapore: Springer Nature Singapore.

[149] Neyrinck AM, Rodriguez J, Taminiau B, Amadieu C, Herpin F, Allaert FA, Cani PD, Daube G, Bindels LB, Delzenne NM. Improvement of gastrointestinal discomfort and inflammatory status by a synbiotic in middle-aged adults: A double-blind randomized placebo-controlled trial. Scientific Report. 2021;11:2627.

[150] Liao JC, Deng JS, Chiu CS, Hou WC, Huang SS, Shie PH, Huang GJ. Anti-inflammatory activities of Cinnamomum cassia constituents in vitro and in vivo. Evidence-Based Complementary and Alternative Medicine. 2012;2012:429320.

[151] Homayouni Rad A, Aghebati-Maleki L, Samadi Kafil H, Gilani N, Abbasi A, Khani N. Postbiotics, as dynamic biomolecules, and their promising role in promoting food safety. Biointerface Res Appl Chem. 2021;11:14529–14544.

[152] Incili GK, Karatepe P, Akgöl M, Güngören A, Koluman A, Ilhak OI. Characterization of lactic acid bacteria postbiotics, evaluation in-vitro antibacterial effect, microbial and chemical quality on chicken drumsticks. Food Microbiol. 2022;104:104001.

[153] Chaney WE, Naqvi SA, Gutierrez M, Gernat A, Johnson TJ, Petry D. Dietary inclusion of a *Saccharomyces cerevisiae*-derived postbiotic is associated with lower *Salmonella enterica* burden in broiler chickens on a commercial farm in honduras. Microorganisms. 2022;10:544.

[154] De Lima AL, Guerra CA, Costa LM, De Oliveira VS, Lemos Junior WJF, Luchese RH. A natural technology for vacuum-packaged cooked sausage preservation with potentially postbiotic-containing preservative. Fermentation. 2022;8:106.

[155] Moradi M, Kousheh SA, Almasi H, Alizadeh A, Guimarães JT, Yilmaz N. Postbiotics produced by lactic acid bacteria: The next frontier in food safety. Comprehensive Reviews in Food Science and Food Safety. 2020;19:3390–3415.

[156] Yildirim S, Röcker B, Pettersen MK, Nilsen-Nygaard J, Ayhan Z, Rutkaite R, Radusin T, Suminska P, Marcos B, Coma V. Active packaging applications for food. Comprehensive Reviews in Food Science and Food Safety. 2018;17:165–199.

[157] Hosseini SA, Abbasi A, Sabahi S, Khani N. Application of postbiotics produced by lactic acid bacteria in the development of active food packaging. Biointerface Res Appl Chem. 2021;12:6164–6183.

[158] Darwish MS, Qiu L, Taher MA, Zaki AA, Abou-Zeid NA, Dawood DH. Health benefits of postbiotics produced by E. coli nissle 1917 in functional yogurt enriched with cape gooseberry (Physalis peruviana L.). Fermentation. 2022;8:128.

[159] Ozma MA, Abbasi A, Akrami S, Lahouty M, Shahbazi N, Ganbarov K, Pagliano P, Sabahi S, Köse Ş, Yousefi M, Dao S, Asgharzadeh M, Hosseini H, Kafil HS. Postbiotics as the key mediators of the gut microbiota-host interactions. Infezioni in Medicina. 2022;30:180–193.

[160] Garnier L, Mounier J, Lê S, Pawtowski A, Pinon N, Camier B. Development of antifungal ingredients for dairy products: From in vitro screening to pilot scale application. Food Microbiology. 2019;81:97–107.

[161] Andrade JC, João AL, De Sousa Alonso C, Barreto AS, Henriques AR. Genetic subtyping, biofilm-forming ability and biocide susceptibility of Listeria monocytogenes strains isolated from a ready-to-eat food industry. Antibiotics. 2020;9:416.

[162] Przekwas J, Wiktorczyk N, Budzyńska A, Wałecka-Zacharska E, Gospodarek-Komkowska E. Ascorbic acid changes growth of food- borne pathogens in the early stage of biofilm formation. Microorganisms. 2020;8:553.

[163] Messaoudi M, Lalonde R, Violle N, Javelot H, Desor D, Nejdi A, Bisson JF, Rougeot C, Pichelin M, Cazaubiel M, Cazaubiel JM. Assessment of psychotropic-like properties of a probiotic formulation (*Lactobacillus helveticus* R0052 and *Bifidobacterium longum* R0175) in rats and human subjects. British Journal of Nutrition. 2011;105:755–764.

[164] Binda S, Hill C, Johansen E, Obis D, Pot B, Sanders ME, Tremblay A, Ouwehand AC. Criteria to qualify microorganisms as "probiotic" in foods and dietary supplements. Frontiers in Microbiology. 2020;11:1662.

[165] Akter S, Park JH, Jung HK. Potential health-promoting benefits of paraprobiotics, inactivated probiotic cells. Journal of Microbiology and Biotechnology. 2020;30:477.

[166] Damián MR, Cortes-Perez NG, Quintana ET, Ortiz-Moreno A, Garfias Noguez C, Cruceño-Casarrubias CE, Sánchez Pardo ME, Bermúdez-Humarán LG. Functional foods, nutraceuticals and probiotics: A focus on human health. Microorganisms. 2022;1065.

[167] Ayad AA, El-Rab DG, Williams LL. Probiotics, prebiotics, and postbiotics have a positive impact. In: Probiotics, Prebiotics, and Postbiotics in Human Health and Sustainable Food Systems. 2025, (vol. 2011, pp. 49).

[168] Soemarie YB, Milanda T, Barliana MI. Fermented foods as probiotics: A review. Journal of Advanced Pharmaceutical Technology & Research. 2021;12:335–339.

[169] Rodrigo L. ed. Immunology of the GI Tract: Recent Advances.

[170] Guli M, Winarsih S, Barlianto W, Illiandri O, Sumarno SP. Mechanism of Lactobacillus reuteri probiotic in increasing intestinal mucosal immune system. Oamjms. 2021;9:784–793.

[171] Hill C, Guarner F, Reid G, Gibson GR, Merenstein DJ, Pot B, Morelli L, Canani RB, Flint HJ, Salminen S, Calder PC. Activity of cecropin P1 and FA-LL-37 against urogenital microflora. Nature Reviews Gastroenterology and Hepatology. 2014;11,506.

[172] Cong J, Zhou P, Zhang R. Intestinal microbiota-derived short chain fatty acids in host health and disease. Nutrients. 2022;14:1977.

[173] Mekonnen SA, Merenstein D, Fraser CM, Marco ML. Molecular mechanisms of probiotic prevention of antibiotic-associated diarrhoea. Current Opinion in Biotechnology. 2020;61:226–234.

[174] Liu Y, Wang J, Wu C. Modulation of gut microbiota and immune system by probiotics, prebiotics, and postbiotics. Frontiers in Nutrition. 2022;8:634897.

[175] Maldonado Galdeano C, Cazorla SI, Lemme Dumit JM, Vélez E, Perdigón G. Beneficial effects of probiotic consumption on the immune system. Annals of Nutrition and Metabolism. 2019;74:115–124.

[176] Wan ML, Forsythe SJ, El-Nezami H. Probiotics interaction with foodborne pathogens: A potential alternative to antibiotics and future challenges. Critical Reviews in Food Science and Nutrition. 2019;59:3320–3333.

[177] Rose EC, Odle J, Blikslager AT, Ziegler AL. Probiotics, prebiotics and epithelial tight junctions: A promising approach to modulate intestinal barrier function. International Journal of Molecular Sciences. 2021;22:6729.

[178] Parada Venegas D, De La Fuente MK, Landskron G, González MJ, Quera R, Dijkstra G, Harmsen HJ, Faber KN, Hermoso MA. Short chain fatty acids (SCFAs)-mediated gut epithelial and immune regulation and its relevance for inflammatory bowel diseases. Frontiers in Immunology. 2019;10:277.

[179] Varvara RA, Vodnar DC. Probiotic-driven advancement: Exploring the intricacies of mineral absorption in the human body. Food Chemistry: X. 2024;21:101067.

[180] Schade DS, Shey L, Eaton RP. Cholesterol review: A metabolically important molecule. Endocrine Practice. 2020;26:1514–1523.

[181] Di Ciaula A, Garruti G, Baccetto RL, Molina-Molina E, Bonfrate L, Portincasa P, Wang DQ. Bile acid physiology. Annals of Hepatology. 2018;16:4–14.

[182] Przerwa F, Kukowka A, Kotrych K. Probiotics in prevention and treatment of cardiovascular diseases. Herba Polonica. 2021;67:77–85.

[183] Olas B. Probiotics, prebiotics and synbiotics – A promising strategy in prevention and treatment of cardiovascular diseases? 2020, 9737.

[184] Hossain TJ, Nafiz IH, Ali F, Mozumder HA, Islam S, Rahman N, Ferdouse J, Khan MS. Antipathogenic action and antibiotic sensitivity pattern of the "Borhani"-associated lactic acid bacterium *Weissella confusa* LAB-11. The Journal of Microbiology, Biotechnology and Food Sciences. 2023;13:e9964.

[185] Shimizu M, Hashiguchi M, Shiga T, Tamura HO, Mochizuki M. Meta-analysis: Effects of probiotic supplementation on lipid profiles in normal to mildly hypercholesterolemic individuals. PloS One. 2015;10:e0139795.

[186] Park S, Kang J, Choi S, Park H, Hwang E, Kang Y, Kim A, Holzapfel W, Ji Y. Cholesterol-lowering effect of *Lactobacillus rhamnosus* BFE5264 and its influence on the gut microbiome and propionate level in a murine model. PLoS One. 2018;13:e0203150.

[187] Aoun A, Darwish F, Hamod N. The influence of the gut microbiome on obesity in adults and the role of probiotics, prebiotics, and synbiotics for weight loss. Preventive Nutrition and Food Science. 2020;25:113.

[188] Obayomi OV, Olaniran AF, Olawoyin DC, Falade OV, Osemwegie OO, Owa SO. Role of enteric dysbiosis in the development of central obesity: A review. Scientific African. 2024;e02204.

[189] Morrison DJ, Preston T. Formation of short chain fatty acids by the gut microbiota and their impact on human metabolism. Gut Microbes. 2016;7:189–200.

[190] Jang HR, Park HJ, Kang D, Chung H, Nam MH, Lee Y, Park JH, Lee HY. A protective mechanism of probiotic Lactobacillus against hepatic steatosis via reducing host intestinal fatty acid absorption. Experimental and Molecular Medicine. 2019;51:1–4.

[191] Barathikannan K, Chelliah R, Rubab M, Daliri EB, Elahi F, Kim DH, Agastian P, Oh SY, Oh DH. Gut microbiome modulation based on probiotic application for anti-obesity: A review on efficacy and validation. Microorganisms. 2019;7:456.

[192] Kim YA, Keogh JB, Clifton PM. Probiotics, prebiotics, synbiotics and insulin sensitivity. Nutrition Research Reviews. 2018;31:35–51.

[193] Singh S, Sarma DK, Verma V, Nagpal R, Kumar M. Unveiling the future of metabolic medicine: Omics technologies driving personalized solutions for precision treatment of metabolic disorders. Biochemical and Biophysical Research Communications. 2023;682:1–20.

[194] Pizarroso NA, Fuciños P, Gonçalves C, Pastrana L, Amado IR. A review on the role of food-derived bioactive molecules and the microbiota–gut–brain axis in satiety regulation. Nutrients. 2021;13:632.

[195] Yu AQ, Li L. The potential role of probiotics in cancer prevention and treatment. Nutrition and Cancer. 2016;68:535–544.

[196] Bedada TL, Feto TK, Awoke KS, Garedew AD, Yifat FT, Birri DJ. Probiotics for cancer alternative prevention and treatment. Biomedicine and Pharmacotherapy. 2020;129:110409.

[197] Sehrawat N, Yadav M, Singh M, Kumar V, Sharma VR, Sharma AK. Probiotics in microbiome ecological balance providing a therapeutic window against cancer. In: Seminars in Cancer Biology. 2021 May 1, (vol. 70, pp. 24–36), Academic Press.

[198] Thananimit S, Pahumunto N, Teanpaisan R. Characterization of short chain fatty acids produced by selected potential probiotic lactobacillus strains. Biomolecules. 2022;12:1829.

[199] Gopalakrishnan V, Spencer CN, Nezi L, Reuben A, Andrews MC, Karpinets TV, Prieto PA, Vicente D, Hoffman K, Wei SC, Cogdill AP. Gut microbiome modulates response to anti–PD-1 immunotherapy in melanoma patients. Science. 2018;359:97–103.

[200] Donaldson MS. Nutrition and cancer: A review of the evidence for an anti-cancer diet. Nutrition Journal. 2004;3:19.

[201] Appleton J. The gut-brain axis: Influence of microbiota on mood and mental health. IMCJ. 2018;17:28.

[202] Sasso JM, Ammar RM, Tenchov R, Lemmel S, Kelber O, Grieswelle M, Zhou QA. Gut microbiome–brain alliance: A landscape view into mental and gastrointestinal health and disorders. ACS Chemical Neuroscience. 2023;14:1717–1763.

[203] Frankiensztajn LM, Elliott E, Koren O. The microbiota and the hypothalamus-pituitary-adrenocortical (HPA) axis, implications for anxiety and stress disorders. Current Opinion in Neurobiology. 2020;62:76–82.

[204] Bruun CF, Haldor Hansen T, Vinberg M, Kessing LV, Coello K. Associations between short-chain fatty acid levels and mood disorder symptoms: A systematic review. Nutritional Neuroscience. 2024;27:899–912.

[205] Feng T, Wang J. Oxidative stress tolerance and antioxidant capacity of lactic acid bacteria as probiotic: A systematic review. Gut Microbes. 2020;12:1801944.

[206] Hsieh TH, Kuo CW, Hsieh KH, Shieh MJ, Peng CW, Chen YC, Chang YL, Huang YZ, Chen CC, Chang PK, Chen KY. Probiotics alleviate the progressive deterioration of motor functions in a mouse model of Parkinson's disease. Brain Sciences. 2020;10:206.

[207] Rudzki L, Stone TW, Maes M, Misiak B, Samochowiec J, Szulc A. Gut microbiota-derived vitamins–underrated powers of a multipotent ally in psychiatric health and disease. Progress in Neuro-Psychopharmacology & Biological Psychiatry. 2021;107:110240

[208] Liu T, Huang Z. Evidence-based analysis of neurotransmitter modulation by gut microbiota. In: Health Information Science: 8th International Conference, HIS 2019. October 18–20, 2019, Proceedings 8 2019 pp. 238–249, Xi'an, China: Springer International Publishing.

[209] Wang J, Ji H, Wang S, Liu H, Zhang W, Zhang D, Wang Y. Probiotic Lactobacillus plantarum promotes intestinal barrier function by strengthening the epithelium and modulating gut microbiota. Frontiers in Microbiology. 2018;9:1953.

[210] Malomo AA, Adeniran H, Balogun D, Oke A, Iyiola O, Olaniran A, Alakija O, Abiose S. Influence of pre-treatment on the microbiological and biochemical properties of wine produced from overripe plantain: Production of Agadagidi. The Journal of Microbiology, Biotechnology and Food Sciences. 2023;12:e8258.

[211] Suganya K, Koo BS. Gut–brain axis: Role of gut microbiota on neurological disorders and how probiotics/prebiotics beneficially modulate microbial and immune pathways to improve brain functions. International Journal of Molecular Sciences. 2020;21:7551.

Priti, Kumkum Verma, Soniya Goyal, Praveen, Charulata,
Bhupesh Gupta, Poonam Bansal*

Chapter 4
Probiotics and Gut Microbiota: A Comprehensive Study of Their Interactions and Benefits

Abstract: Probiotics play a vital role in maintaining a healthy gut microbiome, which enhances digestion and supports immune health. The human gut microbiome is essential for overall well-being and balance. Comprising a diverse array of micro-organisms within the digestive system, it significantly impacts metabolism immune response and vulnerability to diseases. Probiotics, which are live microorganisms, that provide health benefits when consumed in sufficient amount have attracted considerable attention for their ability to influence gut flora and address various health issues. In this chapter, we examine how probiotic supplementation affects the composition and functionality of gut microbiota emphasizing recent developments and scientific insights. We discuss the mechanism through which probiotics promote microbial balance including their influence on intestinal permeability immune function and metabolic activities. Additionally, we discuss the clinical use of probiotics for treating gastrointestinal disorders and their potential advantages for overall health enhancement. This chapter seeks to provide a thorough understanding of how probiotics can be effectively utilized to optimize gut microbiota and improve health outcomes. It also highlights both the potential benefits and challenges of probiotic therapy, underscoring the importance of personalized strategies based on individual microbiome characteristics and specific health needs.

Keywords: probiotics, gut, microflora, antibiotics, human health

*Corresponding author: Poonam Bansal, Department of Biosciences and Technology, MMDU University, Ambala, Haryana, India, e-mail: poonambansal.biochem@gmail.com
Priti, KL Mehta Dayanand College for Women, Faridabad, Haryana, India
Kumkum Verma, IARI, Regional Station, Karnal, Haryana, India
Soniya Goyal, Department of Biosciences and Technology, MMDU University, Ambala, Haryana, India
Praveen, Faculty of Agriculture, Guru Kashi University, Talwandi Sabo, Bathinda, Punjab, India
Charulata, ICAR-Indian Institute of Wheat and Barley Research, Shimla, Himachal Pradesh, India
Bhupesh Gupta, Department of Computer Sciences, MMDU University, Ambala, Haryana, India

https://doi.org/10.1515/9783112205150-004

4.1 Introduction

The human gastrointestinal system hosts trillion of microorganisms collectively known as gut microflora. This intricate ecosystem is vital for maintaining human health, impacting essential physiological processes like digestion, metabolism, and immune function [1]. Within this diverse community's probiotics, beneficial live bacteria have gained significant attention for their potential health benefits. These microorganisms are prioritized by their capacity to provide health benefits when taken in adequate amount [2] and can enhance both the composition and functioning of the gut microbiome. Recent development in metagenomic sequencing and bioinformatics have significantly enhanced our comprehension of gut microbiota composition and its interactions with the host. Studies revealed that maintaining a balance microbiome is crucial for overall health while imbalances in microbial diversity can contribute to a range of health problems such as obesity, diabetes, and inflammatory bowel disease (IBD) [3]. Probiotics have emerged as a potential solution for restoring this balance, various studies demonstrating their effectiveness in modifying gut microflora and improving gut health.

Probiotic exerts their health benefits via different complex mechanisms. They can strengthen the gut barrier, inhibit harmful microorganisms, and influence immune response [4]. For example, various *Lactobacillus* and *Bifidobacterium* species are known to produce short-chain fatty acids that are essential for gut health as they provide energy to colon cells and act as signaling molecules for immune cells [5]. Additionally, probiotics compete with pathogenic bacteria for nutrients and attachment sites, thereby lowering the risk of infections [6]. In addition to their benefits in gastrointestinal system, probiotics are also associated with the range of effects beyond the gut. Various ongoing research demonstrated that microbiota influences brain's activity via the gut-brain axis, potentially affecting mood and cognitive function [7]. Probiotics also prevent allergies and autoimmune disease by regulating immune response [8]. They are essential in sustaining and improving gut microflora that is crucial for overall health. Probiotics preserve gut homeostasis by encouraging good bacteria and suppressing harmful pathogen. They also maintain digestive balance by preventing conditions like irritable bowel syndrome and IBD [9]. Additionally, they reinforce the intestinal barrier and regulate immunological responses to improve the immune system [10]. Recent finding suggests that they improve mental health by interacting with the gut-brain axis [11]. Probiotic presents promising advantages for gut health and overall wellness. While probiotics offer several potential health benefits, there are still challenges associated with their use. The efficacy of probiotic supplementation varies based on several factors like specific strain of the dosage, the length of the time taken, and individual host characteristics such as genetics and the current composition of the gut microbiota. It is essential to understand the unique functions of different probiotic strains in influencing gut microflora to optimize their application in clinical environments.

This chapter offers a comprehensive overview of the current state of probiotics and get microflora. We will explore recent discoveries regarding the mechanisms by which probiotics exert their effects on gut health and related diseases, as well as the challenges encountered in their use. By reviewing the recent literature, we aim to clarify the importance of probiotics in maintaining a balanced gut microbiome and their potential therapeutic roles.

4.2 Healthy Human Gut Microbiome

The human gut microbiome is a complex ecosystem composed of millions of microorganisms residing in the colon essential for maintaining the body's equilibrium. This community includes various strengths of bacteria and yeast that typically engage in a mutually beneficial relationship with the host. Microbial diversity is not uniform across the gastrointestinal tract; the colon harbors the highest concentration of microbes while the stomach and small intestine contain fewer bacterial species as represented in Figure 4.1. Importantly, around 99% of the bacteria in the gut are anaerobic, highlighting the distinctive conditions present in this environment [12].

Research on the human microbiome has identified 2,172 unique microbial spaces that can be classified into 12 different phyla. The main bacterial strains found in the human gut are predominantly from five major phyla: Firmicutes, Bacteroidetes, Actinobacteria, Proteobacteria, and Verrucomicrobia [13]. Several environmental factors, such as diet antibiotic uses and stress, play a significant role in shaping the gut microbiota [14]. Ki metabolites produced by these microorganisms include short-chain fatty acids, bile acids, amino acids, trimethylamine-N-oxide, tryptophan, and indole derivatives [15]. Given the close interplay between gut microbiota and the host, changes in this microbial community are often associated with various diseases. The density of commensal microbes in the gastrointestinal tract measured in colony-forming units increases progressively from the stomach to the jejunum and ultimately to the colon. Disruptions in beneficial microbiota can lead to several gastrointestinal disorders such as irritable bowel syndrome, IBD, celiac disease, and colorectal cancer (CRC). Furthermore, imbalances in gut microbial communities and their metabolites have been linked to dysfunctions or diseases in other parts of the body.

Recent research has enhanced our understanding of the composition and functionality of a healthy gut microbiome. In one study, researchers examined tool samples from 242 healthy young adult using 16S rRNA gene PCR sequencing and whole metagenomic sequencing to assess the microbiomes' structure and functions [16]. The biological diversity and richness in the distal intestine were found to be significantly greater than those in microbiomes from other parts of the body, such as the skin or oral cavity, both in terms of microbial taxa and genetic variation. Additionally, different body sites exhibited distinct predominant taxa.

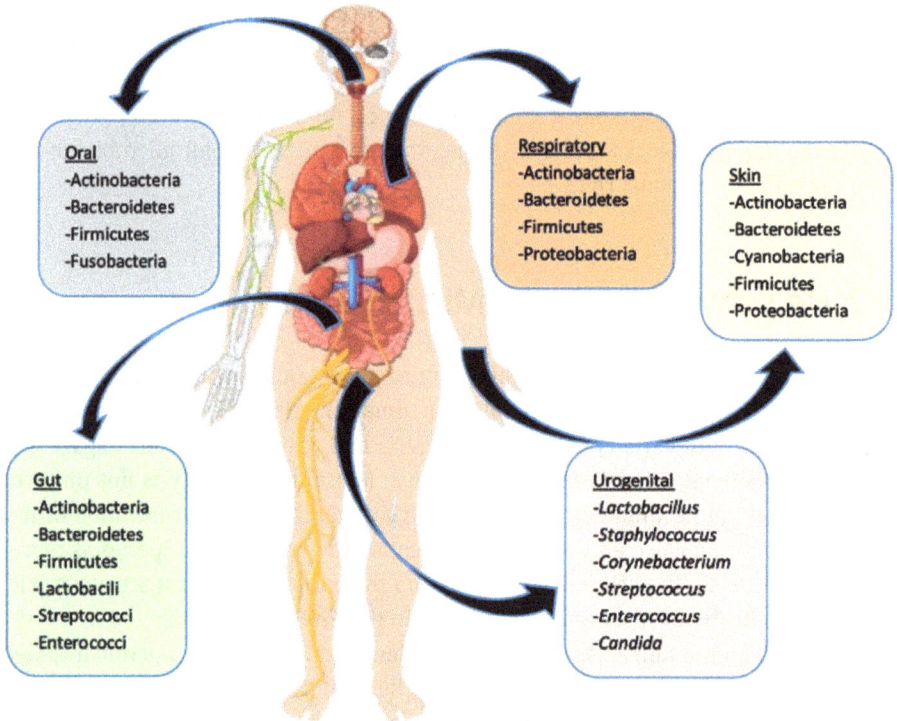

Figure 4.1: Microbial diversity generally associated with human body.

4.3 Diet and Its Impact on the Gut Microbiota

Diet influences the gut microbiota has been a subject of scientific enquiry since the 1960s. Recent studies have employed animal models, and examined intestinal microbiota and metagenomes to investigate how various dietary factors shape the composition and function of the gut microbiome. Dietary supplements can significantly impact the microbiome, resulting in changes in chemical processes in the intestinal lumen. Understanding the connection between diet and gut microbiota is crucial for enhancing overall health. The relationship between diet and its effects on the gut microbiota is summarized in Table 4.1.

Table 4.1: Diet and its effects on the gut microbiota.

Topic	Details	References
Influence of diet on gut microbiota	The relationship between diet and gut microbiota has been studied since the 1960s, with a recent research utilizing animal models and metagenomic analysis	[17]
Impact of diet on microbial composition	High-fat, high-sugar diets can lead to rapid changes in microbial community structure in germ-free mice, increasing Firmicutes and decreasing Bacteroidetes	[17]
Metabolic changes	Changes in diet affect the biochemical reactions in the gut; nutrients are converted into metabolites by intestinal microbes, which may have biological activity	[18]
Endotoxin levels	High-fat diets may increase plasma endotoxin levels, indicating a possible link between diet, gut microbiome perturbations, and health status	[18]
Enterotypes	Fecal metagenomes can be classified into three enterotypes based on dominant genera such as *Bacteroides*, *Prevotella*, and *Ruminococcus*, which are independent of demographic factors	[19]
Dietary associations	Enterotypes are associated with long-term dietary patterns: *Bacteroides* with animal proteins and fats, and *Prevotella* with carbohydrate-rich diets	[20]
Health implications	While the causal relationship between microbial community alterations and health is unclear, dietary patterns may influence enterotype status and potentially disease susceptibility	[20]

4.4 The Role of Microbiota Homeostasis in Supporting Intestinal Health

The human gut microbiota comprises complex and diverse microbial societies that play a vital role in conserving intestinal health [21]. It is assessed that approximately 100 trillion microbial cells exist inside the gut provide a range of metabolic functions that are beneficial to their host [22]. A crucial function of these intestinal microbes is the fermentation of complex carbohydrates from plant sources, in which the host can't digest because of the absence of enzymes needed to break down the structural polysaccharides [23]. The large intestine primarily comprises strict anaerobes categorized as either beneficial or pathogenic [24]. The gut microbiota has the ability of breaking down various substances, and producing organic acid and short-chain fatty acids, for example, propionate acetate and butyrate, which play a vital role in modulating microbial range. Additionally, these microorganisms contribute to several host functions including intestinal development maintenance of homeostasis and defense against harmful bacteria.

Research has shown that an inequity in the intestinal microbiota known as dysbiosis is connected with the progress of metabolic disorders, such as obesity and diabetes, as well as a variety of gastrointestinal issues, including antibiotic-associated diarrhea (AAD), IBD, Crohn's disease (CD), ulcerative colitis (UC), and CRC [25, 26]. Maintaining a balanced gut microbial community is needed for fostering a beneficial relationship between the host and its microbiotaas represented in Figure 4.2. Any disturbance in this balance can lead to detrimental amendments in host metabolism, which are associated with chronic conditions such as IBD, cancer, cardiovascular issues, and metabolic syndrome [27]. Moreover, an imbalanced microbiota has been linked with the onset of asthma [28]. Consequently, understanding the numerous aspects that impact the gut microbiome is vital for avoiding and managing these health concerns.

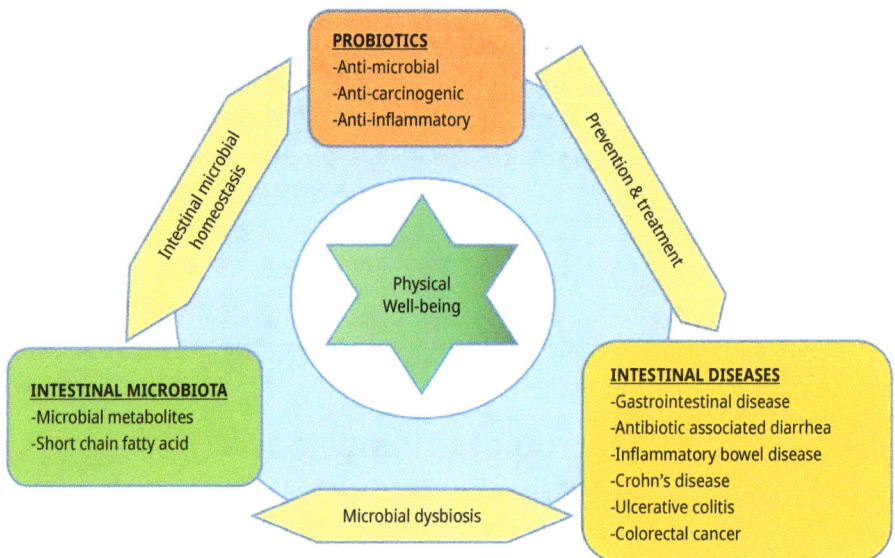

Figure 4.2: The role of probiotics in restoration of gut microbiota for management of gut dysbiosis and intestinal disorders.

4.5 Role of Probiotics in Addressing Gut Microbiome-Related Health Issues

Probiotics are live microorganisms that provide health benefits when ingested in sufficient quantities playing an important role in modulating the gut microbiome and addressing various health concerns (Figure 4.3). Studies have shown that probiotics can enhance the diversity and balance of gut bacteria, which are essential for proper digestive health and immune response. They have been associated with the disease conditions such as irritable bowel syndrome, IBD, and AAD by helping to restore microbial balance. Additionally, probiotics may positively influence metabolic health by affecting gut-brain interactions and reducing inflammation thus offering a supportive option alongside traditional treatments for gut-related issues. In summary, integrating probiotics into dietary practices holds promise for enhancing gut health and alleviating related health problems.

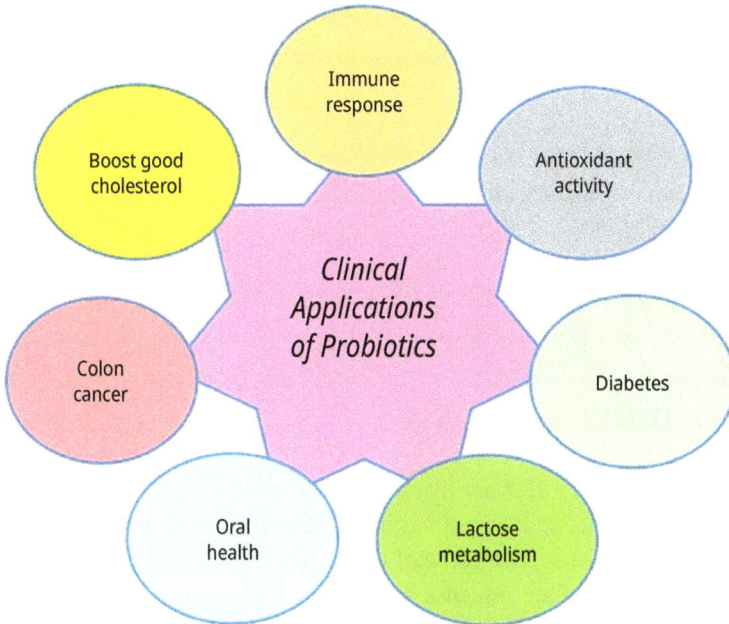

Figure 4.3: Various potent therapeutic and clinical applications of probiotics.

4.6 Dysbiosis of the Human Gut Microbiota

Microbial dysbiosis refers to an imbalance in the composition and function of the gut microbiota [29]. This condition has become increasingly prevalent in contemporary society, often resulting from factors like bacterial infections due to the changes in diet and the use of antibiotics. Recent research has shown that dysbiosis is associated with various intestinal disorders including IBD, irritable bowel syndrome, and celiac disease. Beneficial gut bacteria are essential for preventing the growth of harmful bacteria by competing for resources and space [30]. Probiotics are particularly important for restoring microbial balance and preventing infections especially in individuals who have received antibiotic treatment that may disrupt the gut microbiota.

4.7 Antibiotic-Associated Diarrhea

AAD is a common complication of antibiotic treatment arising from disturbance in the gut microbiota. A primary cause of AAD is the pathogenic bacterium *Clostridium difficile* that can affect the colon when antibiotic-induced changes weaken resistance [31]. Probiotics have emerged as a promising strategy for preventing AAD supported by various randomized controlled trials. For example, found that probiotics can effectively reduce the incidence of *C. defficile*-associated diarrhea in both adults and children [32, 33]. Moreover, a review and meta-analysis suggested that probiotics might act as adjunctive therapy lowering the risk of AAD in our patients of all ages by up to 51% without a significant rise in adverse effects [31]. Notably, *Lactobacillus rhamnosus* and *Saccharomyces boulardii* what identified as particularly effective in offering protection against AAD are not fully understood, underscoring the necessity for further research to elucidate the role of bacterial infections in this condition.

4.8 Inflammatory Bowel Disease

IBD is a long-term inflammatory disorder that affects the gastrointestinal tract including CD, UC, and indeterminate colitis with varying patterns of inflammation to outdoor digestive system [34]. The precise cause of IBD is still unknown but it is thought to arise from an abnormal immune response. Factors such as poor dietary habits and stress may contribute to the onset of IBD. Research has indicated that the gut microbiome might play a significant role in the onset of disease. Studies have shown distinct differences in the microbiota composition between healthy individuals and those suffering from IBD [35]. Maintaining a healthy balance of intestinal microbiota is believed to be essential for IBD prevention [36]. In recent years, probiotics have gained attention as a potential therapeutic strategy to modify the microbiome and improve IBD outcomes. For example, they have been used to achieve remission and as a maintenance treatment for UC [37]. Nevertheless, recent finding suggests that while probiotics may be beneficial as an adjunctive treatment for UC, their effectiveness in treating CD remains unclear [34]. Consequently, there is insufficient evidence to make generalized recommendations for probiotic use among patients with CD.

4.9 Crohn's Disease (CD)

CD is a form of IBD that can impact any part of the gastrointestinal tract. Common symptoms include abdominal discomfort, diarrhea, fever, fatigue, and unintended weight loss. The underlying cause of CD is not fully understood but it is believed that a mix of microbiological genetic and environmental factors contributes to its onset [38]. While there is no known cure for CD, various strategies are available to help manage it symptoms. Corticosteroids may be prescribed to reduce the inflammation in the intestine and immunosuppressants can help lower immune response [39]. Probiotics have also been exploring as a supplementary treatment option alongside standard therapy. A study indicated that probiotic use after surgery was beneficial for patients with CD compare to those who started supplementation later [40]. However, another recent investigation found that multistrain probiotic treatment did not significantly enhance intestinal inflammation in CD patients after probiotic therapy [34]. Therefore, given the varied outcomes of clinical trials regarding probiotics in CD management, further research is needed to better understand their potential role in treatment.

4.10 Colorectal Cancer (CRC)

CRC, often referred to as bowel or colon cancer, is a malignant tumor that can arise in any part of the large intestine including the colon and rectum. The incidence of colon cancer is increasing in developed regions such as Europe, the United States, and Australia [41]. Common symptoms include the presence of blood in stool and unintentional weight loss. Key contributors to the development of this cancer include a zing lifestyle factors and to a lesser degree of genetic factors. In areas where access to regular health screening age limited, younger individuals face a heightened risk of colon cancer [36]. Additionally, individuals with IBD, such as UC or CD, are at an elevated risk for developing CRC [42]. Recent studies have explored the potential role of beneficial bacteria in preventing CRC by altering gut microbiota and enhancing the body's immune response [43].

Although the efficacy of probiotics has been demonstrated in numerous clinical trials and animals' studies, there is still a significant amount of knowledge to be gained. Research indicates that the consumption of beneficial bacteria and fiber can positively alter gut microbiota in individuals with colon cancer promoting the growth of beneficial bacteria such as *Lactobacillus* [44]. In patients who underwent partial colon resection, improvement in the intestinal lining work observed, enhancing its ability to defend against harmful substances. Synbiotics, which combine probiotics and prebiotics, can also impact metabolic processes in the intestine. This body of research suggests that beneficial bacteria may play a role in treating and preventing gastrointestinal disorders in both animals and humans, including the colon cancer [27].

4.11 Probiotics and Obesity

Obesity has become a major global health concern closely associated with imbalance in the intestinal microbiota [45]. The gut microbiome is crucial in reducing local inflammation in adipose tissue by improving immune responses and preventing inflammation. Several factors contribute to the onset of fatty liver disease including inflammation and dysfunction of adipose tissue, dysbiosis in gut microbiota, and changes in gut barrier integrity that affect metabolic and inflammatory pathways in the liver. Consequently, probiotics and prebiotics may provide therapeutic benefits for fatty liver disease by positively influencing the gut microbiome [46]. Research indicates that genetic analysis of vehicle samples from obese individuals show a significant reduction in Bacteroidetes and an increase in Firmicutes compared to those from lean individuals [47]. In a noteworthy study, the transplantation of normal fecal microbiota into adult germ-free mice led to a remarkable 60% increase in insulin resistance and body fat after just 2 weeks, despite a lower food intake [48].

4.12 Probiotics and Pharmacologic Therapies

The gut microbiota significantly influences drug absorption, hepatic metabolism, and the production of active metabolites that the liver does not produce [49]. Certain probiotics can also affect patients' responses to medications and the side effects they experience. For example, enzyme such as beta-glucuronidase produced by bacteria like *Escherichia coli, Bacteroides vulgatus*, and *Clostridium ramosum* can convert irinotecan from its inactive glucuronide form back to its active form, leading to severe gastrointestinal toxicity [50, 51].

Furthermore, the composition of gut microbiota can impact the effectiveness of anti-PD-1 (anti-programmed death-1) immunotherapy in patients with advanced melanoma. A higher abundance of the Ruminococcaceae family and *Faecalibacterium* in fecal samples is associated with better progression-free survival outcomes [52]. Dysbiosis caused by antibiotic use has been linked to diminished response to anti-PD-1 therapy while increased level of Akkermansia *muciniphila* was correlated with enhanced therapeutic effects [53]. Studies have shown that transpiercing fecal microbiota from cancer patients who responded positively to anti-PD-1 therapy into antibiotic-treated mice improved the efficacy of immunotherapy, demonstrating a positive relationship with *A. muciniphila* abundance [53]. In this research, mice that initially exhibited poor responses after fecal microbiota transplantation work treated with *A. muciniphila*, which restored the effectiveness of the anti-PD-1 therapy.

Additionally, the *E. coli* strain has been shown to enhance the bioavailability of amiodarone in rats, whereas probiotics strains like *Lactobacillus casei* seemed to slow down the absorption of amiodarone, though the effect was not significant [54, 55].

Moreover, a study revealed that administration of *Lactobacillus reuteri* k8 reduced the area under the curve for acetaminophen in control mice [56].

Multidrug-resistant bacteria, such as vancomycin-resistant *Enterococcus* (VRE), carbapenemase-producing Enterobacteriaceae (CPE), and strains with extended-spectrum beta-lactamase (ESBL), represent a serious public health concern due to their correlation with high mortality rates [57]. To address this challenge, employing probiotics may be a viable strategy to prevent the colonization of harmful bacteria within the gut microbiota.

The temporary presence of these resistant bacteria can facilitate the transfer of antibiotic-resistance genes to both commensal and pathogenic microorganisms, resulting in the long-term maintenance of these genes in the microbiota. This persistence can increase the risk of severe infections, particularly when there are delays in providing effective antibiotic treatment [58]. Clinical case studies have demonstrated that fecal microbiota transplantation can effectively remove ESBL-carrying and naturally resistant bacterial strains from the gut [59–61].

Moreover, the composition of the microbiota in hospitalized patients affects their vulnerability to colonization by multidrug-resistant bacteria. Probiotic strains like *Lactobacillus plantarum* and *Lactobacillus fermentum* have been associated with reduced colonization by resistant pathogens such as *Acinetobacter baumannii*, *Pseudomonas aeruginosa*, and *Candida albicans* [62, 63]. In laboratory studies, culture supernatants from *Clostridium butyricum*, *C. difficile*, *Clostridium perfringens*, *Enterococcus faecium*, and *L. plantarum* have shown the ability to inhibit the growth and gene transfer of ESBL-carrying bacteria and CPE [64].

Interestingly, VRE appears to be less adapted to the gut environment and can be more easily decolonized compared to other multidrug-resistant organisms. In patients with hematological malignancies, the presence of *Barnesiella* has been linked to a lower frequency of VRE colonization [65]. Treatment with *Barnesiella* or *Lactobacillus paracasei* CNCM I-3,689 has been shown to reduce VRE colonization in mouse models [66, 67]. Additionally, there are clinical reports of successful VRE decolonization following fecal microbiota transplantation in patients with *C. difficile* colitis [68].

4.13 Effects of Probiotics on Intestinal Homeostasis

Maintaining a dynamic balance of intestinal microbiota is essential for a stable ecological environment in the gut [69]. Probiotics contribute to this balance by increasing beneficial bacteria populations through their own growth and by promoting the proliferation of desirable indigenous microbes [69]. All the mechanisms through which probiotics help to maintain intestinal homeostasis are explained in Table 4.2.

Table 4.2: Effects of probiotics on intestinal homeostasis.

Mechanism	Description	Example/study	References
Promotion of beneficial bacteria	Probiotics enhance the growth of beneficial microbes, increasing their population in the intestine	*Lactobacillus* and *Bifidobacterium* promoting beneficial bacteria	[69]
Competitive exclusion	Probiotics outcompete pathogenic bacteria for nutrients and ecological niches, preventing their colonization	*Lactobacillus* strains competing with pathogens like *Salmonella*	[70, 71]
Nutritional competition	Bacterial biopolymers produced by probiotics serve as carbon sources, inhibiting the growth of harmful bacteria	Biopolymers from *Bifidobacterium* and *Lactobacillus* antagonizing pathogens	[72]
Metabolite secretion	Probiotics secrete metabolites (e.g., lactic acid) that lower pH, inhibiting pathogen growth and promoting gut health	Lactic acid production reducing *Vibrio cholerae* in infant mice	[73, 74]
Microbiota adaptation	Probiotic intake helps adapt the gut microbiome for better competition against harmful microbes through genetic changes	SNP adaptation in the gut microbiome related to carbon-source competition	[75]
Biofilm inhibition	Certain probiotics inhibit biofilm formation of pathogens, thereby reducing their colonization and effects	*E. coli* Nissle 1917 disrupting biofilm formation of *Pseudomonas aeruginosa*	[76]

4.14 Probiotic and Cholesterol Metabolism

The rising prevalence of multifactorial disease with neurological conditions and liver-related issues in contemporary society can often be attributed to poor dietary habits. Many of these health problems arise from metabolic dysfunctions with obesity, hypertension, and heart disease are frequently associated with disruptions in lipid metabolism. Probiotics that are beneficial live bacteria offer considerable health benefits and are commonly included in dietary supplements aimed at supporting weight management. Research involving multistrain probiotic supplementation has demonstrated a significant reduction in tumor necrosis factor-alphaand low-density lipoprotein cholesterol levels in obese children while also increasing high-density lipoprotein cholesterol levels [77]. Probiotics from various genera, including *Streptococcus, Saccharomyces, Bacillus, Clostridium, Candida, Lactobacillus*, and *Bifidobacterium*, can modify gut microbiota composition promoting lipid oxidation in the liver and reducing overall fat storage in the body [78].

Additionally, probiotics affect key enzymes involved in cholesterol metabolism including phenyl transferase and HMG-CoA synthase (3-hydroxy-3-methylglutaryl coenzyme A) [79]. They play an essential role in regulating de novo lipogenesis and lipid oxidation; both are crucial in controlling fat accumulation in the liver. Furthermore, probiotics enhance the expression of PPAR-alpha and SREBP1, which leads to increase lipid oxidation and de novo lipogenesis [80].

4.15 Conclusion and Future Prospects

The gut microbiota can be effectively influenced by altering the interactions between the host and microorganisms through methods such as probiotics and personalized nutrition as complementary therapies. Probiotics play a vital role in counteracting these pathogens by strengthening the epithelial barrier producing antimicrobial compounds competing for adhesion sides and limiting nutrient availability to harmful organisms. Several factors can compromise the diversity and complexity of the gut microbiota leading to various health issues. Utilizing probiotics can help sustain health and alleviate other related problems. In conclusion, the role of probiotic in modulating gut microflora offers a promising frontier for enhancing overall health and well-being. Research has increasingly demonstrated the complex interactions between gut microbiota and variety of physiological processes revealing how these microorganisms influence digestion immune function and even mental health. The potential for probiotics to prevent and manage conditions, such as gastrointestinal disorder, obesity, and anxiety, is becoming clearer, underscoring the need for further exploration in this field.

Future studies should prioritize understanding the specific mechanism by which probiotics exert their beneficial effects. This includes identifying the most effective strains and dosages tailored to various health conditions. Additionally, research into the impact of dietary factors and lifestyle on probiotic efficacy will be crucial. Personalized approaches to probiotic supplementation, which consider individual microbiome's profile and genetic background, may lead to more targeted and effective health interventions. Moreover, expanding our knowledge of the interactions between probiotics and other elements of the diet such as prebiotics could unlock new strategies for promoting a healthy gut environment. As we continue to unravel the complexities of the gut microbiome, the integration of probiotics into preventive and therapeutic healthcare strategies has the potential to significantly improve patient's outcomes and enhance the quality of life. Emphasizing interdisciplinary collaboration in research will be the key to losing the full potential of probiotics in public health.

Conflicts of interest: The authors do not declare any conflicts of interest.

References

[1]   Sender R, Fuchs S, Milo R. Revised estimates for the number of human and bacteria cells in the body. PLoS Biology. 2016;14(8):e1002533.

[2]   Hill C, Guarner F, Reid G, Gibson GR, Merenstein DJ, Pot B, Sanders ME. Expert consensus document: The International Scientific Association for Probiotics and Prebiotics consensus statement on the scope and appropriate use of the term probiotic. Nature Reviews Gastroenterology & Hepatology. 2014;11:506–514.

[3]   Jethwani P, Grover K. Gut microbiota in health and diseases – A review. International Journal of Current Microbiology and Applied Sciences. 2019;8(8):1586–1599.

[4]   Kumar S, Nyodu R, Maurya VK, Saxena SK. Host immune response and immunobiology of human SARS-CoV-2 infection. Coronavirus Disease 2019 (COVID-19) Epidemiology, Pathogenesis, Diagnosis, and Therapeutics. 2020; 43–53.

[5]   Koh A, De Vadder F, Kovatcheva-Datchary P, Backhed F. From dietary fiber to host physiology: Short-chain fatty acids as key bacterial metabolites. Cell. 2016;165(6):1332–1345.

[6]   Sekhon BS, Jairath S. Prebiotics, probiotics and synbiotics: An overview. Journal of Pharmacy Education and Research. 2010;1(2):13-36.

[7]   Dinan TG, Cryan JF. The microbiome-gut-brain axis in health and disease. Gastroenterology Clinics. 2017;46(1):77–89.

[8]   Polak E, Stępień AE, Gol O, Tabarkiewicz J. Potential immunomodulatory effects from consumption of nutrients in whole Foods and supplements on the frequency and course of infection: Preliminary results. Nutrients. 2021;13(4):1157.

[9]   Maberry K, Nguyen H, Quinter C, Magalona J. The" Pros" of probiotics: A review of probiotics place in therapy. T-Med. 2024;3(3):113–116.

[10]   Sadanov AK, Gavrilova NN, Ratnikova IA, Orazymbet SE, Protasiuk LE, Massirbaeva AD. Technology for the production of Lyophilizate of an association of Lactic acid bacteria included in the Medicinal product AS-Probionorm. Rjpt. 2023;16(11):5334–5340.

[11]   Sharma H, Bajwa J. Approach of probiotics in mental health as a psychobiotics. Archives of Microbiology. 2022;204(1):30.

[12]   Thursby E, Juge N. Introduction to the human gut microbiota. Biochemical Journal. 2017;474 (11):1823–1836.

[13]   Rinninella E, Raoul P, Cintoni M, Franceschi F, Miggiano GAD, Gasbarrini A, Mele MC. What is the healthy gut microbiota composition? A changing ecosystem across age, environment, diet, and diseases. Microorganisms. 2019;7(1):14.

[14]   Hollins SL, Odgson DM. Stress, microbiota, and immunity. Current Opinion in Behavioral Sciences. 2019;28:66–71.

[15]   Agus A, Clément K, Sokol H. Gut microbiota-derived metabolites as central regulators in metabolic disorders. Gut. 2021;70(6):1174–1182.

[16]   Gevers D, Knight R, Petrosino JF, Huang K, McGuire AL, Birren BW, Huttenhower C. The human microbiome project: A community resource for the healthy human microbiome. PLoS Biology. 2012;10(8):e10013.

[17]   Goodman AL, Kallstrom G, Faith JJ, Reyes A, Moore A, Dantas G, Gordon JI. Extensive personal human gut microbiota culture collections characterized and manipulated in gnotobiotic mice. Proceedings of the National Academy of Sciences. 2011;108(15):6252–6257.

[18]   Pendyala S, Walker JM, Holt PR. A high-fat diet is associated with endotoxemia that originates from the gut. Gastroenterology. 2012;142(5):1100–1101.

[19]   Arumugam M, Raes J, Pelletier E, Le Paslier D, Yamada T, Mende DR, Bork P. Enterotypes of the human gut microbiome. Nature. 2011;473(7346):174–180.

[20] Wu GD, Chen J, Hoffmann C, Bittinger K, Chen YY, Keilbaugh SA, Bewtra M, Knights D, Walters WA, Knight R, Sinha R. Linking long-term dietary patterns with gut microbial enterotypes. Science. 2011 Oct 7;334(6052):105–108.

[21] Clemente JC, Ursell LK, Parfrey LW, Knight R. The impact of the gut microbiota on human health: An integrative view. Cell. 2012;148(6):1258–1270.

[22] Ley RE, Peterson DA, Gordon JI. Ecological and evolutionary forces shaping microbial diversity in the human intestine. Cell. 2006;124(4):837–848.

[23] Flint HJ, Scott KP, Duncan SH, Louis P, Forano E. Microbial degradation of complex carbohydrates in the gut. Gut Microbes. 2012;3(4):289–306.

[24] Apajalahti J. Comparative gut microflora, metabolic challenges, and potential opportunities. Journal of Applied Animal Research. 2005;14(2):444–453.

[25] Qin J, Li R, Raes J, Arumugam M, Burgdorf KS, Manichanh C, Wang J. A human gut microbial gene catalogue established by metagenomic sequencing. Nature. 2010;464(7285):59–65.

[26] Wen H, Yin X, Yuan Z, Wang X, Su S. Comparative Analysis of Gut Microbial Communities in Children under 5 Years Old with Diarrhea. 2018.

[27] Marteau P, Seksik P, Jian R. Probiotics and intestinal health effects: A clinical perspective. British Journal of Nutrition. 2002;88(S1):s51–s57.

[28] Hilty M, Burke C, Pedro H, Cardenas P, Bush A, Bossley C, Cookson WO. Disordered microbial communities in asthmatic airways. PloS One. 2010;5(1):e8578.

[29] Carding S, Verbeke K, Vipond DT, Corfe BM, Owen LJ. Dysbiosis of the gut microbiota in disease. Microbial Ecology in Health and Disease. 2015;26(1):26191.

[30] Ouwehand AC, Kirjavainen PV, Grönlund MM, Isolauri E, Salminen SJ. Adhesion of probiotic micro-organisms to intestinal mucus. International Dairy Journal. 1999;9(9):623–630.

[31] Blaabjerg S, Artzi DM, Aabenhus R. Probiotics for the prevention of antibiotic-associated diarrhea in outpatients – A systematic review and meta-analysis. Antibiotics. 2017;6(4):21.

[32] Conway S, Hart A, Clark A, Harvey I. Does eating yogurt prevent antibiotic-associated diarrhea: A placebo-controlled randomized controlled trial in general practice. Bjgp. 2007;57(545):953–959.

[33] Vanderhoof JA, Whitney DB, Antonson DL, Hanner TL, Lupo JV, Young RJ. *Lactobacillus* GG in the prevention of antibiotic-associated diarrhea in children. Journal of Pediatrics. 1999;135(5):564–568.

[34] Bjarnason I, Sission G, Hayee BH. A randomised, double-blind, placebo-controlled trial of a multi-strain probiotic in patients with asymptomatic ulcerative colitis and Crohn's disease. Inflammopharmacology. 2019;27:465–473.

[35] Shadnoush M, Hosseini RS, Khalilnezhad A, Navai L, Goudarzi H, Vaezjalali M. Effects of probiotics on gut microbiota in patients with inflammatory bowel disease: A double-blind, placebo-controlled clinical trial. Korean Journal of Gastroenterology. 2015;65(4):215–221.

[36] Yang J, Yu J. The association of diet, gut microbiota and colorectal cancer: What we eat may imply what we get. Protein & Cell. 2018;9(5):474–487.

[37] Furrie E, Macfarlane S, Kennedy A, Cummings JH, Walsh SV, O'neil D A, Macfarlane GT. Synbiotic therapy (Bifidobacterium longum/Synergy 1) initiates resolution of inflammation in patients with active ulcerative colitis: A randomized controlled pilot trial. Gut. 2005;54(2):242–249.

[38] Baumgart DC, Carding SR. Inflammatory bowel disease: Cause and immunobiology. The Lancet. 2007;369(9573):1627–1640.

[39] Cowan DC, Cowan JO, Palmay R, Williamson A, Taylor DR. Effects of steroid therapy on inflammatory cell subtypes in asthma. Thorax. 2010;65(5):384–390.

[40] Fedorak RN, Feagan BG, Hotte N, Leddin D, Dieleman LA, Petrunia DM, Madsen K. The probiotic VSL 3 has anti-inflammatory effects and could reduce endoscopic recurrence after surgery for Crohn's disease. Clinical Gastroenterology and Hepatology. 2015;13(5):928–935.

[41] Center MM, Jemal A, Smith RA, Ward E. Worldwide variations in colorectal cancer. CA: A Cancer Journal for Clinicians. 2009;59(6):366–378.

[42] Triantafillidis JK, Nasioulas G, Kosmidis PA. Colorectal cancer and inflammatory bowel disease: Epidemiology, risk factors, mechanisms of carcinogenesis and prevention strategies. Anticancer Research. 2009;29(7):2727–2737.

[43] Liong MT. Roles of probiotics and prebiotics in colon cancer prevention: Postulated mechanisms and in-vivo evidence. International Journal of Molecular Sciences. 2008;9(5):854–863.

[44] Krebs B. Prebiotic and synbiotic treatment before colorectal surgery-randomized double-blind trial. Collegium Antropologicum. 2016;40(1):35–40.

[45] Chandrasekaran P, Weiskirchen R. The role of obesity in type 2 diabetes mellitus – An overview. International Journal of Molecular Sciences. 2024;25(3):1882.

[46] Chandrasekaran P, Weiskirchen R. The pivotal role of the membrane-bound O-acyltransferase domain containing 7 in non-alcoholic fatty liver disease. Livers. 2023;4(1):1–14.

[47] Davis CD. The gut microbiome and its role in obesity. Nutrition Today. 2016;51(4):167–s174.

[48] Liu BN, Liu XT, Liang ZH, Wang JH. Gut microbiota in obesity. World Journal of Gastroenterology. 2021;27(25):3837.

[49] Spanogiannopoulos P, Bess EN, Carmody RN, Turnbaugh PJ. The microbial pharmacists within us: A metagenomic view of xenobiotic metabolism. Nature Reviews Microbiology. 2016;14(5):273–287.

[50] Wallace BD, Wang H, Lane KT, Scott JE, Orans J, Koo JS, Redinbo MR. Alleviating cancer drug toxicity by inhibiting a bacterial enzyme. Science. 2010;330(6005):831–835.

[51] Guthrie L, Gupta S, Daily J, Kelly L. Human microbiome signatures of differential colorectal cancer drug metabolism. NPJ Biofilms and Microbiomes. 2017;3(1):27.

[52] Gopalakrishnan V, Spencer CN, Nezi L, Reuben A, Andrews MC, Karpinets TV, Wargo J. Gut microbiome modulates response to anti-PD-1 immunotherapy in melanoma patients. Science. 2018;359(6371):97–103.

[53] Routy B, Le Chatelier E, Derosa L, Duong CP, Alou MT, Daillère R, Zitvogel L. Gut microbiome influences efficacy of PD-1-based immunotherapy against epithelial tumors. Science. 2018;359 (6371):91–97.

[54] Matuskova Z, Anzenbacherova E, Vecera R, Tlaskalova-Hogenova H, Kolar M, Anzenbacher P. Administration of a probiotic can change drug pharmacokinetics: Effect of *E. coli* Nissle 1917 on amidarone absorption in rats. PloS One. 2014;9(2):e87150.

[55] Matuskova Z, Anzenbacher P, Vecera R, Siller M, Tlaskalova-Hogenova H, Strojil J, Anzenbacherova E. Effect of *Lactobacillus casei* on the pharmacokinetics of amiodarone in male Wistar rats. European Journal of Drug Metabolism and Pharmacokinetics. 2017;42:29–36.

[56] Kim JK, Choi MS, Jeong JJ, Lim SM, Kim IS, Yoo HH, Kim DH. Effect of probiotics on pharmacokinetics of orally administered acetaminophen in mice. Drug Metabolism and Disposition. 2018;46 (2):122–130.

[57] Caballero S, Kim S, Carter RA, Leiner IM, Susac B, Miller L, Pamer EG. Cooperating commensals restore colonization resistance to vancomycin-resistant *Enterococcus faecium*. Ch&m. 2017;21 (5):592–602.

[58] Kaushik A, Ammerman NC, Parrish NM, Nuermberger EL. New β-lactamase inhibitors nacubactam and zidebactam improve the in vitro activity of β-lactam antibiotics against *Mycobacterium abscessus* complex clinical isolates. Antimicrobial Agents and Chemotherapy. 2019;63(9):10–1128.

[59] Singh N, Mishra SK, Sachdev V, Sharma H, Upadhyay AD, Arora I, Saraya A. Effect of oral glutamine supplementation on gut permeability and endotoxemia in patients with severe acute pancreatitis: A randomized controlled trial. Pancreas. 2014;43(6):867–873.

[60] Crum-Cianflone NF, Sullivan E, Ballon-Landa G. Fecal microbiota transplantation and successful resolution of multidrug-resistant-organism colonization. Journal of Clinical Microbiology. 2015;53 (6):1986–1989.

[61]   Millan B, Park H, Hotte N, Mathieu O, Burguiere P, Tompkins TA, Madsen KL. Fecal microbial transplants reduce antibiotic-resistant genes in patients with recurrent *Clostridium difficile* infection. Clinical Infectious Diseases. 2016;62(12):1479–1486.

[62]   Singhi SC, Kumar S. Probiotics in critically ill children. F1000Research. 2015;5.

[63]   Dallal MS, Zamaniahari S, Davoodabadi A, Hosseini M, Rajabi Z. Identification and characterization of probiotic lactic acid bacteria isolated from traditional Persian pickled vegetables. GMS Hygiene and Infection Control. 2017;12.

[64]   Kunishima H, Ishibashi N, Wada K, Oka K, Takahashi M, Yamasaki Y, Kaku M. The effect of gut microbiota and probiotic organisms on the properties of extended spectrum beta-lactamase producing and carbapenem resistant Enterobacteriaceae including growth, beta-lactamase activity and gene transmissibility. Journal of Infection and Chemotherapy. 2019;25(11):894–900.

[65]   Ubeda C, Bucci V, Caballero S, Djukovic A, Toussaint NC, Equinda M, Pamer EG. Intestinal microbiota containing *Barnesiella* species cures vancomycin-resistant *Enterococcus faecium* colonization. Infection and Immunity. 2013;81(3):965–973.

[66]   Tannock GW, Munro K, Harmsen HJM, Welling GW, Smart J, Gopal PK. Analysis of the fecal microflora of human subjects consuming a probiotic product containing *Lactobacillus rhamnosus* DR20. Applied and Environmental Microbiology. 2000;66(6):2578–2588.

[67]   Crouzet L, Derrien M, Cherbuy C, Plancade S, Foulon M, Chalin B, Serror P. *Lactobacillus paracasei* CNCM I-3689 reduces vancomycin-resistant *Enterococcus* persistence and promotes Bacteroidetes resilience in the gut following antibiotic challenge. Scientific Reports. 2018;8(1):5098.

[68]   Stripling J, Kumar R, Baddley JW, Nellore A, Dixon P, Howard D, Rodriguez JM. Loss of vancomycin-resistant *Enterococcus* fecal dominance in an organ transplant patient with *Clostridium difficile* colitis after fecal microbiota transplant. Open Forum Infectious Diseases. 2016;2(2):ofv078. Oxford University Press.

[69]   Fassarella M, Blaak EE, Penders J, Nauta A, Smid H, Zoetendal EG. Gut microbiome stability and resilience: Elucidating the response to perturbations in order to modulate gut health. Gut. 2021;70 (3):595–605.

[70]   Monteagudo-Mera A, Rastall RA, Gibson GR, Charalampopoulos D, Chatzifragkou A. Adhesion mechanisms mediated by probiotics and prebiotics and their potential impact on human health. Applied Microbiology and Biotechnology. 2019;103;6463–6472.

[71]   Litvak Y, Mon KK, Nguyen H, Chanthavixay G, Liou M, Velazquez EM, Bäumler AJ. Commensal Enterobacteriaceae protect against Salmonella colonization through oxygen competition. Cell Host & Microbe. 2019;25(1):128–139.

[72]   Castro-Bravo N, Wells JM, Margolles A, Ruas-Madiedo P. Interactions of surface exopolysaccharides from *Bifidobacterium* and *Lactobacillus* within the intestinal environment. Frontiers in Microbiology. 2018;9:2426.

[73]   Shao Y, Evers SS, Shin JH, Ramakrishnan SK, Bozadjieva-Kramer N, Yao Q, Seeley RJ. Vertical sleeve gastrectomy increases duodenal *Lactobacillus* spp. richness associated with the activation of intestinal HIF2α signaling and metabolic benefits. Molecular Metabolism. 2022;57:101432.

[74]   Mao N, Cubillos-Ruiz A, Cameron DE, Collins JJ. Probiotic strains detect and suppress cholera in mice. Science Translational Medicine. 2018;10(445):eaao2586.

[75]   Ma C, Zhang C, Chen D, Jiang S, Shen S, Huo D, Zhang J. Probiotic consumption influences universal adaptive mutations in indigenous human and mouse gut microbiota. Communications Biology. 2021;4(1):1198.

[76]   Aljohani AM, El-Chami C, Alhubail M, Ledder RG, O'Neill CA, McBain AJ. *Escherichia coli* Nissle 1917 inhibits biofilm formation and mitigates virulence in *Pseudomonas aeruginosa*. Frontiers in Microbiology. 2023;14:1108273.

[77]   Lin K, Zhu L, Yang L. Gut and obesity/metabolic disease: Focus on microbiota metabolites. Medical Communications. 2022;3(3):e171.

[78]   Fawad JA, Luzader DH, Hanson GF, Moutinho TJ Jr, McKinney CA, Mitchell PG, Moore SR. Histone deacetylase inhibition by gut microbe-generated short-chain fatty acids entrains intestinal epithelial circadian rhythms. Gastroenterology. 2022;163(5):1377–1390.

[79]   Howard EJ, Lam TK, Duca FA. The gut microbiome: Connecting diet, glucose homeostasis, and disease. Annual Review of Medicine. 2022;73(1):469–481.

[80]   Wu J, Yang K, Fan H, Wei M, Xiong Q. Targeting the gut microbiota and its metabolites for type 2 diabetes mellitus. Frontiers in Endocrinology. 2023;14:1114424.

Poonam Bansal, Priti, Mahiti Gupta, Kumkum Verma, Bhupesh Gupta, Soniya Goyal*

Chapter 5
Health Benefits and Mechanism of Action Associated with Biotics: Prebiotics and Synbiotics

Abstract: Over the past few decades, synbiotics, probiotics, and prebiotics have been extensively researched for their various health benefits. Often classified as functional foods, they are known to influence, modify, or restore the microbial flora in the human body, contributing to the proper functioning of the immune system. Probiotics are live microbial strains that, when consumed, enhance intestinal health and function. In contrast, prebiotics are indigestible or partially digestible food components typically included in the regular diet. Synbiotics, a combination of probiotics and prebiotics, have gained significant attention in recent years for their potential to enhance various aspects of human health. Prebiotics are nondigestible substances that promote the growth and activity of probiotics. Traditionally, synbiotics have been used to support gut health and manage gastrointestinal disorders. However, their health benefits extend beyond the digestive system, with research exploring their role in addressing conditions such as allergies, boosting immune function, and improving metabolic health. Numerous clinical studies have highlighted the effectiveness of synbiotics in improving gastrointestinal health and managing various illnesses. Randomized controlled trials have shown that synbiotics can help alleviate symptoms of irritable bowel syndrome and inflammatory bowel disease and may also reduce the incidence of infectious diarrhea. Additionally, observational and interventional studies suggest that synbiotics could aid in the prevention and treatment of allergies, autoimmune disorders, and metabolic conditions. However, the mechanisms behind these potential benefits re-

Acknowledgment: The authors (SG, PB, and MG) acknowledge the help and support by the head of the Department of Biosciences and Technology, Maharishi Markandeshwar (Deemed to be University), Mullana-Ambala, India.

*Corresponding author: Soniya Goyal**, Department of Biosciences and Technology, Maharishi Markandeshwar (Deemed to be University), Mullana-Ambala, Haryana, India,
e-mail: soni.goyal48@gmail.com
Poonam Bansal,Mahiti Gupta, Department of Biosciences and Technology, Maharishi Markandeshwar (Deemed to be University), Mullana-Ambala, Haryana, India
Priti, K. L. Mehta Dayanand College for Women, Faridabad, Haryana, India
Kumkum Verma, IARI, Regional Station, Karnal, Haryana, India
Bhupesh Gupta, Department of Computer Sciences, Maharishi Markandeshwar (Deemed to be University), Mullana-Ambala, Haryana, India

https://doi.org/10.1515/9783112205150-005

main unclear, and further research is needed to determine optimal dosages and formulations for specific health conditions. This chapter reviews the existing literature on the impact of prebiotics, probiotics, and synbiotics in enhancing human health, highlighting current trends and advancements in the field.

Keywords: prebiotics, probiotics, synbiotics, safety aspects, health benefits

5.1 Introduction

At present, beyond the fundamental function of nutrition in providing essential nutrients for growth and development, additional factors are gaining importance, such as promoting health and preventing disease. In a world dominated by highly processed foods, there is heightened focus on the composition and safety of what we consume. Food quality is crucial in twenty-first-century epidemics due to issues like food poisoning, allergies, obesity, cardiovascular disease, and cancer. Scientific studies highlight the health benefits associated with prebiotics and probiotics in individuals' diet [1]. Synbiotics combine properties of both prebiotics and probiotics and developed to address potential challenges related to probiotic survival in the gastrointestinal (GI) tract [2]. By integration of these components into a single product, the synergistic interaction is expected to produce superior advantages compared to the effects of prebiotics or probiotics used individually. This review aims to explore the current understanding of probiotics, prebiotics, and synbiotics effects on human health, as well as their mechanism of action. To achieve therapeutic effects, it is very crucial to select an appropriate probiotic strains, prebiotics, and dosages, warranting dedicated sections on these topics [3]. It is mandatory to conduct more research to develop novel probiotic strains, optimize the selection of appropriate prebiotics and probiotics for synbiotic formulations, establish appropriate dosing, ensure safety, and document health benefits through clinical trials [4]. These effects should be validated through well-designed clinical studies conducted by independent research institutions.

5.2 Probiotics

The word "probiotics" originates from a Greek phrase that means "for life" and refers to live, nonpathogenic organisms that provide benefits to their hosts. The word was first coined by Vergin, after studying the negative impact of antimicrobial drugs and other microbial agents on gut microbes [5]. He noted that "probiotika" supported healthy gut microflora. Later, Lilly and Stillwell redefined probiotics as "a product synthesized by one microbe which stimulates the growth of another microbe." Fuller further refined the term to describe "non-pathogenic microbes that, when consumed, have a beneficial effect on physiology and health of host" [6]. The current definition, jointly established by the FDA and WHO, defines probiotics as "live microbes which, when delivered in appropriate quantities, provide beneficial effect on health of the host" [7]. The most common sources of probiotics are nutritional supplements and fermented foods such as kefir, yogurt, and sauerkraut. Some of the commonly used probiotic microorganisms include *Lactobacillus reuteri*, *Lactobacillus rhamnosus*, various strains of *Lactobacillus casei*, *Pediococcus acidilactici* and *Bifidobacterium*, members of the, *Bacillus coagulans*, specific enterococci like *Enterococcus faecium* SF68, *Escherichia coli* strain Nissle 1917, *Lactobacillus acidophilus* group, and the yeast *Saccharomyces boulardii* [8]. Such probiotics are typically incorporated to foods, especially fermented dairy products, either individually or in combination. These microorganisms help maintain digestive health, enhance immune function, and may offer other wellness benefits by promoting a balanced microbiome. With ongoing research, new probiotic genera and strains are continually being discovered and synthesized [9].

5.3 Prebiotics

The term "prebiotic" was first defined in 1995 to describe an indigestible component of food. Prebiotics are ingredients that the human body cannot digest; they resist gastric acid, are not fragmented by mammalian enzymes, and are not absorbed in the digestive system [10]. Instead, they undergo fermentation by gut flora and selectively activate some beneficial microbes in the colon, promoting their development and activity to positively influence the health of the host. In 2004, Gibson and colleagues found that prebiotics are components that can be fermented selectively to alter the composition and function of beneficial gut flora. These components are commonly known as "bifidogenic factors" [11]. In 2017, the International Scientific Association for Probiotics and Prebiotics revised the definition, describing prebiotics as substances which can be particularly utilized and transformed by intestinal flora in ways that support host health. The updated definition broadens prebiotics to include noncarbohydrates and extends their site of action beyond the GI tract, with applications not limited to food alone [12].

A compound is considered as a prebiotic on the basis of following criteria: (i) it must be resistant to stomach acidity, be unaffected by mammalian enzymes, and not be absorbed in the digestive tract; (ii) it must be fermentable by gut microbial flora; and (iii) it must particularly facilitate the growth and/or activity of certain intestinal microbes, thereby enhancing the health of the host [13]. While not all prebiotics are derived from carbohydrates, two main characteristics can assist in distinguishing fiber from prebiotics made up of carbohydrates: (i) fibers are carbohydrates that have a degree of polymerization (DP) of 3 or more, and (ii) they cannot be hydrolyzed by intestinal enzymes [14]. It is important to consider that solubility or fermentability of fiber is not a critical factor in this distinction. As noted, prebiotics mainly target lactobacilli and bifidobacteria species, promoting short-chain fatty acid productions and reducing pH levels. An intake of dietary fiber is essential for maintaining the gut's mucosal barrier function. Recent studies have showed that prebiotic fibers may significantly impact microbiota present in gut [15]. Research indicates that consuming inulin supplements may help lessen the detrimental impact of high-fat meals on mucus layer permeability and metabolic processes [4]. Studies have linked the Western diet's low-fiber content to chronic diseases by lowering the intestinal mucus barrier and promoting the growth of microbiota, which can increase a person's susceptibility to infections and inflammation [16]. Additionally, galacto-oligosaccharides by themselves can increase *Lactobacillus*, but when combined with fructo-oligosaccharides, they enhance bifidobacteria and reduce *Clostridium* present in the gut [17]. It has been demonstrated that inulin and arabino-xylooligosaccharides improve immunological response and intestinal barrier function [16].

Prebiotics are subject to stringent regulations in Europe, requiring comprehensive scientific validation and approval from the European Commission and the European Food Safety Authority (EFSA). EFSA must conduct a scientific review of any product for health claims before the approval of the Commission. For example, research demonstrates that chicory inulin increases the frequency of bowel movements, supporting the health claim that "Inulin improves bowel function" [18]. Prebiotics such as FOS, GOS, and inulin, which were consumed prior to 1997, are regarded as safe. However, those produced after 1997, like some human milk oligosaccharides, are considered as novel foods and need to be evaluated for safety. The approval process also takes into account of novel foods' history of safe use in non-EU countries. In the United States, the term "prebiotic" is not officially recognized by the Food and Drug Administration (FDA). Rather, prebiotics are controlled according to the goods they are meant to be used in, such as food items, nutritional supplements, medical foods, medications, cosmetics, or devices. In 2016, the FDA issued guidance on the new dietary ingredient notification process that includes safety and efficacy standards. Furthermore, nondigestible carbohydrates having positive physiological effects were included in the 2014 revisions to fiber labeling standards, which redefined fiber [19]. A prebiotic to be considered as a fiber must have certain properties and the FDA must receive supporting documentation through particular petition procedures. It is expected that the FDA will issue more advice to elucidate prebiotic labeling and regulation.

5.3.1 Prebiotic Types

The term "prebiotic" encompasses a wide variety of products, having distinct structures and modes of action. Prebiotics include diverse structures, such as oligosaccharides and resistant starches, and recent study has increasingly succeeded in identifying and characterizing them. Each prebiotic type offers distinct health benefits [20]. These substances enter the colon as substrates for fermentation by good gut bacteria after not being broken down in the upper respiratory tract [21]. In addition to other potentially health-promoting compounds, this fermentation process produces metabolites like short-chain fatty acids (SCFAs). Prebiotics frequently contain short polysaccharides like inulin, galactofructose, oligofructose, xylo-oligosaccharides, and galacto-oligosaccharides, as well as carbohydrates that are difficult to digest, such as oligosaccharides. There are some types of prebiotics discussed below:

5.3.1.1 Inulin

Dietary fibers that are prebiotics, such as inulin, are members of the fructans class. While many plants naturally produce inulin, chicory root is among the richest sources. Additionally, foods high in inulin include leeks, onions, garlic, asparagus, bananas, and Jerusalem artichokes [22]. Fructose molecules and terminal glucose molecules are joined by $\beta(2-1)$-glycosidic linkages to form inulin. A length of chain varies with shorter chains referred to as oligofructans and larger ones referred as polysaccharides. The reason inulin is regarded as a prebiotic is that it cannot be broken down in the upper digestive tract and enters the colon intact, where it serves as a substrate for fermentation by good bacteria. Notably, gut-health-promoting bacteria such as bifidobacteria and lactobacilli help in the fermentation of inulin. By acting as a substrate for fermentation, inulin promotes the development of beneficial microbes and promotes a more diverse and healthy gut microbiota. Its consumption has been associated with a number of health advantages, including better digestive health, increased immunity, and possibly a lower risk of developing several chronic illnesses [23]. SCFAs such as acetate, butyrate, and propionate are produced during the fermentation of inulin in the colon. These SCFAs have therapeutic properties and provide energy to colonocytes [24]. A study by Mitchell et al [25] suggests that it may improve sensitivity against insulin and reduce the level of postprandial glucose, supporting blood sugar regulation. This effect is thought to be mediated by SCFA production, which may influence signaling pathways of insulin and metabolism of glucose.

5.3.1.2 Fructooligosaccharides

Prebiotic fibers known as fructooligosaccharides (FOSs) are made up of short fructose chains that terminate in a glucose molecule. These are naturally found in certain vegetables, grains, and fruits. FOS molecules are low-molecular-weight oligosaccharides. Their structure includes a terminal glucose and fructose units linked by $\beta(2-1)$-glycosidic bonds [26]. The length of the FOS chain can vary depending on the processing method and source. Like other prebiotic fibers, FOS do not undergo digestion in the upper part of digestive system, comes in the colon where beneficial bacteria use it as a fermentation substrate. Specific bacteria, such as bifidobacteria and lactobacilli, responsible for fermentation of FOS, contribute in an improvement of gut health. Consuming FOS has been linked with numerous health benefits, including enhancement of immune function, improved digestive health, and a potential reduction in the risk of some chronic diseases. According to a study, FOS especially promotes the proliferation of intestinal bifidobacteria, a phenomenon known as the bifidogenic effect [27]. This may help to regulate bowel motions and alleviate constipation by improving the consistency and frequency of stools. Additionally, FOS has demonstrated antimicrobial properties against various pathogenic bacteria, including *Salmonella* spp. and *Escherichia coli* [28].

5.3.1.3 Galactooligosaccharides

Galactooligosaccharides (GOSs) are also belonging to prebiotic fiber, made up of short chains of galactose molecules. These oligosaccharides maintain a balance of microbes in the gut by promoting the growth of *bifidobacteria*. Wang et al. [29] reported that the supplements containing GOS modifies the make- up and activity of the gut microbiota, fostering a diverse and balanced microbial community. This effect may aid in preventing illnesses and disorders related to digestive system, contributing positively to overall gut health. Advances in modern analytical techniques have enabled the identification of numerous structures of GOS, which differ in polymerization degree, branch locations, and glycosidic linkage. Once ingested, GOS modify microbes present in gut, prompting the production of fatty acids of short chain and demonstrating significant biological activity. They precisely increase the proliferation of probiotics, inhibit pathogenic bacterial growth and attachment and help in the management of GI, neurological, metabolic, and allergic conditions. GOS also influence metabolite production and aid in ion storage and absorption. Furthermore, GOS are widely used as food additives since they are stable, safe, highly soluble, and taste clean. They can enhance the texture, flavor, appearance, taste, viscosity, shelf life, rheological properties, and health advantages of food products [29].

5.3.1.4 Starch and Glucose-Derived Oligosaccharides

This is also known as resistant starch (RS), a type of starch that is unable to digest in the upper digestive tract. By producing significant amounts of butyrate, RS offers health benefits and has therefore been proposed as a prebiotic. RS is a kind of starch that travels to the colon, where it serves as a fermentation substrate for gut microorganisms, avoiding digestion in the small intestine. It is composed of amylopectin and amylose molecules, the two primary components of starch. Because of its structure,

RS cannot be broken down by the enzymes found in people's small intestines. Based on its origins and physical characteristics, RS is divided into a number of varieties. RS1, known as physically inaccessible starch, is found in complete or partially ground grains and seeds. RS2 includes high-amylose granular starch, commonly present in green bananas, high-amylose corn and raw potatoes. Retrograded starch (RS3) develops when foods with starch like rice, pasta, and potatoes are cooked and afterwards cooled. RS4 is chemically modified starch, designed to resist digestion [30]. Due to its unique structure, RS is indigestible in the intestinal tract and moves into the colon. SCFAs, such as propionate, acetate, and butyrate, are produced during fermentation and have a number of health advantages. Kim et al [31] investigated the impact of eating a resistant starch snack on post-meal glucose levels and observed a noteworthy decrease in blood glucose. This suggests that RS may help to improve regulation of blood sugar levels. This effect is thought to be mediated by the production of SCFAs, which can influence signaling pathways of insulin and metabolism of glucose. RS is naturally found in many foods, including cooked and cooled potatoes, whole grains, seeds, green bananas, legumes, and certain types of high-amylose corn [32]. Although factors include storage conditions, cooking methods and processing can affect the amount of RS in these foods.

5.3.1.5 Beta-Glucans

Some yeasts, microbes, algae, molds, and cereals have cell walls that contain beta-glucans, which are prebiotic fibers made of glucose molecules joined by β-glycosidic linkages. Although their main purpose is to aid in immune system modulation, recent study evidences that they may also have prebiotic effects [33]. Its' structure varies depending on their molecular weight, branching degree, and source. Common sources include yeast (*Saccharomyces cerevisiae*), barley, oats, and some bacteria (e.g., *Lactobacillus* species) [34]. Oats and barley, particularly in their bran and germ layers, are rich in beta-glucans. Beta-glucan supplements derived from yeast or mushrooms are also available. These fibers are often incorporated as functional ingredients in variety of food including cereal, bread, pasta, and dietary supplements [35].

5.3.1.6 Lactulose

Lactulose is an artificial disaccharide sugar formed from galactose and fructose, linked together with β(1–4)-glycosidic bonds, making it indigestible. Lactulose is frequently used as a laxative to relieve constipation, but it also serves as a prebiotic by promoting the development of good gut bacteria. A study examined lactulose's effects on constipation, showing increased levels of bacteria like *Prevotella*, bifidobacteria, *Anaerostipes*, *Bacillus cereus*, *Bacillus*, *Mogibacterium*, and *Oribacterium* in the constipation group after treatment, while *Anaerotruncus* diminished in a healthy control group. This suggests that lactulose can enhance the intestinal microbes, increase probiotic populations, and alleviate constipation. By providing a substrate for fermentation, lactulose promotes more balanced and diverse gut microbes. It is available in various forms, including tablets, powders, and oral solutions, and serves as both laxative and prebiotic supplement to support the health of gut [36].

5.3.1.7 Mannan Oligosaccharides

Mannan oligosaccharides (MOSs) are a type of prebiotic fiber extracted from yeast cell walls or certain plant sources, such as some seaweeds and specific yeasts [37]. MOS is a fermentation substrate for probiotic bacteria that is made up of short chains of mannose molecules joined by glycosidic linkages. It does not undergo digestion in the upper digestive system and enters the colon intact [38]. Mannans are made up of mannose units joined by $\beta(1–4)$-glycosidic linkages, and depending on the source and processing, they may also contain other sugar side chains, like glucose or galactose. MOS vary in the extent of polymerization, ranging from short-chain oligosaccharides to higher polysaccharides. β-MOS are specifically fermented by gut microbes, encouraging growth of beneficial microbes (probiotics) while either having no effect on or inhibiting the growth of harmful enteric pathogens. This fermentation also leads to the synthesis of useful metabolites, such as SCFAs [39]. β-MOS also possess various bioactive properties and offer multiple health benefits. Utilizing enzymes like β-mannanases for β-MOS production is considered the most efficient and environmentally friendly method.

5.3.1.8 Pectin Oligosaccharides

Pectin – prebiotic fiber – is found in numerous fruits, particularly in apples, peels of berries, and citrus fruits. Its conformation varies based on factors such as fruit type and ripeness [40]. Pectin is a polysaccharide primarily composed of chains of galacturonic acid with side chains containing sugars, such as arabinose, rhamnose and galactose. One of its unique properties is its ability to form a gel-like substance when mixed with water, making it useful as a gelling, stabilizing, and thickening agent in foods like jams, jellies, and fruit preserves [41]. As a heteropolysaccharide, pectin's structure includes various sugar units. The main component of pectin is galacturonic acid involved in the backbone of the molecule, while side chains may include sugars like galactose, arabinose, and rhamnose. Pectin enters the colon intact and serves as a substrate for fermentation by good bacteria since it is resistant to digestion in the upper digestive system. Pectin consumption is linked to a number of health advantages, including as improved digestive health, increased immunity, and possibly a de-creased risk of developing some chronic illnesses [42].

Pectin oligosaccharides (POSs) are emerging as promising prebiotic candidates, particularly for their role in regulating gut microbiota. Studies have showed that POS may be more efficacious as prebiotics than pectin itself. Research indicates that POS can promote beneficial bacterial populations similarly to or even better than FOSs [43]. Additionally, oral administration of pectin-derived acidic oligosaccharides has been shown to significantly influence gut microbial composition and enhance fecal SCFA production, thereby modulating immune response in chronic *P. aeruginosa* lung infections. Currently, POSs are mainly produced through the partial depolymerization of pectin [44].

5.3.2 Production of Prebiotics

Prebiotics are important for human health and are naturally found in a wide range of foods, notably onion, tomato, asparagus, wheat, sugar beet, Jerusalem artichoke, gar-lic, soybean, chicory, banana, honey, beans, barley, peas, rye, and human and cow's milk. Moreover, prebiotics have also been found in microalgae and seaweeds [45]. As prebiotics occur in low concentrations in these foods, they are often produced on an industrial scale. The raw ingredients used to make some prebiotics include starch, su-crose, and lactose. Most industrial prebiotics are classified as GOS and FOS, leading to a significant research into methods for their large-scale production. Prebiotics can be produced through the transgalactosylation activity of specific carbohydrates using ei-ther chemical or enzymatic methods and employing raw materials like xylan, lactose, starch, and sucrose [46].

5.3.2.1 Production of Fructooligosaccharides (FOSs)

FOS is naturally present in around 36,000 plant species; however, the concentration in these sources is insufficient to exert significant prebiotic effects. As a result, FOS often needs to be synthesized. FOS can also serve as an alternative artificial sweet-ener. When fermented, FOS produces various acids and gases, providing some energy to the body. Several methods for producing FOS have been documented by various researchers [47]. FOS can be produced chemically by glycosyltransferase and glycosi-dase enzymes; however, these reactions require hazardous and costly compounds, re-sulting in a low concentration of FOS, which makes industrial-scale production im-practical [48]. Fructosyltransferase (FTase) is the main enzyme involved in the formation of FOS; it transfers one to three fructose molecules from sucrose to produce FOS. FOS can be synthesized chemically, as well as enzymatically. With the aid of gly-cosidase and glycosyltransferase, FOS can be chemically produced. However, these reactions involve hazardous and expensive compounds, and produced low-yield FOS, making large-scale production impractical. A more effective approach for industrial production involves the enzyme FTase, which converts sucrose into FOS by transferring one to three fructose molecules. Sources of enzymes for FOS synthesis fall into two categories: plant sources, such as Jerusalem artichoke, asparagus, onion, and sugar beet; and microbial sources, which include various bacterial and fungal strains like *Aspergillus* sp., *Fusarium* sp., *Saccharomyces cerevisiae*, *Penicillium* sp., *Aureobasidium* sp., *Kluyveromyces*, *Zymomonas mobilis*, *Arthrobacter* sp., *Candida*, and *Bacillus macerans* [49]. Plant-based enzymes have a comparatively low yield of FOS, and production by large-scale enzyme is restricted by seasonal factors. Consequently, production of industrial FOS primarily relies on enzymes from fungus viz. *Aureobasidium* spp. or *Aspergillus niger* [50].

Meiji Seika Co. achieved the first commercial production of FOS in 1984 in Japan (marketed as Neosugar) using enzymes from *Aspergillus niger*, successfully demonstrating its valuable functional properties. Using *Aureobasidium pullulans* immobilized cells; Cheil Foods & Chemicals Co. in Korea recently created an industrial-scale production method [51]. The industrial production of FOS has expanded its demand in the sweetener market, allowing it to be widely used in various foods and feed products and enabling competition with traditional sweeteners like natural sugar and high-fructose corn syrup. For the production of FOS, either whole microbial cells or free enzymes can be utilized. The resultant FOS concentration is influenced by a number of factors. FTases typically produce a maximum yield of 55–60% FOS, depending on the initial concentration of sucrose [52]. Glucose, a by-product of fermentation, inhibits trans-glycosylation, making it essential to remove both glucose and sucrose residues to achieve higher FOS yields. Some researchers have suggested FOS production can be enhanced by using β-fructofuranosidase and glucose oxidase [53]. β-Fructofuranosidase can convert sucrose into FOS, while glucose oxidase converts glucose into gluconic acid. Gluconic acid can then be removed using coagulation with calcium carbonate ($CaCO_3$) and ion-exchange resins. This combined enzyme approach can increase FOS yield as much as 98%. β-fructofuranosidase can be sourced from *Apostichopus japonicus* and glucose oxidase from *A. niger*. Additionally, nanofiltration methods can separate glucose from FOS, further enhance production of FOS production as much as 90%. *Zymomonas mobilis* and *S. cerevisiae* are capable of removing small saccharides like sucrose, fructose, and glucose by converting them into ethanol and carbon dioxide [54]. However, *S. cerevisiae* is incapable of fermenting oligosaccharides that consist of four or more monosaccharide units. During sucrose fermentation, *Z. mobilis* also generates small quantities of FOS and sorbitol.

5.3.2.2 Production of Galactooligosaccharides (GOSs)

These are prebiotic ingredients with a variety of useful qualities that promote the gut microbiota's healthy growth and offer major health advantages. Although, these prebiotics are not absorbed by animals or humans, they contribute positively to various health functions. Due to the growing demand for functional foods, GOS production has garnered considerable attention and can be achieved through enzymatic and chemical methods. The process by which mineral acids can convert monosaccharides into oligosaccharides is known as "reversion." This mechanism, which was initially noticed in the 1950s, explains how oligosaccharides are produced through acidic hydrolysis of lactose [55]. Research has thoroughly explored the conditions favorable for oligosaccharide formation during this process and the resulting oligosaccharide structures. This method produces a composite mixture of di and trisaccharide with various links in both (α and β) anomeric configurations, as well as anhydro-sugars [56]. However, due to requirement of extreme conditions and deficiency of product specificity, production of lactose through acidic hydrolysis is not widely used for GOS production on large scale [57].

The preferred technique for creating GOS is enzymatic catalysis from lactose using glycosyltransferases (EC 2.4) or glycoside hydrolases (EC 3.2.1). The primary enzymes involved in GOS production are galactosidase and galactosyltransferase. Galactosyltransferase is a stereoselective enzyme capable of generating maximum quantities of GOS [58]. However, GOS synthesis using galactosyltransferase is quite expensive due to the need for nucleotide sugars as donors in the reaction. Several strategies, such as the production of globotriose or the use of human milk oligosaccharides have been explored to reduce the costs associated with this process. Glycosyl groups are transferred from a donor sugar to an acceptor sugar with the help of these enzymes. Despite being highly regioselective, stereoselective, and efficient, glycosyltransferases are not widely used for production of industrial GOS because of their limited availability, high cost of enzyme formulations commercially, and particular substrates, like sugar nucleotides, are required [59]. Producing GOS through galactosidase is more cost-effective than using galactosyltransferase; however, galactosidase yields lower quantities of GOS and is less stereospecific. The yield of GOS from galactosidase can be enhanced through several approaches: (i) raising the concentrations of donor and acceptor molecules in the reaction; (ii) decreasing the amount of water in the reaction environment; (iii) driving the equilibrium of the reaction in the direction of product formation by eliminating products from the medium; and (iv) altering the synthesis conditions [60].

β-Galactosidases are derived from various sources, including *Aspergillus oryzae*, bifidobacteria, lactobacilli, and *Sterigmatomyces elviae*. Each source of β-galactosidase produces distinct types of GOS which differ in quantity, DP, and glycosidic linkages. Optimal GOS production requires specific conditions tailored to the source of β-galactosidase: fungal and bacterial enzymes typically require an acidic pH, while yeast sources function best at a neutral pH [61]. Additionally, thermophilic sources necessitate higher temperatures. For GOS production, either whole cells or the natural form of β-galactosidase can be utilized, synthetic variants of the enzyme are also feasible. Using the entire cell is advantageous when isolating β-galactosidase would be cost-prohibitive, as it is more economical due to the presence of natural co-factors within the cell membrane. However, since β-galactosidase primarily requires metal ions as co-factors, whole-cell use is less essential for GOS synthesis [62].

By-products like glucose and galactose, which lack prebiotic properties and reduce GOS yield, can be removed via other metabolic activities in whole-cell processes. For example, *Rasopone minuta*, *Sirobasidium magnum*, and *S. elviae* can utilize glucose as a source of carbon in lactose-based media for production of GOS. Similarly, galactoses can upregulate β-galactosidase expression, while yeast cells use glucose as a carbon source. However, whole-cell use can result in metabolic by-products (e.g., acetic acid, ethanol, and lactic acid) that may interfere with GOS synthesis, necessitating additional removal steps. Temperature is another challenge; while it can enhance GOS yields, high temperatures may harm non-thermophilic cells. Studies have shown that using nonviable or resting cells can overcome these issues, resulting in higher GOS yields [63].

Recombinant β-galactosidases offer advantages include high yield production, easy purification, enhanced stability, and activity due to molecular engineering. *Bacillus subtilis* and *Escherichia coli* are commonly utilized as hosts for recombinant enzyme production. However, *E. coli* may produce endotoxins, have challenges with disulfide bond expression, and generate acetate, which is toxic. *B. subtilis*, while nontoxic, can produce large amounts of proteases (which degrade proteins) and suffer from plasmid instability [64].

5.3.3 Health Benefits of Prebiotics

Numerous studies conducted in the last few decades have demonstrated the health advantages of prebiotics, highlighting their beneficial impact on the GI tract, including immune modulation, pathogen prevention, gut barrier function enhancement, decreased population of pathogenic microbes, and SCFA production [65]. Prebiotics have also been connected to mental health by generating metabolites which affect energy, cognition, and brain function, as well as cardiovascular health by lowering blood cholesterol levels and affecting insulin resistance. They also contribute to bone health by improving the bioavailability of minerals [12]. In terms of GI tract benefits, nondigestible carbohydrates like prebiotics can significantly influence the makeup and function of intestinal microbes [66]. Because they lack certain hydrolyzing enzymes, these prebiotics cannot be digested in the small intestine. Instead, they pass through the colon undigested and serve as a substrate for fermentation by lactobacilli and bifidobacteria, two types of beneficial bacteria [67]. Figure 5.1 illustrates some of the therapeutic benefits associated with prebiotics.

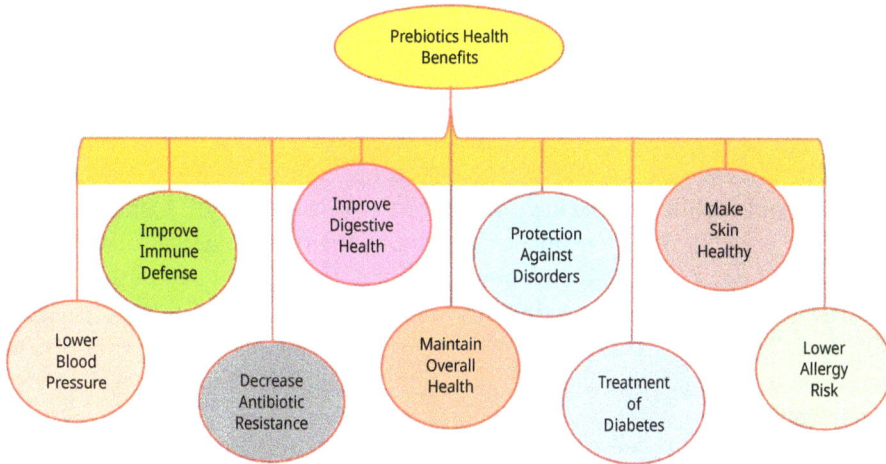

Figure 5.1: Various potent health benefits of prebiotics.

5.3.4 Mechanism of Action of Prebiotics

Since gut microbes are involved in development of several GI disorders [68], there is growing interest in dietary approaches to modulate microbiota composition. Because the gut microbes can metabolize several of these polysaccharides and produce SCFAs, including acetate, butyrate, and propionate, this has led to increasing emphasis on prebiotics. However, impact of prebiotics on microbes residing in the colon remains a topic of debate. Limited human studies have yielded contradictory results on SCFA levels. For example, a study by Liu et al. [69] with healthy volunteers reported a notable decrease in bacteria which produce butyrate and an increase in *Bifidobacterium* levels following administration of FOS and GOS. This shift may be due to elevated lactic acid levels, which can create an environment less favorable for butyrate-producing bacteria. Only a few number of the many possible prebiotics that have been researched, such as inulin, FOS, and GOS, have been proven to work in human trials. Although fructans are recognized as primary substrates for beneficial microbes, GOS and lactulose appear to promote a greater increase in lactobacilli and bifidobacteria compared to inulin.

Kanner et al. [70] demonstrated that gastric acid secretion can promote the oxidation of lipids and other food substances. Their studies suggest that dietary antioxidants, such as inulin, may help prevent lipid peroxidation in the stomach. Generally, dietary supplementation with inulin or oligofructose aids in protecting against oxidative stress, thereby helping to prevent inflammatory reactions linked to oxidative stress. Recent study has focused on the intricate mechanisms through which prebiotics influence microbial composition, immunological responses, pathways of metabolism, and gut barrier functionality. Beneficial microorganisms use prebiotics as a substrate for fermentation, producing metabolites like SCFAs, which are essential for preserving gut equilibrium and have advantages outside the gut [71]. To maximize the benefits of prebiotics, it is essential to understand the specific mechanisms through which they support gut health and lower the risk of GI disorders and related comorbidities. Prebiotics are nondigestible fibers that arrive in the colon intact, such as GOS, inulin, and FOS. These fibers promote the growth and multiplication of beneficial bacteria in the gut microbiota, including probiotics like bifidobacteria and lactobacilli [72].

SCFAs produced through prebiotics fermentation reduce the gut pH from about 6.5 to 5.5, creating an acidic environment that discourages the development of harmful microbes while promoting beneficial bacterial growth, thereby supporting healthy gut [73]. SCFA synthesis was stimulated at higher pH reliant on both the donor and the substrate with inulin enhancing butyrate production at higher pH, thus supporting *butyricimonas*, a butyrate-producing bacterium. Higher colonic pH was also linked to increased abundance of Rikenellaceae, *Phascolarctobacterium*, and *Bacteroides* and increased production of propionate using GOS and FOS. Prebiotics promote the growth of beneficial bacteria that modify immune function by interacting with gut-associated lymphoid tissue, or GALT. This relationship lowers inflammation and promotes a healthy immunological response [74]. SCFAs in the GI tract also help regulate immune responses, promoting tolerance and modulating immune cells, such as T regulatory cells, supporting gut homeostasis and overall immunity. Prebiotics also enhance the function of lymphocytes and phagocytes, essential immune cells involved in identifying and eliminating pathogens. By fostering beneficial bacterial growth and modulating immune responses, prebiotics strengthen the body's prebiotics also help maintain the functioning of gut barrier by promoting bacterial growth, which produces tight junction proteins and mucins that seal gaps between epithelial cells, preventing pathogens from entering the bloodstream defenses against infections and maintain gut health. Prebiotics also indirectly inhibit pathogenic bacterial growth by promoting advantageous bacterial growth, which competes for attachment sites and nutrients, thereby reducing colonization of pathogenic microbes [75]. Mechanism of prebiotics is illustrated in Figure 5.2.

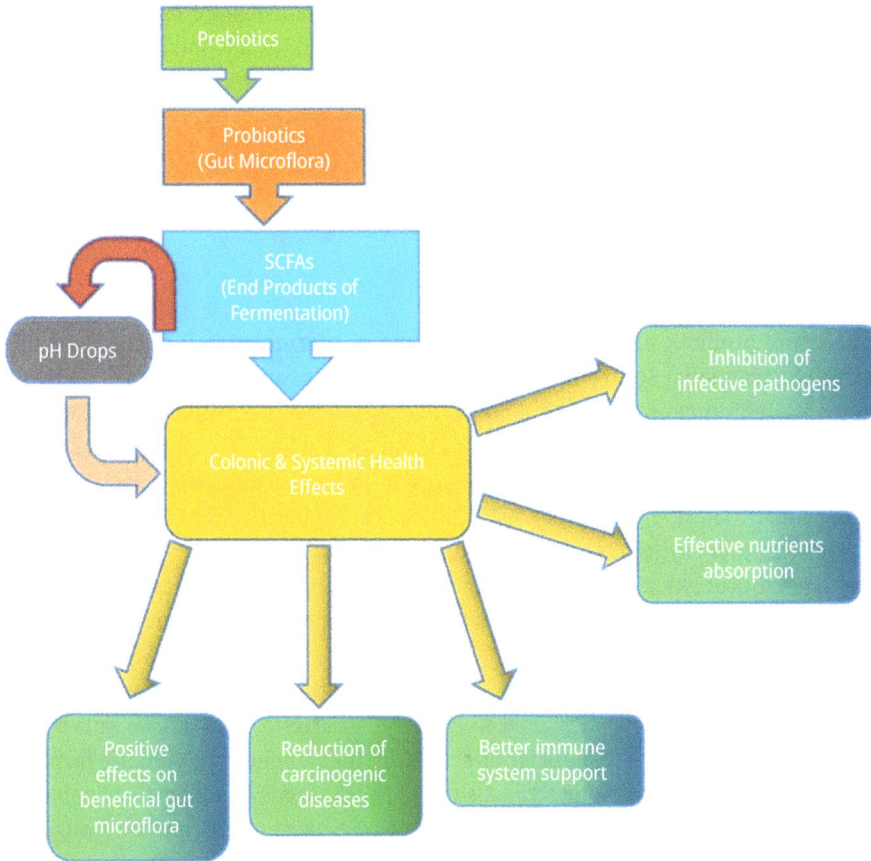

Figure 5.2: Mechanism of action of prebiotics and systemic health effects.

5.3.5 Safety Assessment of Prebiotics

Prebiotics are typically considered to have no serious or potentially fatal side effects. Since digestive enzymes are unable to disintegrate oligosaccharides, as well as polysaccharides, these compounds come in the colon, where they are able to ferment by microbes present in the gut. As a result, negative effects of prebiotics primarily stem from their osmotic properties. Common side effects include cramping, osmotic diarrhea, bloating, and flatulence. The chain length of prebiotics has a significant function in the occurrence of these side effects. Interestingly, shorter-chain prebiotics are more likely to cause side effects, as they are primarily metabolized in the proximal colon and undergo rapid fermentation. In contrast, longer-chain prebiotics are fermented more gradually in distal colon. In addition to chain length, a dosage of prebiotics may influence their safety profile. For instance, low doses (2.5–10 g/day) can lead to flatulence, while high doses (40–50 g/day) may be responsible for osmotic diarrhea. Notably, a regular intake of 2.5–10 g is necessary to achieve advantageous health effects of prebiotics. This implies that prebiotics may have mild to moderate adverse effects, even when used within their recommended dosage range. Most prebiotic products in the market contain amount ranging from 1.5 to 5 g per serving [76].

5.4 Synbiotics

A mixture of probiotic and prebiotic constituents in a formulation is known as synbiotic. The International Scientific Association for Probiotics and Prebiotics currently redefined its definition of synbiotics, although the idea was first proposed 25 years ago. An updated concept of synbiotics describes them as "a make-up of live microbes and substrate(s) specifically utilized microbes present in host, giving well-being benefit to an individual." These preparations may be designed either to complement the host's microbes or to act synergistically, where the prebiotic specifically supports the co-administered probiotics, resulting in one or more health benefits. Numerous studies have shown that prebiotics are more effective when combined with probiotics as component of a synbiotic formulation. The word "synbiotic" terms as a synergistic relationship where prebiotic component is specifically utilized by the live probiotic microbe [77]. This amalgam is designed to enhance an activity and in vivo survival of probiotics, amplifying the beneficial effects of both components. Synbiotic formulations typically include probiotic strains such as *Saccharomyces boulardii*, lactobacilli, *Bacillus coagulans*, and bifidobacteria spp. Prebiotics used often consist of oligosaccharides include FOS, inulin, GOSs, and xylooligosaccharides and prebiotics derived from sources naturally include chicory and yacon roots. Some studies reported health advantages of synbiotic utilization in humans include: (1) enhanced levels of bifidobacteria and lactobacilli, leading to a stable gut microbial composition, (2) refined function of liver in patients with cirrhosis, (3) enhanced immunomodulatory capacity, and (4) preventing bacterial translocation and lowering the frequency of hospital-acquired infections in surgical patients [78].

Recently, the definition of synbiotics has been refined to emphasize preparations that promote synergy between probiotics, which metabolize accompanying prebiotics to restore gut balance and support health of host. Synergistic interactions between prebiotics and probiotics facilitate growth of selective microbes or stimulate specific metabolic pathways via gut flora. A presence of a rapidly fermentable substrate enhances probiotic survival and activity. Additionally, prebiotics can protect probiotics from proteolysis and stomach acidity, potentially through steric hindrance or by forming a protective coating around the probiotic [79]. To maximize their effectiveness, it is crucial to carefully select specific combinations of substrates and microorganisms in synbiotic products, as they can offer superior benefits as compared to products having probiotics or prebiotics alone. Numerous studies have highlighted positive impact of synbiotics in managing conditions such as nonalcoholic fatty liver disease, obesity, necrotizing enterocolitis in very low birth-weight infants, diabetes, and hepatic encephalopathy treatment. Various synbiotic agents include *Lactobacillus* and *Bifidobacterium* strains plus FOS, *Bacillus coagulans* and FOS *Bifidobacterium lactis* B94 plus inulin, *Bifidobacterium breve* strain Yakult, and GOS and *Lactobacillus sporogenes* plus inulin in diminishing several diseases and ailments such as irritable bowel syndrome, gastroenteritis in children, ulcerative colitis, constipation in adult women, and diabetes [80].

5.4.1 Selection Criteria and Mechanism of Action of Synbiotics

A major justification for employing synbiotic is that it is a true probiotic struggle to survive in GI tract without its corresponding prebiotic source. Without this essential nutrient, the probiotic becomes more vulnerable to oxygen exposure, low pH levels, and temperature fluctuations. By combining benefits of both probiotics and prebiotics in synergy, synbiotics significantly elevate the proliferation of advantageous bacteria in the digestive system. The primary consideration when formulating a synbiotic is selecting an appropriate probiotic and prebiotic, each capable of independently providing well-being to the host. Identifying specific characteristics, a prebiotic must possess to positively influence the probiotic is a key approach [81]. An ideal prebiotic should promote the growth of specific beneficial microbes while having little to no stimulatory effect on other microbial populations.

Since probiotics primarily exert their activity in large and small intestines but prebiotics primarily function in the large intestine, their combination can produce a synergistic effect [82]. The primary function of prebiotics is to act as a selective medium that facilitates the development, fermentation, and intestinal transit of probiotic strains. According to published literature, prebiotics help probiotic component to become more adaptable to external factors like temperature, pH, and oxygenation in the host's gut [83]. However, the exact mechanism by which an additional energy source improves tolerance to these factors remains unclear. This combination of components creates microbial dietary supplements that are viable, with favorable environment having positive effect on an individual's health. An action of synbiotics have been recognized by two mechanisms: (1) action via increased probiotic microbe viability; and (2) action via provision of specific health advantages. The synergism of probiotics with prebiotics modulates metabolic activity in intestine while maintaining intestinal biostructure, promoting beneficial microbiota growth, and suppressing harmful microbes in GI tract [84]. Synbiotics help minimize the quantity of harmful metabolites, deactivate nitrosamines and carcinogenic substances, and significantly increase the amount of SCFAs, methyl acetates, ketones, and carbon disulfides contributing in the improvement of host's health [85]. Their therapeutic benefits include anticancer, antibacterial, and anti-allergic properties. In addition to preventing intestinal degradation, constipation, and diarrhea, synbiotics may also effectively aid in preventing osteoporosis, lowering levels of fat and sugar in blood, control the immune system, and managing brain disorders linked to impaired liver function [86].

5.5 Conclusion and Future Prospects

Synbiotic-based supplements are increasingly recognized as valuable tools for promoting health in humans and animals, preventing diseases, and reducing disease-related risks. The significance of synbiotics in preserving mammalian health has been highlighted by studies performed between gut microbiota and immunology relationships in both humans and animals. Evidence suggests that synbiotics positively influence intestinal microbial ecology in both humans and animals, contributing to the alleviation of various health conditions [87]. While most synbiotic studies focus on humans, there is a growing interest in their effects on animals. Research in farm animals has shown that synbiotic supplementation can increase beneficial bacteria and decrease potential pathogens. Despite advances in our knowledge, important information regarding synbiotic effects on the host is still deficient. Identifying optimal formulation of probiotics and prebiotics could further reduce the chances of enteric diseases and address specific bacterial imbalances [88].

The development of synbiotics holds great potential for personalized and precision medicine, as the optimal combinations and type of probiotic strains and prebiotics can be tailored to meet individual needs and health goals. Emerging technologies, such as machine learning and artificial intelligence, could make it easier to optimize and personalize synbiotic compositions according to each person's particular microbiome. What began as a focus on improving gut health has evolved into a broader understanding of the diverse health benefits synbiotics may offer. The future of synbiotics in medicine appears promising, but further research is essential to fully harness their potential and refine their use in preventing and treating a wide range of illnesses [89].

Conflicts of interest: The authors do not declare any conflicts of interest.

References

[1]   Nair R, Paul P, Mahajan S, Maji I, Gupta U, Aalhate M, et al. Emerging era of "biotics": Prebiotics, probiotics, and synbiotics. In: Synbiotics in Human Health: Biology to Drug Delivery. 2024, pp. 319–348, Singapore: Springer Nature Singapore.

[2]   Kumari R, Singh A, Yadav AN, Mishra S, Sachan A, Sachan SG. Probiotics, prebiotics, and synbiotics: Current status and future uses for human health. In: New and Future Developments in Microbial Biotechnology and Bioengineering. 2020, pp. 173–190, Elsevier.

[3]   Al-Habsi N, Al-Khalili M, Haque SA, Elias M, Olqi NA, Al Uraimi T. Health benefits of prebiotics, probiotics, synbiotics, and postbiotics. Nutrients. 2024;16(22):3955.

[4]   Liang D, Wu F, Zhou D, Tan B, Chen T. Commercial probiotic products in public health: Current status and potential limitations. Critical Reviews in Food Science and Nutrition. 2024;64 (19):6455–6476.

[5]   Sanz Y, Portune K, Del Pulgar EG, Benítez-Páez A. Targeting the microbiota: Considerations for developing probiotics as functional foods. In: The Gut-Brain Axis. 2016, pp. 17–30, Academic Press.

[6]   Latif A, Shehzad A, Niazi S, Zahid A, Ashraf W, Iqbal MW, et al. Probiotics: Mechanism of action, health benefits and their application in food industries. Frontiers in Microbiology. 2023;14:1216674.

[7]   Munir A, Javed GA, Javed S, Arshad N. *Levilactobacillus brevis* from carnivores can ameliorate hypercholesterolemia: In vitro and in vivo mechanistic evidence. Journal of Applied Microbiology. 2022;133(3):1725–1742.

[8]   Bansal P, Kumar R, Dhanda S. Characterization of starter cultures and nutritional properties of *Pediococcus acidilactici* NCDC 252: A potential probiotic of dairy origin. Journal of Food Processing and Preservation. 2022;46(10):e16817.

[9]   Kechagia M, Basoulis D, Konstantopoulou S, Dimitriadi D, Gyftopoulou K, Skarmoutsou N, et al. Health benefits of probiotics: A review. ISRN Nutrition. 2013;2013:481651.

[10] Gibson GR, Roberfroid MB. Dietary modulation of the human colonic microbiota: Introducing the concept of prebiotics. Journal of Nutrition. 1995;125(6):1401–1412.

[11] Gibson GR, Probert HM, Van Loo J, Rastall RA, Roberfroid MB. Dietary modulation of the human colonic microbiota: Updating the concept of prebiotics. Nutrition Research Reviews. 2004;17 (2):259–275.

[12] Gibson GR, Hutkins R, Sanders ME, Prescott SL, Reimer RA, Salminen SJ, et al. Expert consensus document: The International Scientific Association for Probiotics and Prebiotics (ISAPP) consensus statement on the definition and scope of prebiotics. Nature Reviews Gastroenterology and Hepatology. 2017;14(8):491–502.

[13] Kuo SM. The interplay between fiber and the intestinal microbiome in the inflammatory response. Advances in Nutrition. 2013;4(1):16–28.

[14] Howlett J, Betteridge V, Champ M, Craig SS, Meheust A, Jones JM. The definition of dietary fiber – discussions at the Ninth Vahouny Fiber Symposium: Building scientific agreement. Food and Nutrition Research. 2010;54(1):5750.

[15] Kumari A, R KG, Sudhakaran VA, Warrier AS, Singh NK. Unveiling the health benefits of prebiotics: A comprehensive review. Indian Journal of Microbiology. 2024;64(2):376-388.

[16] Obayomi OV, Olaniran AF, Owa SO. Unveiling the role of functional foods with emphasis on prebiotics and probiotics in human health: A review. Journal of Functional Foods. 2024;119:106337.

[17] Dinleyici EC, Eren M, Ozen M, Yargic ZA, Vandenplas Y. Effectiveness and safety of *Saccharomyces boulardii* for acute infectious diarrhea. Expert Opinion on Biological Therapy. 2012;12(4):395–410.

[18] EFSA Panel on Dietetic Products, Nutrition and Allergies (NDA). Scientific opinion on the substantiation of a health claim related to "native chicory inulin" and maintenance of normal defecation by increasing stool frequency pursuant to Article 13.5 of Regulation (EC) No 1924/2006. EFSA Journal. 2015;13(1):3951.

[19] Hutkins RW, Krumbeck JA, Bindels LB, Cani PD, Fahey G Jr, Goh YJ, et al. Prebiotics: Why definitions matter. Current Opinion in Biotechnology. 2016;37:1–7.

[20] Rezende ESV, Lima GC, Naves MMV. Dietary fibers as beneficial microbiota modulators: A proposed classification by prebiotic categories. Nutrition. 2021;89:111217.

[21] Rawi MH, Zaman SA, Pa'ee KF, Leong SS, Sarbini SR. Prebiotics metabolism by gut-isolated probiotics. Journal of Food Science and Technology. 2020;57(8):2786–2799.

[22] Mudannayake DC, Jayasena DD, Wimalasiri KM, Ranadheera CS, Ajlouni S. Inulin fructans – Food applications and alternative plant sources: A review. International Journal of Food Science and Technology. 2022;57(9):5764–5780.

[23] Nazzaro F, Fratianni F, De Feo V, Battistelli A, Da Cruz AG, Coppola R. Polyphenols, the new frontiers of prebiotics. Advances in Food and Nutrition Research. 2020;94:35–89.

[24] Huang Z, Boekhorst J, Fogliano V, Capuano E, Wells JM. Distinct effects of fiber and colon segment on microbiota-derived indoles and short-chain fatty acids. Food Chemistry. 2023;398:133801.

[25] Mitchell CM, Davy BM, Ponder MA, McMillan RP, Hughes MD, Hulver MW, et al. Prebiotic inulin supplementation and peripheral insulin sensitivity in adults at elevated risk for type 2 diabetes: A pilot randomized controlled trial. Nutrients. 2021;13(9):3235.

[26] Nobre C, Simões LS, Gonçalves DA, Berni P, Teixeira JA. Fructooligosaccharides production and the health benefits of prebiotics. In: Current Developments in Biotechnology and Bioengineering. 2022, pp. 109–138.

[27] Costa GT, Vasconcelos QD, Aragão GF. Fructooligosaccharides on inflammation, immunomodulation, oxidative stress, and gut immune response: A systematic review. Nutrition Reviews. 2022;80(4):709–722.

[28] Balta I, Linton M, Pinkerton L, Kelly C, Stef L, Pet I, Corcionivoschi N. The effect of natural antimicrobials against *Campylobacter* spp. and its similarities to *Salmonella* spp., *Listeria* spp., *Escherichia coli*, *Vibrio* spp., *Clostridium* spp., and *Staphylococcus* spp. Food Control. 2021;121:107745.

[29] Wang W, Liu F, Xu C, Liu Z, Ma J, Gu L, et al. *Lactobacillus plantarum* 69-2 combined with galacto-oligosaccharides alleviates d-galactose-induced aging by regulating the AMPK/SIRT1 signaling pathway and gut microbiota in mice. Journal of Agricultural and Food Chemistry. 2021;69 (9):2745–2757.

[30] Tian S, Sun Y. Influencing factor of resistant starch formation and application in cereal products: A review. International Journal of Biological Macromolecules. 2020;149:424–431.

[31] Kim HK, Nanba T, Ozaki M, Chijiki H, Takahashi M, Fukazawa M, et al. Effect of the intake of a snack containing dietary fiber on postprandial glucose levels. Foods. 2020;9(10):1500.

[32] Artavia G, Cortés-Herrera C, Granados-Chinchilla F. Total and resistant starch from foodstuff for animal and human consumption in Costa Rica. Current Research in Food Science. 2020;3:275–283.

[33] Xin Y, Ji H, Cho E, Roh KB, You J, Park D, Jung E. Immune-enhancing effect of water-soluble beta-glucan derived from enzymatic hydrolysis of yeast glucan. Biochemistry and Biophysics Reports. 2022;30:101256.

[34] Singla A, Gupta OP, Sagwal V, Kumar A, Patwa N, Mohan N, et al. Beta-glucan as a soluble dietary fiber source: Origins, biosynthesis, extraction, purification, structural characteristics, bioavailability, biofunctional attributes, industrial utilization, and global trade. Nutrients. 2024;16(6):900.

[35] Lante A, Canazza E, Tessari P. Beta-glucans of cereals: Functional and technological properties. Nutrients. 2023;15(9):2124.

[36] Biscarrat P, Cassandre BF, Philippe L, Claire C. Pulses: A way to encourage sustainable fiber consumption. Trends in Food Science & Technology. 2023;143:104281.

[37] Liu X, Li X, Bai Y, Zhou X, Chen L, Qiu C, et al. Natural antimicrobial oligosaccharides in the food industry. International Journal of Food Microbiology. 2023;386:110021.

[38] Singh V, Shaida B. Probiotics, prebiotics, and synbiotics: A potential source for a healthy gut. In: The Gut Microbiota in Health and Disease. 2023, pp. 217–230.

[39]   Rana M, Jassal S, Yadav R, Sharma A, Puri N, Mazumder K, Gupta N. Functional β-mannooligosaccharides: Sources, enzymatic production and application as prebiotics. Critical Reviews in Food Science and Nutrition. 2023;64(28):10221–10238.

[40]   Gamonpilas C, Buathongjan C, Kirdsawasd T, Rattanaprasert M, Klomtun M, Phonsatta N, Methacanon P. Pomelo pectin and fiber: Some perspectives and applications in food industry. Food Hydrocolloids. 2021;120:106981.

[41]   Alam M, Pant K, Brar DS, Dar BN, Nanda V. Exploring the versatility of diverse hydrocolloids to transform techno-functional, rheological, and nutritional attributes of food fillings. Food Hydrocolloids. 2024;146:109275.

[42]   Wu D, Ye X, Linhardt RJ, Liu X, Zhu K, Yu C, et al. Dietary pectic substances enhance gut health by its polycomponent: A review. Comprehensive Reviews in Food Science and Food Safety. 2021;20 (2):2015–2039.

[43]   Gomez B, Gullón B, Yáñez R, Schols H, Alonso JL. Prebiotic potential of pectins and pectic oligosaccharides derived from lemon peel wastes and sugar beet pulp: A comparative evaluation. Journal of Functional Foods. 2016;20:108–121.

[44]   Zhang S, Hu H, Wang L, Liu F, Pan S. Preparation and prebiotic potential of pectin oligosaccharides obtained from citrus peel pectin. Food Chemistry. 2018;244:232–237.

[45]   Varzakas T, Kandylis P, Dimitrellou D, Salamoura C, Zakynthinos G, Proestos C. Preparation and processing of religious and cultural foods. In: Innovative and Fortified Food: Probiotics, Prebiotics, GMOs, and Superfood. 2018, pp. 67–129, London: Elsevier.

[46]   Panesar PS, Kumari S, Panesar R. Biotechnological approaches for the production of prebiotics and their potential applications. Critical Reviews in Biotechnology. 2013;33(4):345–364.

[47]   Sangeetha PT, Ramesh MN, Prapulla SG. Recent trends in the microbial production, analysis and application of fructooligosaccharides. Trends in Food Science & Technology. 2005;16(10):442–457.

[48]   Barreteau H, Delattre C, Michaud P. Production of oligosaccharides as promising new food additive generation. Food Technology and Biotechnology. 2006;44(3):323-333.

[49]   Mohkam M, Nezafat N, Berenjian A, Negahdaripour M, Behfar A, Ghasemi Y. Role of Bacillus genus in the production of value-added compounds. In: Bacilli and Agrobiotechnology. 2016, pp. 1–33.

[50]   Maiorano AE, Piccoli RM, Da Silva ES, De Andrade Rodrigues MF. Microbial production of fructosyltransferases for synthesis of pre-biotics. Biotechnology Letters. 2008;30:1867–1877.

[51]   Davani-Davari D, Negahdaripour M, Karimzadeh I, Seifan M, Mohkam M, Masoumi SJ, et al. Prebiotics: Definition, types, sources, mechanisms, and clinical applications. Foods. 2019;8(3):92.

[52]   Kumar CG, Sripada S, Poornachandra Y. Status and future prospects of fructooligosaccharides as nutraceuticals. In: Role of Materials Science in Food Bioengineering. 2018, pp. 451–503.

[53]   Sheu DC, Duan KJ, Cheng CY, Bi JL, Chen JY. Continuous production of high-content fructooligosaccharides by a complex cell system. Biotechnology Progress. 2002;18(6):1282–1286.

[54]   Lin TJ, Lee YC. High-content fructooligosaccharides production using two immobilized microorganisms in an internal-loop airlift bioreactor. Journal of the Chinese Institute of Chemical Engineers. 2008;39(3):211–217.

[55]   Aronson M. Trans galactosidation during lactose hydrolysis. 1952.

[56]   Huh KT, Toba T, Adachi S. Oligosaccharide structures formed during acid hydrolysis of lactose. Food Chemistry. 1991;39(1):39–49.

[57]   Torres DP, Gonçalves MDPF, Teixeira JA, Rodrigues LR. Galacto-oligosaccharides: Production, properties, applications, and significance as prebiotics. Comprehensive Reviews in Food Science and Food Safety. 2010;9(5):438–454.

[58]   Weijers CA, Franssen MC, Visser GM. Glycosyltransferase-catalyzed synthesis of bioactive oligosaccharides. Biotechnology Advanced. 2008;26(5):436–458.

[59]   De Roode BM, Franssen MC, Van der Padt A, Boom RM. Perspectives for the industrial enzymatic production of glycosides. Biotechnology Progress. 2003;19:1391–1402.

[60] Iqbal MW, Riaz T, Mahmood S, Liaqat H, Mushtaq A, Khan S, et al. Recent advances in the production, analysis, and application of galacto-oligosaccharides. Food Reviews International. 2023;39:5814–5843.

[61] Osman A, Tzortzis G, Rastall RA, Charalampopoulos D. Bbgiv is an important bifidobacterium β-galactosidase for the synthesis of prebiotic galactooligosaccharides at high temperatures. Journal of Agricultural and Food Chemistry. 2012;60:740–748.

[62] Burton SG, Cowan DA, Woodley JM. The search for the ideal biocatalyst. Nature Biotechnology. 2002;20:37–45.

[63] Onishi N, Tanaka T. Purification and characterization of galacto-oligosaccharide-producing β-galactosidase from Sirobasidium magnum. Letters in Applied Microbiology. 1997;24:82–86.

[64] Demain AL, Vaishnav P. Production of recombinant proteins by microbes and higher organisms. Biotechnology Advanced. 2009;27:297–306.

[65] Slavin J. Fiber and prebiotics: Mechanisms and health benefits. Nutrients. 2013;5:1417–1435.

[66] Walker AW, Ince J, Duncan SH, Webster LM, Holtrop G, Ze X, et al. Dominant and diet-responsive groups of bacteria within the human colonic microbiota. ISME Journal. 2011;5:220–230.

[67] Guarino MPL, Altomare A, Emerenziani S, Di Rosa C, Ribolsi M, Balestrieri P, et al. Mechanisms of action of prebiotics and their effects on gastro-intestinal disorders in adults. Nutrients. 2020;12:1037.

[68] Lo Presti A, Zorzi F, Del Chierico F, Altomare A, Cocca S, Avola A, et al. Fecal and mucosal microbiota profiling in irritable bowel syndrome and inflammatory bowel disease. Frontiers in Microbiology. 2019;10:1655.

[69] Liu F, Li P, Chen M, Luo Y, Prabhakar M, Zheng H, et al. Fructooligosaccharide (FOS) and galactooligosaccharide (GOS) increase Bifidobacterium but reduce butyrate producing bacteria with adverse glycemic metabolism in healthy young population. Scientific Reports. 2017;7:11789.

[70] Kanner J, Lapidot T. The stomach as a bioreactor: Dietary lipid peroxidation in the gastric fluid and the effects of plant-derived antioxidants. Free Radical Biology and Medicine. 2001;31:1388–1395.

[71] Marnpae M, Balmori V, Kamonsuwan K, Nungarlee U, Charoensiddhi S, Thilavech T, et al. Modulation of the gut microbiota and short-chain fatty acid production by gac fruit juice and its fermentation in in vitro colonic fermentation. Food and Function. 2024;15:3640–3652.

[72] Limbu D, Sarkar BR, Adhikari MD. Role of probiotics and prebiotics in animal nutrition. Sustainable Agriculture Reviews. 2024;4:173–204.

[73] Roupar D, González A, Martins JT, Gonçalves DA, Teixeira JA, Botelho C, et al. Modulation of designed gut bacterial communities by prebiotics and the impact of their metabolites on intestinal cells. Foods. 2023;12:4216.

[74] Zhou P, Chen C, Patil S, Dong S. Unveiling the therapeutic symphony of probiotics, prebiotics, and postbiotics in gut-immune harmony. Frontiers in Nutrition. 2024;11:1355542.

[75] Rousseaux A, Brosseau C, Bodinier M. Immunomodulation of B lymphocytes by prebiotics, probiotics and synbiotics: Application in pathologies. Nutrients. 2023;15:269.

[76] Svensson UK, Håkansson J. Safety of food and beverages: safety of probiotics and prebiotics. 2014.

[77] Gomez Quintero DF, Kok CR, Hutkins R. The future of synbiotics: Rational formulation and design. Frontiers in Microbiology. 2022;13:919725.

[78] Roy S, Dhaneshwar S. Role of prebiotics, probiotics, and synbiotics in management of inflammatory bowel disease: Current perspectives. World Journal of Gastroenterology. 2023;29:2078.

[79] Sekhon BS, Jairath S. Prebiotics, probiotics and synbiotics: An overview. Journal of Pharmaceutical Education and Research. 2010;1.

[80] Palai S, Derecho CMP, Kesh SS, Egbuna C, Onyeike PC. Prebiotics, probiotics, synbiotics and its importance in the management of diseases. In: Functional Foods and Nutraceuticals: Bioactive Components, Formulations and Innovations. 2020, pp. 173–196.

[81] Scavuzzi BM, Henrique FC, Miglioranza LHS, Simão ANC, Dichi I. Impact of prebiotics, probiotics and synbiotics on components of the metabolic syndrome. Annals of Nutrition Disorders and Therapy. 2014;1:1009.

[82] Hamasalim HJ. Synbiotic as feed additives relating to animal health and performance. Advances in Microbiology. 2016;6:288–302.

[83] Sekhon BS, Jairath S. Prebiotics, probiotics and synbiotics: An overview. Journal of Pharmaceutical Education and Research. 2010;1.

[84] De Vrese M, Schrezenmeir J, Stahl U, Donalies UE, Nevoigt E. Food biotechnology. Advances in Biochemical Engineering and Biotechnology. 2008.

[85] Manigandan T, Mangaiyarkarasi SP, Hemalatha R, Hemalatha VT, Murali NP. Probiotics, prebiotics and synbiotics-a review. Biomedical and Pharmacology Journal. 2012;5:295.

[86] Mounir M, Ibijbijen A, Farih K, Rabetafika HN, Razafindralambo HL. Synbiotics and their antioxidant properties, mechanisms, and benefits on human and animal health: A narrative review. Biomolecules. 2022;12:1443.

[87] Yadav M, Sehrawat N, Sharma AK, Kumar S, Singh R, Kumar A, et al. Synbiotics as potent functional food: Recent updates on therapeutic potential and mechanistic insight. Journal of Food Science and Technology. 2024;61:1–15.

[88] Gomte SS, Rout B, Agnihotri TG, Peddinti V, Jain A. Future perspective and safety issues of synbiotics in different diseases. In: Synbiotics in Human Health: Biology to Drug Delivery. 2024, pp. 281–307, Singapore: Springer Nature Singapore.

[89] Abdelkawi A, Martinez JPP, Pathak S, Pathak Y. Synbiotics: Traditional approach, present status, and future outlook. In: Anxiety, Gut Microbiome, and Nutraceuticals. 2024, pp. 333–344, CRC Press.

Mukesh Yadav, Sunil Kumar, Namita Ashish Singh, Sushil Kumar
Upadhyay, Amit Kumar, Nirmala Sehrawat*

Chapter 6
Synbiotics in Gut Microbiota Balance and Human Health: Dietary Interventions, Therapeutic Potential, and Future Perspectives

Abstract: The human gut microbiota has emerged as a central regulator of health and illnesses, influencing immunity, metabolism, and neurobehavioral functions. Dietary interventions aimed at modulating gut microbiota have gained considerable attention, with probiotics and prebiotics being extensively studied. Synbiotics, which combine probiotics and prebiotics, represent an advanced strategy to synergistically enhance microbial balance and host health. Emerging evidence suggests that synbiotics exert beneficial effects by selectively stimulating beneficial microbes, improving intestinal barrier integrity, enhancing synthesis of short-chain fatty acids (SCFAs), and modulating systemic inflammation. Present review summarizes the mechanistic insights into synbiotic action; their dietary formulations and therapeutic potential to treat gut-associated human illnesses as part of precision medicine according to the available literature. This also includes the current challenges, regulatory perspectives, and the growing importance of personalized synbiotic interventions tailored to individual microbiome profiles. Synbiotics are currently more viable future therapeutic approaches for maintaining healthy microbiota within the host gut, which has a direct impact on the onset or development of linked disorders or diseases.

Acknowledgment: The authors (MY, NS, and SKU) are thankful to the head of the Department of Bio-Sciences and Technology, MMEC, Maharishi Markandeshwar (Deemed to be University), Mullana-Ambala (Haryana), India, for providing necessary support and facilities.

*Corresponding author: Nirmala Sehrawat, Department of Bio-Sciences and Technology, MMEC, Maharishi Markandeshwar (Deemed to be University), Mullana-Ambala 133207, Haryana, India, e-mail: nirmalasehrawat@gmail.com
Mukesh Yadav, Sushil Kumar Upadhyay, Department of Bio-Sciences and Technology, MMEC, Maharishi Markandeshwar (Deemed to be University), Mullana-Ambala 133207, Haryana, India
Sunil Kumar, Department of Microbiology, Graphic Era University, Dehradun, Uttarakhand, India
Namita Ashish Singh, Department of Microbiology, Mohanlal Sukhadia University, Udaipur 313001, Rajasthan, India
Amit Kumar, Department of Biotechnology, Sharda School of Engineering and Technology, Sharda University, Greater Noida, Uttar Pradesh, India

https://doi.org/10.1515/9783112205150-006

Keywords: gut microbiota, dysbiosis, synbiotics, dietary interventions, therapeutic potential

6.1 Introduction

The human gastrointestinal (GI) system supports trillions of microorganisms known as the gut microbiota, a diverse and dynamic population. The majority of these microorganisms are bacteria, mostly from the phyla Proteobacteria, Actinobacteria, Bacteroidetes, and Firmicutes [1, 2]. Hence, this varied ecology is essential for immunological regulation, food metabolism, and pathogen defense. Three clusters of the gut microbiota have been identified based on how they affect the human body: (1) Beneficial microorganisms, such as *Bifidobacterium* and *Lactobacillus*, that detoxify the human gut, generate advantageous metabolites, and protect the intestinal tract. (2) Neutral microorganisms (two traits): good for human health under normal development conditions, but can cause varying degrees of disease when they surpass a particular growth standard or spread to other bodily areas, such as *Enterococcus*. (3) Pathogenic microbes secrete toxins in the intestinal tract and thus harm human health, e.g., *Salmonella* and *Helicobacter pylori* (Figure 6.1). Many factors, including genetics, age, diet, antibiotic use, lifestyle, and mode of birth delivery, affect the gut microbiota's makeup [3]. Among all factors, dietary factors are regarded to be the most direct and effective route to therapeutic intervention since they significantly shape and regulate gut microbiota [4]. The human GI tract's commensal microbiota load, expressed in colony-forming unit (cfu)/mL, progressively rises from the stomach to the jejunum to the colon, as illustrated in Figure 6.1 [5].

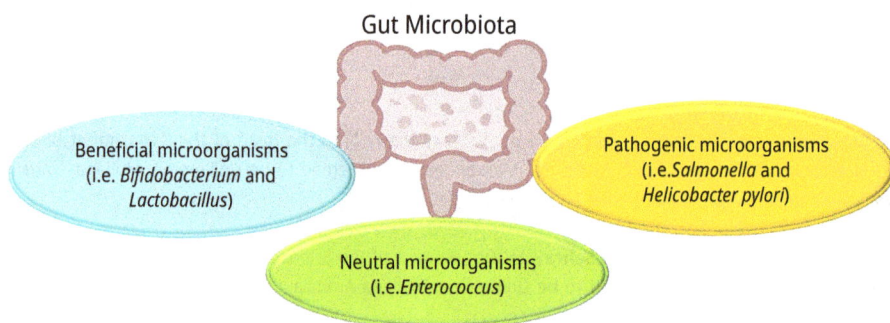

Figure 6.1: Types of microbial clusters inhabiting the intestinal gut.

6.2 Gut Dysbiosis and Onset of Human Illnesses

Gut microbiota (commensals as well as pathogenic) influences various aspects of physiological processes and cellular signaling that affect human health. It is also a key regulator of different gut-associated axes [6]. The gut microbiota is stable, resilient, and interacts symbiotically with the host when it is healthy [7]. In contrast, the loss of this equilibrium affected gut-associated axis that causes severe human illnesses including onset or progression of chronic, metabolism-associated, and neurodegenerative diseases or disorders. Disturbance in gut microbiota's diversity, composition, or functional capacities exerts negative health impacts on the host, referred as gut dysbiosis. Dysbiosis, or imbalance in microbial composition has wide-ranging and complex health effects as it is linked to a variety of humans illnesses, such as autoimmune disorders, GI disorders, metabolic syndromes or diseases (diabetes, obesity), allergies, neurological diseases, IBDs (inflammatory bowel diseases), and IBS (irritable bowel syndrome), and even cancer [8, 9]. Several mechanisms are responsible for development of gut-associated diseases or disorders as a result of dysbiosis such as (a) gut barrier dysfunction, (b) immune dysregulation, (c) metabolite imbalance, and (4) altered neuroimmune interactions. The balance of the physiological microbiota community in the gut, or eubiosis, is crucial for preserving an individual's health. Accordingly, strategies aimed at modulating the gut microbiome have attracted scientific and clinical interest. Increasing evidence suggests that dietary interventions can beneficially alter the gut microbiome to maintain homeostasis. Among them, synbiotics are becoming more well-known due to their potential to restore homeostasis and modify gut microbiota, which may have therapeutic advantages [10, 11].

6.3 Synbiotics: An Efficient and Emerging Gut Microbiota Modulator

Synbiotics are a combination of probiotics (live beneficial bacteria) and prebiotics (substrates that selectively feed beneficial microbes) designed to act synergistically to improve both microbial survival and host benefit. The Greek words "syn" and "biotic" both mean "together" and "life," respectively. In 1995, Gibson and Roberfroid defined "synbiotics" as "a mixture of probiotics and prebiotics that beneficially affect the host by improving the survival and implantation of live microbial dietary supplements in the GI tract, by selectively stimulating the growth and/or activating the metabolism of one or a limited number of health-promoting bacteria, thus improving host welfare" [12]. Synbiotics are described as "a mixture comprising live microorganisms and substrate(s) selectively utilized by host microorganisms that confers a health benefit on the host" according to the International Scientific Association for Probiotics and Prebiotics (ISAPP). This definition accommodates both complimentary synbiotics, which combine separate probiotics and prebiotics, and synergistic synbiotics, in which the prebiotic directly improves the survival or function of the selected probiotic strain [13]. This formulation emphasizes that the substrate can target either the co-administered (allochthonous) microbes or the resident (autochthonous) microbiota, which opened the door to intentionally designed pairings rather than ad hoc probiotic + fiber mixtures. Probiotics are live bacteria that give health advantages when taken in suitable proportions. Prebiotics are specifically fermented substrates that promote the growth of beneficial microbes, have been widely applied. However, limitations exist in their individual effectiveness due to variability in colonization, substrate specificity, and host response [13, 14]. Different selection criteria that help in efficient symbiotic formulation are presented in Table 6.1.

Table 6.1: Selection criterion for synbiotics.

Criterion	Significance
Efficacy	Greater efficacy compared to probiotics or prebiotics alone, research showing better outcomes in particular health issues
Safety	Consider the safety of both probiotic strains and prebiotic fibers, and watch for unpleasant gastrointestinal effects
Selection of specific microbial strain	To ensure compatibility and efficacy of both probiotic strains and prebiotic fibers, choose strains known for their synergy
Accessibility	Ensure the availability of both probiotic and prebiotic components, and consider accessibility in various forms to accommodate individual preferences
Selection	Choose products with established synergy of probiotics and prebiotics, and assess compatibility with additional dietary or medicinal therapies

6.3.1 Probiotics

Probiotics are "live microorganisms which, when administered in adequate amounts, confer a health benefit on the host," according to the FAO/WHO (Food and Agriculture Organization of the United Nations/World Health Organization). *Saccharomyces boulardii*, *Bifidobacterium*, and *Lactobacillus* are common probiotic strains. Probiotics exert health effects by preventing the growth of infections, bolstering the intestinal barrier, regulating immunological responses, and generating antimicrobial chemicals. Probiotics are also implicated in the regulation of bile acid (BA) metabolism, lactic acid production, short-chain fatty acid (SCFA) production, and the suppression of alpha-glucosidase activity [9]. Because probiotics occupy host tissue and stop harmful bacteria from colonizing, they are highly recommended for preserving gut microbial balance and resolving microbial dysbiosis. However, the efficacy of probiotics depends upon dose type, mode of delivery, and host [15, 16]. Probiotics exert their significant effects in reshaping the gut microbiota (GM) composition via several mechanisms including: (1) modulation of gut microbiota composition, (2) strengthen intestinal barrier and its function, (3) competitive exclusion of harmful microbes, (4) antimicrobial compounds production, (5) alteration of host immune system, and (6) regulatory effects on host inflammatory responses [8, 17]. Modern DNA technology has advanced to the point that it is now feasible to design microbes to treat diseases. Genetic engineering technology has produced "smart probiotics," which give probiotic use new life. It works better than probiotics that are naturally occurring [6]. Earlier studies demonstrated promising role of probiotics intake in maintaining gut microbiota balance and prevention of several diseases including GI disorders. But, when probiotics are utilized for therapeutic purposes on particular individuals, it is crucial to employ the right dosages [18, 19].

6.3.2 Prebiotics

Prebiotics are indigestible food elements that specifically stimulate the growth and function of beneficial gut bacteria (i.e., bifidobacteria and lactobacilli) that support healthy gut microbiome and confer various health benefits as per the ISAPP [20-22]. Glenn Gibson and Marcel Roberfroid introduced the term "prebiotic" in 1995 as "non-digestible food ingredients that beneficially affect the host by selectively stimulating the growth and/or activity of one or a limited number of bacteria in the colon, to improve host health" [23–25]. Prebiotics act as food for the host's natural microbiota, as compared to the probiotics, that introduce live beneficial bacteria. Although the benefits of prebiotics vary depending on the strain and species, they generally promote the native gut flora. Traditional and emergent prebiotics are the two categories of prebiotics that are distinguished by their chemical makeup and capacity to specifically promote the growth of good gut bacteria. Inulin, resistant starches, galacto-oligosaccharides (GOS), and fructo-oligosaccharides (FOS) are few examples [26]. Prebiotics promote a balanced gut ecology that is essential to human health by being fermented by good bacteria in the gut. During the fermentation process, several SCFAs are created, which significantly increases the proliferation, activity, and diversity of beneficial bacteria. These microbes in turn suppress the pathogens by decreasing colonic pH, competing for nutrients, and producing antimicrobial compounds. Improved gut health and function exerts broader physiological benefits [27]. The SCFAs strengthen the barrier's integrity and generate a protective mucus layer, which decreases the translocation of dangerous chemicals into the bloodstream. Additionally, they have anti-inflammatory qualities and support systemic health advantages. Prebiotics along with probiotics are gaining more attention recently in modulating gut microbiome to manage or treat severe human illnesses [22, 28, 29].

Individual responses toward prebiotics depend upon person's genetic background, personalized nutrition, and differences in composition of the gut microbiota. Different protein sources can affect the makeup and actions of our gut microflora, which can have different impacts on our health as changing the metabolites produced by gut microbes and controlling the quantity of vital intestinal flora [4]. Researchers have begun to emphasize the advantages of plant-based protein in recent years, and but for maximum nutritional value, they advise a dual-protein diet that incorporates both plant- and animal-based protein sources [30]. Although we are aware that dietary proteins interact with the gut microbiota to control our general health, but still the complete understanding of the mechanisms involved is unclear. Earlier studies demonstrated the significant prebiotics potential in improving gut health and modulating host immune response to treat or manage GI tract-associated cellular processes or molecular events that regulate onset and progression of gut-associated disorders or chronic diseases [31]. Recent developments in discovery of novel prebiotic, creative delivery methods, regulatory issues, and their use to personalized nutrition have been summarized by Chakraborty et al. [25] and Ali et al. [29].

6.4 Mechanistic Insights of Synbiotics and Gut Microbiata Modulation

Synbiotics are dietary supplements that contain both probiotics (beneficial living microorganisms) and prebiotics (nondigestible food elements which stimulate the growth and survival of useful microbes). This synergistic combination optimizes gut microbial balance, enhances host health, and offers therapeutic potential in GI and systemic diseases [10, 32]. Synbiotics are categorized into two groups: **(1) complementary synbiotics** contain probiotics and prebiotics that act independently, offering parallel benefits (i.e., *Lactobacillus rhamnosus* with inulin) and (2) **synergistic synbiotics** include specific prebiotics that improve the growth and metabolic activities of the co-administered probiotics (i.e., *Bifidobacterium longum* with GOS). Synbiotic-mediated gut microbiota regulation occurs through modulation of microbial composition, strengthening of intestinal barriers, and modulation of host immune responses. Synbiotics regulate gut microbiota through several mechanisms that involves multiple host-microbe interactions, including: (a) enhancing colonization and survival of beneficial bacteria by providing selective substrates, (b) suppressing pathogenic microorganisms through competitive exclusion and antimicrobial production, (c) promoting the SCFAs production that helps in maintaining pH in GI tract and strengthen barrier integrity, and host metabolism, (d) anti-inflammatory cytokine induction and pro-inflammatory pathway repression, (e) modulating BA metabolism and nutrient absorption, and (f) regulation of immune responses via modulation of gut-associated lymphoid tissue [10, 11] (Table 6.2). Recent mechanistic research has improved our understanding of synbiotic activity by developing selected substrates. For example, selecting prebiotic structures (certain oligosaccharides, resistant starches, and polyphenol conjugates) that favor the metabolic pathways of paired bacteria (e.g., butyrogenic vs. propionigenic routes), resulting in predictable SCFA profiles [11, 33].

Table 6.2: Summary of mechanisms of synbiotics in gut microbiota regulation.

Synbiotic mechanism	Impact on gut microbiota and its significance
SCFA production	Prebiotics fermentation by the probiotics in the gut generates acetate, propionate, and butyrate. These regulate intestinal pH, provide energy to colonocytes, and exert anti-inflammatory effects
Immune modulation	Synbiotics enhance anti-inflammatory cytokines, suppress pro-inflammatory responses, and regulate gut-associated lymphoid tissue (GALT)
Barrier integrity	Stimulation of mucin secretion and tight junction proteins strengthens the gut epithelial barrier, preventing pathogen entry
Pathogen suppression	Competitive exclusion, antimicrobial compound production, and reduced pathogen adhesion to intestinal mucosa

6.5 Dietary Interventions and Formulations of Synbiotics

Health and illness are significantly influenced by the human gut flora. Dietary elements have emerged as significant gut microbiota regulators. Synbiotics are nutritional supplements that contain both probiotics (useful living microorganisms) and prebiotics (nondigestible food elements that promote the growth of helpful microbes). This synergistic combination optimizes gut microbial balance, enhances host health, and offers therapeutic potential in GI and systemic diseases including cancer [11, 34]. Dietary intake of synbiotics can be done in any form such as: (a) traditional fermented foods (yogurt, kefir, kimchi, and sauerkraut) enriched with prebiotic fibers, (b) functional foods fortified with synbiotics (milk, cereals, and infant formula), and (c) nutraceutical supplements (capsules and powders containing selected strains + fibers). Individual responses toward synbiotic therapies can be affected by various factors such as genetics, strain-specific probiotic properties, nature and dose of prebiotics, overall health, gut microbiota composition at the time of intervention, food matrix, and delivery vehicle [35]. Excessive dosages of synbiotic may also cause transient GI distress (i.e., stomach discomfort, gas, and bloating) and changes in bowel habits (i.e., constipation or diarrhea) in the host. Hence, the daily dietary intake of synbiotics might be a useful tactic for enhancing gut health and general well-being of a person [32, 33]. Emerging evidence highlights the variability of synbiotic efficacy among individuals due to differences in micro-biome composition, genetics, and diet. Advances in metagenomics, metabolomics, and nutrigenomics are paving the way for personalized synbiotic formulations tailored to patient-specific needs. One of the fastest technical advances is in delivering viable microbes together with their substrates through the GI tract and into food products either in the form of microbeads and microencapsulation that protect probiotics during processing, storage, and gastric transit while releasing both microbes and substrate in the intestine. This may also improves viability in food matrices and clinical dosing. Another method is use of synbiotic as food formats (yogurts, dairy, and nondairy fermented foods) that retain probiotic viability during shelf life and deliver specific prebiotic substrates is now well documented in randomized studies [10, 36, 37].

6.6 Therapeutic Potential of Synbiotics in Human Health

Synbiotics are compositions that combine prebiotics and probiotics. Prebiotics and probiotics are combined to generate "synbiotics," in which certain prebiotic components promote probiotic bacteria growth. The fundamental goal of using synbiotics is to maintain the amount of useful microbiome that aids in gut health by adjusting metabolism according to the body's requirements. Probiotics, prebiotics, and synbiotics constitute potent triad biotics that exert significant health outcomes via modulation of gut microflora and host immune system which in turn regulate gut-associated cellular processes [11, 38]. Different sources of these biotics along with their health significance are depicted in Table 6.3. Synbiotics may be the safest and most effective therapeutic option for preventing gut pathogen colonization and infection, particularly bacterial infections. Synbiotics enhance immune system performance by generating a harmonious bacteria composition within the GI tract. They help improve effective digestion and nutrient absorption [39, 40]. Synbiotics can have two effects: they can increase the growth of beneficial microflora, such as *Bifidobacteria*, after taking only prebiotics, or they can improve the host's health after consuming both prebiotics and probiotics. Synbiotics have been found to have a wide range of pharmacological characteristics, including antidepression, antiparkinson, anti-inflammatory, anticancer, neuroprotective, and so on. Synbiotics are indicated as complements to existing treatments for a variety of chronic conditions, respiratory tract infections and caries prevention [41, 42]. Gomez Quintero et al. [36] examined several specific examples of complimentary and synergistic synbiotics, as well as the logic behind their development. Authors also discussed the therapeutic significance of synbiotic approach for well-being of humans in future [36]. The therapeutic significance of synbiotics is given in Table 6.4 and Figure 6.2.

Table 6.3: Different sources and therapeutic significance of powerful triad of biotics (probiotics, prebiotics, and synbiotics).

Biotic types	Sources	Beneficial microbes involved	Types of bioactive compounds produced	Health significance	References
Probiotics	Yoghurt, miso, and kefir; sauerkraut (cabbage), cheese, pickles, and tempeh (soybeans)	Lactococcus sp. Bifidobacteria Lactobacillus sp.	Bacteriocins, SCFAs, exopolysaccharides, immunomodulatory, compounds, amino acids, vitamins, and oligosaccharides	Antimicrobial use, food preservative, lactose intolerance, diarrheal disease, ulcer treatment, immune stimulation, and colon cancer	[24, 43]
Prebiotics	Chicory roots, soy sauce, honey, miso, onions, asparagus, garlic, Jerusalem antichoke, oat, wheat, carbohydrate-based products (biscuits), protein-based products (yoghurt)	Bifidobacteria Bacteriodes Lactobacillus sp.	Flavanols Fructo-oligosaccharides Omalto-oligosaccharides Xylo-oligosaccharides Lactulose Lactosucrose Oligosaccharides	Produced dietary fibers: Inulin, beta-glucan, lactulose, fructo-oligosaccharides, xylo-oligosaccharides, galacto-oligosaccharide, isomalto-oligosaccharides, resistant starches, and arabino oligosaccharides	[29, 44, 45]
Synbiotics	Combined application of probiotics and prebiotics as green banana, chicory roots, onions, oats, acacia, gum, garlic, asparugus, yoghurt, whole grains, cheese, and sauerkraut	Lactococcus sp., Bifidobacteria, Lactobacillus sp.	Bioactive compound (SCFAs) production and other compounds (bacteriocins) to inhibit pathogens	– Emerging functional food as therapeutics for human well-being – Influences: Metabolism, gastrointestinal tract (GIT) stability, immune function, neurological effects, psychological effects for mental well-being, cardiovascular risk reduction, and reduce metabolic syndrome	[40, 46]

6.7 Recent Advances in Synbiotics as Therapeutics and Associated Challenges

Literature supported that synbiotics provide superior functional benefits over probiotics or prebiotics alone by delivering live beneficial microbes and their selective substrates at the same time, such as improving colonization efficiency, resilience to GI stressors, boosting SCFA synthesis, altering cytokine profiles, regulating the immunological response, and long-term modulation of the gut microbial ecosystem. Their possible medical uses include enhanced mucosal immunity, GI problems, metabolic syndrome, and recovery from antibiotic-induced dysbiosis. These alterations have been demonstrated to increase intestinal barrier integrity, restore microbial balance (eubiosis), and reduce systemic inflammation, which collectively exert significant host health impacts [14, 47]. Ngoc et al. [48] investigated the effectiveness of synbiotics as a treatment for Alzheimer's disease (AD) that outperforms typical probiotic or prebiotic therapies. The authors discovered that synbiotics can be utilized not only to address cognitive decline but also to minimize AD-related psychosocial stress, enhancing the overall quality of life of Alzheimer's patients [48]. Oti et al. [49] explored synbiotics' potential as an adjuvant therapy for COVID-19. The authors also summarized how synbiotics alter the gut axis, manage GI symptoms, modulate the immunological response, and may reduce the risk of acute respiratory distress syndrome and other consequences [49].

Lv et al. [50] studied the impact of synbiotics as a supplement on cardiovascular risk factors in nonalcoholic fatty liver disease (NAFLD) patients. After controlling for publication bias, the authors found that synbiotic supplementation had no significant effect on NAFLD patients' blood pressure, anthropometric indices, or lipid profile parameters. Also, the synbiotic supplementation reduces C-reactive protein levels but does not affect tumor necrosis factor-alpha [50]. Mok et al. [51] examined the impact of *Limosilactobacillus reuteri* KUB-AC5 (10^8 cfu) and powder of *Wolffia globosa* (6 g/day) on continuous human GI model under in vitro conditions. The author's findings underlined the ability of synbiotic supplementation to raise beneficial bacterial populations, increase in microbial diversity, and improve metabolic health to manage obesity issues [51]. Parhi et al. [44] investigated how prebiotics affect the growth and viability of probiotic microbes in food matrices. Authors also noted the increased interest in synbiotics as functional dietary components. The authors also indicated that, despite synbiotics' potential benefits for gut health and overall well-being, further research is required to completely understand their mechanisms of action and maximize their use [44].

Consumers' rising concern about their health has led to an increase in the consumption of functional foods and beverages. Many merchants now sell functional foods that have considerable health benefits. Changes in lifestyle and diet are also driving up demand for these items. Synbiotics have been loosely characterized in recent decades as "probiotics and prebiotics that benefit the host." Synbiotics selectively increase the development and metabolism of the selective microbes in the colon, ultimately improving host health. As per future market insights, Parhi et al. [44] discussed that the global synbiotic products market is anticipated to be worth $638.2 million in 2023. Sales will grow at a 7.5% compound annual rate throughout the forecast period, reaching US$1116.9 million by 2033. The increased fortification of functional foods is predicted to drive growth in the synbiotic product market [44]. Synbiotics are gaining more attention today in scientific community as effective gut microbiota modulators for human well-being. Dietary intake of synbiotics is generally safe in healthy populations in optimal dose but caution should be followed in immune-compromised individuals due to risk of bacteremia or fungemia. However, there is a lack of standardized formulations and dosages across clinical trials, uncertain microbial strain-specific effects along with limited large-scale and long-term human studies regarding synbiotics applicability. Hence consumer awareness about necessity of good gut health should be raised and stable formulations of synbiotics with proven efficacy should be designed in near future for therapeutic applications. Hence there is great demand of designer synbiotics, microbiome-informed personalized nutrition, and integration with digital health monitoring and artificial intelligence-driven microbiome diagnostics in present era [6, 36, 52].

6.8 Conclusion

Synbiotics represent a promising nutritional and therapeutic strategy, bridging the gap between diet and clinical therapy. By combining probiotics and prebiotics, they offer synergistic benefits in modulating gut microbiota, enhancing immunity, and preventing or managing chronic diseases. While current evidence supports their potential in GI, metabolic, cardiovascular, and neurological health, further high-quality clinical trials are required to establish standardized guidelines. Future of synbiotics lies in precision nutrition and their integration into preventive healthcare systems.

Table 6.4: Therapeutic potential of synbiotics in management of gut-associated human diseases.

Type of illness	Disease/disorder	Mechanism involved	Recorded effects of synbiotics	References
Gastrointestinal disorders	Antibiotic-associated diarrhea, IBS, IBD, and *C. difficile* infection	Modulation of gut microbiota, SCFA production, improved gut barrier integrity, and reduced inflammatory cytokines	Reduction of bloating, abdominal pain, stool irregularities in IBS, improved mucosal healing in IBD, and decrease in recurrence of *C. difficile* infection	[53, 54]

Table 6.4 (continued)

Type of illness	Disease/disorder	Mechanism involved	Recorded effects of synbiotics	References
Immune system alteration	Allergies (eczema and asthma), respiratory infections, and vaccine response	Enhanced epithelial barrier, decrease in anti-inflammatory cytokines (IL-10), and modulation of dendritic and T cells	Reduction of allergy incidence in children, improved vaccine efficacy, and reduction in frequency of respiratory infections	[39, 55]
Cardiovascular health problems	Hyperlipidemia and hypertension	Bile salt deconjugation, improved lipid metabolism, and SCFA-mediated endothelial function	Reduction of cholesterol and triglycerides and improved blood pressure regulation	[40, 50, 56]
Metabolic disorders	Type 2 diabetes, obesity, and NAFLD	Improved insulin sensitivity, SCFA-mediated appetite regulation and energy balance, and reduction of endotoxemia	Improved glycemic control, reduced body fat and BMI, and reduced hepatic steatosis	[46, 51, 57, 58]
Neurological and mental health problems	Alzheimer's disease, aging, depression, anxiety, cognitive decline, and autism spectrum disorders	Gut-brain axis modulation, neurotransmitter (serotonin and GABA) regulation, and reduces systemic inflammation	Improved mood, cognition, and anxiety and early evidence for behavioral improvements in ASD	[48, 59, 60]
Cancer prevention and adjunct therapy	Colon cancer and chemotherapy support	Reduced carcinogen-producing bacteria, enhanced immune surveillance, and improved treatment tolerance	Reduction of chemotherapy side effects; improved quality of life and reduced tumor-promoting microbiota	[9, 61]

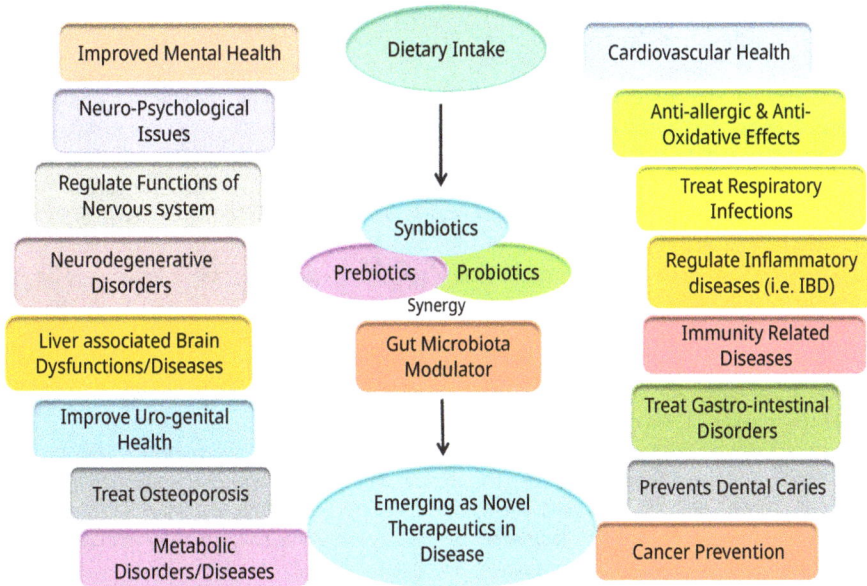

Figure 6.2: Potent therapeutic applications of synbiotics to treat gut-associated disease or disorders for human well-being.

Conflicts of interest: The authors do not declare any conflicts of interest.

References

[1]   Pitocco D, Di Leo M, Tartaglione L, De Leva F, Petruzziello C, Saviano A, Pontecorvi A, Ojetti V. The role of gut microbiota in mediating obesity and diabetes mellitus. European Review for Medical and Pharmacological Sciences. 2020 Feb 1;24(3):1548-1562.

[2]   Agus A, Clément K, Sokol H. Gut microbiota-derived metabolites as central regulators in metabolic disorders. Gut. 2021 Jun 1;70(6):1174–1182.

[3]   Akagawa S, Akagawa Y, Yamanouchi S, Kimata T, Tsuji S, Kaneko K. Development of the gut microbiota and dysbiosis in children. Bioscience of Microbiota, Food and Health. 2021;40(1):12–18.

[4]   Bartlett A, Kleiner M. Dietary protein and the intestinal microbiota: An understudied relationship. IScience. 2022 Nov 18;25(11):105313.

[5]   Chandrasekaran P, Weiskirchen S, Weiskirchen R. Effects of probiotics on gut microbiota: An overview. International Journal of Molecular Sciences. 2024 May 30;25(11):6022.

[6]   Hou K, Wu ZX, Chen XY, Wang JQ, Zhang D, Xiao C, Zhu D, Koya JB, Wei L, Li J, Chen ZS. Microbiota in health and diseases. Signal Transduction and Targeted Therapy. 2022 Apr 23;7(1):135.

[7]   Fan Y, Pedersen O. Gut microbiota in human metabolic health and disease. Nature Reviews Microbiology. 2021 Jan;19(1):55–71.

[8]   Sharma VR, Singh M, Kumar V, Yadav M, Sehrawat N, Sharma DK, Sharma AK. Microbiome dysbiosis in cancer: Exploring therapeutic strategies to counter the disease. In: Seminars in Cancer Biology. 2021a May 1, (vol. 70, pp. 61–70). Academic Press.

[9]   Fakharian F, Thirugnanam S, Welsh DA, Kim WK, Rappaport J, Bittinger K, Rout N. The role of gut dysbiosis in the loss of intestinal immune cell functions and viral pathogenesis. Microorganisms. 2023 Jul 21;11(7):1849.

[10]   Nair R, Paul P, Mahajan S, Maji I, Gupta U, Aalhate M, Singh SK, Singh PK. Correction to: Emerging Era of "Biotics": Prebiotics, Probiotics, and Synbiotics. In: Synbiotics in Human Health: Biology to Drug Delivery 2024 Aug 1, pp. C1–C1. Singapore: Springer Nature Singapore.

[11]   Yadav M, Sehrawat N, Sharma AK, Kumar S, Singh R, Kumar A, Kumar A. Synbiotics as potent functional food: Recent updates on therapeutic potential and mechanistic insight. Journal of Food Science and Technology. 2024 Jan;61(1):1–5.

[12]   Gibson GR, Roberfroid MB. Dietary modulation of the human colonic microbiota: Introducing the concept of prebiotics. The Journal of Nutrition. 1995 Jun 1;125(6):1401–1412.

[13]   Swanson KS, Gibson GR, Hutkins R, Reimer RA, Reid G, Verbeke K, Scott KP, Holscher HD, Azad MB, Delzenne NM, Sanders ME. The International Scientific Association for Probiotics and Prebiotics (ISAPP) consensus statement on the definition and scope of synbiotics. Nature Reviews Gastroenterology & Hepatology. 2020 Nov;17(11):687–701.

[14]   Yadav MK, Kumari I, Singh B, Sharma KK, Tiwari SK. Probiotics, prebiotics and synbiotics: Safe options for next-generation therapeutics. Applied Microbiology and Biotechnology. 2022 Jan;106(2):505–521.

[15]   Kim SK, Guevarra RB, Kim YT, Kwon J, Kim H, Cho JH, Kim HB, Lee JH. Role of probiotics in human gut microbiome-associated diseases. Journal of Microbiology and Biotechnology. 2019;29(9):1335–1340.

[16]   Bhatia A, Sharma D, Mehta J, Kumarasamy V, Begum MY, Siddiqua A, Sekar M, Subramaniyan V, Wong LS, Mat Rani NN. Probiotics and synbiotics: Applications, benefits, and mechanisms for the improvement of human and ecological health. Journal of Multidisciplinary Healthcare. 2025 Dec 31;18:1493–1510.

[17]   Sehrawat N, Yadav M, Singh M, Kumar V, Sharma VR, Sharma AK. Probiotics in microbiome ecological balance providing a therapeutic window against cancer. In: Seminars in Cancer Biology. 2021 May 1, (vol. 70, pp. 24–36). Academic Press.

[18]   Ma T, Shen X, Shi X, Sakandar HA, Quan K, Li Y, Jin H, Kwok LY, Zhang H, Sun Z. Targeting gut microbiota and metabolism as the major probiotic mechanism-An evidence-based review. Trends in Food Science & Technology. 2023 Aug 1;138:178–198.

[19]   Patra D. Synthetic biology-enabled engineering of probiotics for precision and targeted therapeutic delivery applications. Exon. 2024 Nov 17;1(2):54–66.

[20]   Bevilacqua A, Campaniello D, Speranza B, Racioppo A, Sinigaglia M, Corbo MR. An update on prebiotics and on their health effects. Foods. 2024;13(3):446.

[21]   Bisht D, Pal D, Shrestha R. Introduction to Probiotics, Prebiotics, and Synbiotics: A Holistic Approach, Probiotics. 2024, pp. 1–28, CRC Press.

[22]   Yoo S, Jung S-C, Kwak K, Kim J-S. The role of prebiotics in modulating gut microbiota: Implications for human health. International Journal of Molecular Sciences. 2024;25(9):4834.

[23]   Chavan AR, Singh AK, Gupta RK, Nakhate SP, Poddar BJ, Gujar VV, Purohit HJ, Khardenavis AA. Recent trends in the biotechnology of functional non-digestible oligosaccharides with prebiotic potential. Biotechnology and Genetic Engineering Reviews. 2023;39:1–46.

[24]   Maftei NM, Raileanu CR, Balta AA, Ambrose L, Boev M, Marin DB, Lisa EL. The potential impact of probiotics on human health: An update on their health-promoting properties. Microorganisms. 2024 Jan 23;12(2):234.

[25]   Chakraborty M, Budhwar S, Kumar S. Probiotics, Prebiotics, and Synbiotics as Dietary Supplements: A Healthy Approach, the Functional Foods. 2025, pp. 227–246, Apple Academic Press.

[26]   Liu Y, Wang J, Wu C. Modulation of gut microbiota and immune system by probiotics, pre-biotics, and post-biotics. Front. Nutr. 2022;8:634897.

[27]   You S, Ma Y, Yan B, Pei W, Wu Q, Ding C, Huang C. The promotion mechanism of prebiotics for probiotics: A review. Front. Nutr. 2022;9:1000517.

[28] Sharma V, Sharma N, Sheikh I, Kumar V, Sehrawat N, Yadav M, Ram G, Sankhyan A, Sharma AK. Probiotics and prebiotics having broad spectrum anticancer therapeutic potential: Recent trends and future perspectives. Current Pharmacology Reports. 2021b Apr;7(2):67–79.

[29] Ali S, Hamayun M, Siraj M, Khan SA, Kim HY, Lee B. Recent Advances in Prebiotics: Classification. Mechanisms, and Health Applications. Future Foods. 2025 Jun 9;12:100680.

[30] Huang Y, Zhang K, Zhang L, Qiu J, Fu L, Yin T, Wang J, Qin R, Zhang J, Dong X, Wang G. Dosage of dual-protein nutrition differentially impacts the formation of atherosclerosis in ApoE−/− mice. Nutrients. 2022 Feb 18;14(4):855.

[31] Kok CR, Rose D, Hutkins R. Predicting personalized responses to dietary fiber interventions: Opportunities for modulation of the gut microbiome to improve health. Annual Review of Food Science and Technology. 2023;14(1):157–182.

[32] Cosier DJ, Lambert K, Neale EP, Probst Y, Charlton K. The effect of oral synbiotics on the gut microbiota and inflammatory biomarkers in healthy adults: A systematic review and meta-analysis. Nutrition Reviews. 2025 Feb;83(2):e4–24.

[33] Jiang H, Cai M, Shen B, Wang Q, Zhang T, Zhou X. Synbiotics and gut microbiota: New perspectives in the treatment of type 2 diabetes mellitus. Foods. 2022 Aug 13;11(16):2438.

[34] Lauw S, Kei N, Chan PL, Yau TK, Ma KL, Szeto CY, Lin JS, Wong SH, Cheung PC, Kwan HS. Effects of synbiotic supplementation on metabolic syndrome traits and gut microbial profile among overweight and obese Hong Kong Chinese individuals: A randomized trial. Nutrients. 2023 Oct 2;15 (19):4248.

[35] Shang Z, Pai L, Patil S. Unveiling the dynamics of gut microbial interactions: A review of dietary impact and precision nutrition in gastrointestinal health. Frontiers in Nutrition. 2024 May 30;11:1395664.

[36] Gomez Quintero DF, Kok CR, Hutkins R. The future of synbiotics: Rational formulation and design. Frontiers in Microbiology. 2022 Jul 22;13:919725.

[37] Guo L, Yang L, Wang K, Liu W, Wang S, Zhang Y, Li R, Wu Z, Chen C. Synbiotic microencapsulation of Lactiplantibacillus plantarum-lentinan for enhanced growth in broilers. AMB Express. 2025 Aug 31;15(1):128.

[38] Hu Q, Liu Y, Fei Y, Zhang J, Yin S, Zou H, Zhu F. Efficacy of probiotic, prebiotic, and synbiotics supplements in individuals with anemia: A systematic review and meta-analysis of randomized controlled trials. BMC Gastroenterology. 2024 Dec 23;24(1):472.

[39] Cicero AF, Fogacci F, Bove M, Giovannini M, Borghi C. Impact of a short-term synbiotic supplementation on metabolic syndrome and systemic inflammation in elderly patients: A randomized placebo-controlled clinical trial. European Journal of Nutrition. 2021 Mar;60(2):655–663.

[40] Liu X, Tong Y, Qin J, Zhao Y. Efficacy and safety of probiotic and synbiotic supplementation in metabolic syndrome: A systematic review and meta-analysis. Nutrition, Metabolism and Cardiovascular Diseases. 2025 Apr;25:104100.

[41] Bijle MN, Ekambaram M, Lo EC, Yiu CK. Synbiotics in caries prevention: A scoping review. PloS One. 2020 Aug 12;15(8):e0237547.

[42] Singh L, Kaur H, Bhatti R. Therapeutic Potential of Synbiotics in Management of Various Disorders. In: Dua K. (ed.) Synbiotics in Human Health: Biology to Drug Delivery. 2024, Singapore: Springer, https://doi.org/10.1007/978-981-99-5575-6_16.

[43] Markowiak P, Śliżewska K. Effects of probiotics, prebiotics, and synbiotics on human health. Nutrients. 2017 Sep;9(9):1021.

[44] Parhi P, Liu SQ, Choo WS. Synbiotics: Effects of prebiotics on the growth and viability of probiotics in food matrices. Bioactive Carbohydrates and Dietary Fibre. 2024 Dec 7;32:100462.

[45] Davani-Davari D, Negahdaripour M, Karimzadeh I, Seifan M, Mohkam M, Masoumi SJ, Berenjian A, Ghasemi Y. Prebiotics: Definition, types, sources, mechanisms, and clinical applications. Foods. 2019 Mar;8(3):92.

[46] Niu X, Zhang Q, Liu J, Zhao Y, Shang N, Li S, Liu Y, Xiong W, Sun E, Zhang Y, Zhao H. Effect of synbiotic supplementation on obesity and gut microbiota in obese adults: A double-blind randomized controlled trial. Frontiers in Nutrition. 2024 Nov 27;11:1510318.

[47] Polanía AM, García A, Londoño L. Advancement in the research and development of synbiotic products. Microbial Bioreactors in Industrial and Molecular Applications. 2023;1:55–79.

[48] Ngoc AP, Zahoor A, Kim DG, Yang SH. Using synbiotics as a therapy to protect mental health in Alzheimer's disease. Journal of Microbiology and Biotechnology. 2024 Jun 21;34(9):1739.

[49] Oti VB, Adah OF, Dzator J, Omoloye NA, Kandagor B. Unravelling the potential of synbiotics in ameliorating COVID-19 complications: A concise review. One Health Bulletin. 2024 Jun 1;4(2):47–54.

[50] Lv M, Shafagh G, Yu S. Effect of synbiotics on the cardiovascular risk factors in patients with non-alcoholic fatty liver: A GRADE assessed systematic review and meta-analysis. BMC Gastroenterology. 2025 May 26;25(1):407.

[51] Mok K, Tomtong P, Ogawa T, Nagai K, Torrungruang P, Charoensiddhi S, Nakayama J, Wanikorn B, Nitisinprasert S, Vongsangnak W, Nakphaichit M. Synbiotic-driven modulation of the gut microbiota and metabolic functions related to obesity: Insights from a human gastrointestinal model. BMC Microbiology. 2025 Apr 27;25(1):250.

[52] Chen T, Wang J, Liu Z, Gao F. Effect of supplementation with probiotics or synbiotics on cardiovascular risk factors in patients with metabolic syndrome: A systematic review and meta-analysis of randomized clinical trials. Frontiers in Endocrinology. 2024 Jan 8;14:1282699.

[53] Yassine F, Najm A, Bilen M. The role of probiotics, prebiotics, and synbiotics in the treatment of inflammatory bowel diseases: An overview of recent clinical trials. Frontiers in Systems Biology. 2025 Apr 16;5:1561047.

[54] Smolinska S, Popescu FD, Zemelka-Wiacek M. A Review of the Influence of Prebiotics, Probiotics, Synbiotics, and Postbiotics on the Human Gut Microbiome and Intestinal Integrity. Journal of Clinical Medicine. 2025 May 23;14(11):3673.

[55] Zhang Y, Hong J, Zhang Y, Gao Y, Liang L. The Effects of Synbiotics Surpass Prebiotics in Improving Inflammatory Biomarkers in Children and Adults: A Systematic Review, Meta-analysis, and Meta-evidence of Data from 5,207 Participants in 90 Randomized Controlled Trials. Pharmacological Research. 2025 Jun 26;218:107832.

[56] Salamat S, Jahan-Mihan A, Tabandeh MR, Mansoori A. Randomized clinical trial evaluating the efficacy of synbiotic supplementation on serum endotoxin and trimethylamine N-oxide levels in patients with dyslipidaemia. Archives of Medical Sciences. Atherosclerotic Diseases. 2024 Feb 1;9:e18.

[57] Zolghadrpour MA, Jowshan MR, Heidari Seyedmahalleh M, Karimpour F, Imani H, Asghari S. The effect of a new developed synbiotic yogurt consumption on metabolic syndrome components in adults with metabolic syndrome: A randomized controlled clinical trial. Nutrition & Diabetes. 2024 Dec 18;14(1):97.

[58] Rong L, D C, Jia P, Tsoi KK, Wong SH, Sung JJ. Use of probiotics, prebiotics, and synbiotics in non-alcoholic fatty liver disease: A systematic review and meta-analysis. Journal of Gastroenterology and Hepatology. 2023 Oct;38(10):1682–1694.

[59] Wang W, Liu F, Xu C, Liu Z, Ma J, Gu L, Jiang Z, Hou J. Lactobacillus plantarum 69–2 combined with galacto-oligosaccharides alleviates d-galactose-induced aging by regulating the AMPK/SIRT1 signaling pathway and gut microbiota in mice. Journal of Agricultural and Food Chemistry. 2021 Feb 10;69(9):2745–2757.

[60] Pasinetti GM. Synbiotic-derived metabolites reduce neuroinflammatory symptoms of Alzheimer's disease. Current Developments in Nutrition. 2020;4:nzaa062_035.

[61] Alam Z, Shang X, Effat K, Kanwal F, He X, Li Y, Xu C, Niu W, War AR, Zhang Y. The potential role of prebiotics, probiotics, and synbiotics in adjuvant cancer therapy especially colorectal cancer. Journal of Food Biochemistry. 2022 Oct;46(10):e14302.

Nafees Ahmed, Harsh M. Prajapati, Mohit R. Chauhan, Vikram H. Raval,
Nirmala Sehrawat, Namita Ashish Singh*

Chapter 7
Functional Foods and Management of Human Gut Microbiome: Recent Advances and Future Prospective

Abstract: Functional foods are natural/processed nutritional staples that deliver essential nutrients and offer beneficial effects on the host's health, such as disease prevention by enhancing the immune system's ability to prevent infections from pathogens. Human gut microbiome has garnered budding interest in extensive research connected to health and disease. Gut microbiota constitutes a complex ecosystem, which plays a series of fundamental functions such as nutrient absorption, immune modulation, mental health, as well as protection against pathogens. The microbial diversity of the human gut varies considerably across the digestive tract. This chapter covers the composition, factors affecting the gut microbiome, i.e., heavy metals, nanoparticles, endocrine disrupting chemicals, etc., and role of gut microbiota in body process. The chapter also highlights the major bioactive components, their mechanism of action, metabolic products and their effects in functional foods such as probiotics, prebiotics, and synbiotics in regulating the gut microbiome, and preventing illness. Current developments in functional food research, including technological advancements and individualized nutrition, as well as potential future perspective and challenges, are discussed.

Keywords: functional food, gut microbiome, disease prevention, microbial diversity

7.1 Introduction

The gut microbiota is a complex ecosystem of bacteria, fungi, viruses, and parasites that interacts with the host. With over 100 trillion bacteria, this ecosystem performs a

---

*Corresponding author: Namita Ashish Singh**, Department of Microbiology, Mohanlal Sukhadia
University, Udaipur 313001, Rajasthan, India, e-mail: namita.singh@mlsu.ac.in
Nafees Ahmed, Department of Microbiology, Mohanlal Sukhadia University, Udaipur 313001, Rajasthan,
India
Harsh M. Prajapati, Mohit R. Chauhan, Vikram H. Raval, Department of Microbiology and
Biotechnology, University School of Sciences, Gujarat University, Ahmedabad 380009, Gujarat, India
Nirmala Sehrawat, Department of Bio-Sciences and Technology, MMEC, Maharishi Markandeshwar
(Deemed to be University), Mullana-Ambala 133207, Haryana, India

https://doi.org/10.1515/9783112205150-007

crucial function in digestion, immunology, and pathogen prevention. Over the past decade, culture-independent techniques have been used to study the intestinal microbiome, including external and internal factors, "normal" and "abnormal" variations, and their relationship with health and disease [1].

This microbial network is dominated by commensal bacteria, whose genes are 100 times more numerous than those found in the human genome. These bacteria also outnumber mammalian cells by approximately ten-fold [2]. Numerous physiological functions, such as cellular metabolism, digestion, nutrition absorption, and immune system development, depend on their contributions. The biggest pool of immune cells and over 70% of the body's immune system are found in the gastrointestinal (GI) tract, making it a crucial immunological organ [3]. Tightly controlled microbiota-host interactions are essential for immune system priming, education, and development, especially for preserving T-cell homeostasis [4].

The gut microbiota plays a crucial role in human metabolism by producing enzymes which are not found in the human genome. These enzymes help break down polysaccharides, polyphenols, and synthesize vitamins [5]. Comparative studies in germ-free and conventional microbiota, or human microbiota-associated animals, as well as in vitro studies using human feces or continuous culture gut models, provide evidence for the microbiota's role in dietary metabolism and its impact on health [6].

Furthermore, observational studies comparing the fecal microbiota of healthy subjects to those of patients strongly suggest that the gut microbiota plays an important role in the etiology and/or development of a variety of GI diseases and conditions, including inflammatory bowel disease (IBD), irritable bowel syndrome (IBS), colon cancer, and antibiotic-associated diarrhea. Recent data suggests that the microbiome may possibly play a role in obesity and diabetes [7, 8].

An estimated three million genes makeup the genome of gut microbes, which is 150 times greater than that of humans [4]. The gut microbiota can affect systemic processes, as well as local gut health because of its wide genetic repertoire. Numerous illnesses, such as obesity, depression, chronic ileal inflammation, liver disorders, and cardiovascular diseases (CVDs), have been linked to dysbiosis, or an imbalance in the microbial makeup [9]. Additionally, the involvement of the gut microbiota in disease etiology and its potential for therapeutic regulation have been proven via fecal transfer trials.

The construction of the gut microbiota is quite dynamic and affected by a number of variables, including illness conditions, age, diet, and changes in lifestyle. It has been referred to as an "acquired organ" because of its capacity to adjust to host and environmental needs. Asthma, allergies, IBD, and viral diseases have all been associated with changes in its composition [10]. Furthermore, new research indicates that metabolic disorders such obesity, diabetes mellitus (DM), metabolic syndrome, and cardiovascular disease may be exacerbated by dysbiosis [11].

In order to shape the gut microbiota and create physiological states that affect disease-risk later in life, early nutrition is essential. Early nutritional interventions can have a lasting impact on metabolic processes and health outcomes, according to experimental models. Public health nutrition initiatives that seek to lower long-term disease risks by implementing early dietary changes have been influenced by these findings [12].

In diseases like prediabetes and type II diabetic mellitus (T2DM), the connection between nutrition and metabolic health is more noticeable. About 37% of adults in the US alone suffer from prediabetes, a condition marked by consistently high blood glucose levels. Up to 70% of people with prediabetes develop T2DM if they don't receive treatment [13]. Other elements of the metabolic syndrome, such as obesity, hypertension, cardiovascular disease, hypertriglyceridemia, and nonalcoholic fatty liver disease (NAFLD), are intimately linked to this disorder. Therefore, controlling blood glucose levels and halting the progression of disease depend heavily on dietary decisions that control postprandial glycemic reactions [14].

Probiotic research continues to improve, with new discoveries and technologies emerging throughout time. The potential health advantages of probiotics are continually being investigated as the field evolves. This chapter covers the composition, roles, and determinants of the gut microbiome. It also highlighted the role of functional foods, such as probiotics, prebiotics, synbiotics, and their bioactive substances, in regulating the gut microbiome and averting illness. Current developments in functional food research are discussed, including technological advancements and individualized nutrition, as well as potential future directions and obstacles in gut health optimization.

7.2 The Human Gut Microbiome: Structure and Function

Microbial cell populations attain their peak density in the intestinal compartment, where they collectively constitute a complex microbial community referred to as the gut microbiota [15]. It is believed that there are more than 10^4 different kinds of microorganisms living in the human intestines [16]. Firmicutes, Bacteroidetes, Actinobacteria, Probacteria, Fusobacteria and Verrucomicrobia are the most common gut microbial phyla, with Firmicutes and Bacteroidetes accounting for 90% of the microbiota [17]. The Firmicutes phylum is made up of about 200 species, including *Lactobacillus, Bacillus, Clostridium, Enterococcus*, and *Ruminococcus. Clostridium* genera makeup 95% of the Firmicutes phylum. Bacteroidetes includes prominent genera like *Bacteroides* and *Prevotella*. The Actinobacteria phylum is proportionally less abundant with the *Bifidobacterium* genus accounting for the majority [1]. There is no single ideal gut microbiota composition because it varies for everyone.

7.2.1 Composition of Gut Microbiota

The microbial diversity of the human gut varying considerably across the digestive tract. Although the stomach and small intestine host relatively few bacterial species, the colon is highly inhabited with up to 10^{12} cells per gram of intestinal material, representing between 300 and 1,000 species. Bacteria account for up to 60% of the feces' dry bulk. In addition to bacteria, fungi, protists, archaea, and viruses exist, yet their roles are poorly understood [18].

7.2.1.1 Spatial Variation

Microbial composition changes greatly throughout the GI system. Figure 7.1 depicts the distribution of prominent bacterial species in several parts of the human digestive tract.

a. Oral cavity
The oral cavity is the opening to the GI tract, where food enters and mixes with saliva. It includes several microenvironments, such as tonsils, teeth, gums, tongue, cheeks, and hard and soft palates. According to Human Oral Microbiome Database (HOMD), the oral cavity harbors more than more than 700 oral bacterial species [19]. Among them six major phyla dominate: Firmicutes, Bacteroidetes, Proteobacteria, Actinobacteria, Spirochaetes, and Fusobacteria. In healthy individuals' saliva, the most common genera are *Gemella, Veillonella, Neisseria, Fusobacterium, Streptococcus, Prevotella, Pseudomonas*, and *Actinomyces* [20].

b. Throat and distal esophagus
These regions have similar kind of microbial communities dominated by *Streptococcus, Prevotella, Actinomyces, Gemella, Rothia, Granulicatella, Haemophilus*, and *Veillonella. Streptococcus* is the predominant genus [21].

c. Stomach
Helicobacter pylori determines microbial diversity in the stomach. *Streptococcus, Actinomyces, Prevotella*, and *Gemella* are prevalent in *H. pylori* – negative stomachs, implying that they are transitory inhabitants swallowed from throat. *H. pylori*, when present, dominates the bacterial phyla, showing that it has adapted to the stomach environment [22].

d. Small intestine
Enterococci, *Escherichia coli, Klebsiella*, Lactobacilli, Staphylococci, and Streptococci are frequently identified. Aerobic and facultative anaerobic microorganisms are more common in this location. The small intestine contains aerobic *Enterococcus* groups,

lactobacilli, streptococci, and Gammaproteobacteria, but the large intestine is dominated by anaerobes [23].

e.   Cecum and recto-sigmoidal colon

The microbiota present in cecal is more complicated than those in jejunum and ileum. It comprises of key fecal bacteria groups such as *Clostridium leptum, C. coccoides,* and *Bacteroides, Lactobacillus, Enterococcus,* and *E. coli* make up most of the cecal microflora, with facultative aerobes taking the lead. In contrast, the recto-sigmoidal colon has a separate and more diversified microbial community, dominated by stringent anaerobes such as *Bacteroides, C. coccoides,* and *C. leptum.* Individual microbial composition varies, as does that of healthy and old populations. Gammaproteobacteria is abundant in both group's fecal microbiota, while it is uncommon in healthy people's feces. This phylum is also present in the jejunum, ileum, and cecum, usually in smaller numbers [23].

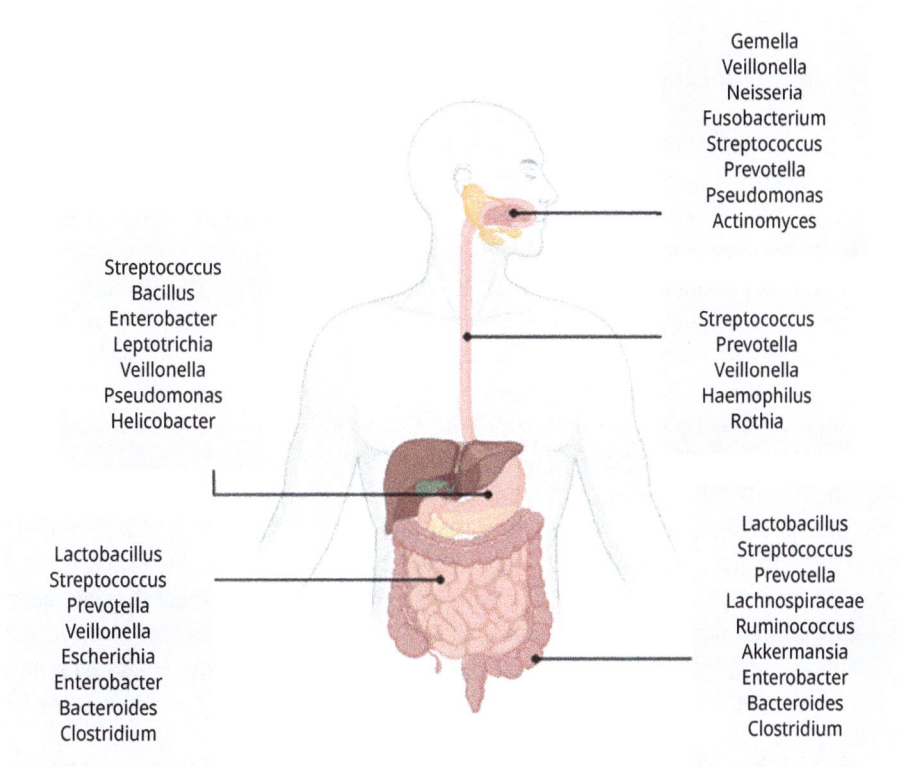

Gemella
Veillonella
Neisseria
Fusobacterium
Streptococcus
Prevotella
Pseudomonas
Actinomyces

Streptococcus
Bacillus
Enterobacter
Leptotrichia
Veillonella
Pseudomonas
Helicobacter

Streptococcus
Prevotella
Veillonella
Haemophilus
Rothia

Lactobacillus
Streptococcus
Prevotella
Veillonella
Escherichia
Enterobacter
Bacteroides
Clostridium

Lactobacillus
Streptococcus
Prevotella
Lachnospiraceae
Ruminococcus
Akkermansia
Enterobacter
Bacteroides
Clostridium

Figure 7.1: Variability of microbial genera in the human gastrointestinal (GI) tract (predominant bacterial genera in oral cavity, esophagus, stomach, small intestine, and colon are delineated in this figure).

Firmicutes, Bacteroidetes, Actinobacteria, and Proteobacteria are the chief bacterial phyla in human gut and the most common bacterial genera are *Bacteroides, Clostridium, Peptococcus, Bifidobacterium, Eubacterium, Ruminococcus, Faecalibacterium,* and *Peptostreptococcus* [1]. Among them, *Bacteroides* is the most common genus accounting for over 30% of gut bacteria and playing an important role in host physiology [18].

7.2.2 Factors Influencing Gut Microbiota

The gut microbiota is the aggregate term for the billions of microorganisms that inhabit the digestive system. This vast microbial community, which makes up around 10^{12} colony-forming units/mL in the colon, affects not only the function of distant organs but also the GI physiology and the host's vulnerability to disease [24]. Despite having trillions of microorganisms, the adult gut microbiota is primarily dominated by four phyla: Firmicutes, Bacteroidetes, Actinobacteria, and Proteobacteria [11].

7.2.2.1 Impact of Heavy Metals' on the Gut Microbiota

The gut microbiota has been demonstrated to be significantly impacted by heavy metal exposures thus far. Following 4 weeks of exposure to 10 ppm arsenic in the drinking water, the gut microbiota and metabolome profiles of 6-week-old female C57Bl/6 mice were assessed. Several gut flora modifications brought on by arsenic exposure caused treated versus control people to cluster differently in terms of β-diversity. Additionally, metabolomic research showed that a number of metabolites were changed after exposure to arsenic, and these changes were linked to changes in the gut microbiota. For example, this group's decreased levels of indole lactic acid were negatively correlated with Clostridiales Family XIII Incertae Sedis but positively correlated with alterations in Erysipelotrichaceae [25].

The Flint Drinking Water Crisis has brought lead (Pb), another heavy metal, back into the spotlight. Another study found that the chemical changed the phylogenetic diversity and gut microbiome trajectory of adult C57Bl/6 female mice exposed to 10 ppm PbCl2 in drinking water for 13 weeks at a concentration of about 2 mg/kg body weight/day. Fecal samples taken at 4 and 13 weeks after exposure also showed gut metabolic disruptions. When wild populations are exposed to heavy metal pollution on a regular basis in both aquatic and terrestrial settings, the gut microbiota may be affected [25].

Heavy metals were tested by comparing the profiles of two populations of Mongolian toads, one of which lived in an area that was heavily contaminated by heavy metals and the other of which lived in an area that was comparatively unpolluted. Bacteroidetes were overrepresented in the polluted area, while Tenericutes were overrepresented in the unpolluted area. The proportion of beneficial bacteria in the gut microbiome and the Firmicutes/Bacteroidetes ratio were lower in polluted toads than in un-polluted toads. Toads exposed to heavy metal contamination also showed decreased species diversity and OTU (other taxonomic unit) proportion [25].

7.2.2.2 Effect of Particulate Matter

Wild type (WT)129/SvEv mice were tested to ascertain the effects of particulate matter (PM), a major pollutant in ambient air, on the intestines and gut flora. In order to evaluate the long-term effects of exposure, IL10 deficient (−/−) mice were given the same treatment for 35 days WT after being exposed to urban PM10 (EHC-93: 18 µg/g/day) for 7 or 14 days. Mice exposed to PM10 for a brief period of time showed changes in immune gene expression, increased production of pro-inflammatory cytokines into the small intestine, increased permeability (leakiness) of the gut, and splenocytes that were hyporesponsive to the PM. Pro-inflammatory cytokine expression was elevated in the colon of IL10$^{-/-}$ mice exposed for an extended period of time, and the relative abundances of Bacteroidetes spp., Firmicutes spp., and Verrucomicrobia spp. were significantly altered in these animals [25].

7.2.2.3 Impact of Nanoparticles on the Gut Microbiota

One study investigated the effects of 48 h of exposure to silver nanoparticles (AgNPs) (25, 100, and 200 mg/L) on a defined bacterial community created from a healthy human donor in order to ascertain whether AgNPs may alter the composition of the human gut microbiome. Based on changes in fatty acid methyl ester profiles and gas generation, their results show that these particles caused a shift to more pathogenic bacterial species. Additionally, after being exposed to different concentrations of AgNPs, the bacterial communities of *Bacteroides ovatus*, *Roseburia faecalis*, *Eubacterium rectale*, *Roseburia intestinalis*, and *Ruminococcus torques* drastically decreased, while *Raoultella* spp. and *E. coli* increased [25].

The mortality rate for redworms increased by 35% when zinc nanoparticles (ZnNPs) at a concentration of 1,000 mg/kg were added to the substrate soil. The worm's gut microbiome was similarly impacted by ZnNP exposure, as evidenced by a decrease in α-diversity (303 OTUs in controls vs to 78 OTUs in treated individuals). While ZnNP-treated worms showed a decrease in Firmicutes, exposed individuals showed overgrowths of Proteobacteria, mainly as a result of increases in *Verminephrobacter* spp. and *Ochrobactrum* spp. [25].

7.2.2.4 Endocrine-Disrupting Chemical's Impact on the Gut Microbiome

The gut microbiome of canines, zebrafish, and rodent models can be impacted by adult and developmental exposure to bisphenol A (BPA), estradiol (E2), or ethinyl estradiol (EE, estrogen in birth control pills). The intestinal flora was restructured, with a significant increase in the CKC4 phylum, and the hepatic expression of vitellogenin, a biomarker of estrogen exposure in male fish, increased after 5 weeks of exposure to BPA (200 or 2,000 µg/L) or E2 (500 ng/L or 2,000 ng/L) in adult male zebrafish [25].

Microbiome changes were observed by adolescence in Sprague-Dawley female rats that were chronically exposed from birth to adulthood to diethyl phthalate (DEP: 0.1735 mg/kg bodyweight), methylparaben (MPB: 0.1050 mg/kg bodyweight), triclosan (TCS: 0.05 mg/kg body weight), or a combination of these three chemicals. However, many of these changes were less pronounced by adulthood. Adolescent alterations included a decrease in Firmicutes (bacilli spp.) and an increase in Bacteroidetes (*Prevotella* spp.) relative abundance across all treatment groups. Adolescents who received DEP or MPB treatment had lower body weights [25].

To date, studies have looked at how the EDCs BPA and phthalates affect the gut microbiota in zebrafish, California mice, CD1 mice, dogs, and Sprague-Dawley rats. The combined research shows some gut microbes that are impacted across species, even though the bacteria that are impacted vary from study to study. BPA-exposed CD1 mice and rats exposed to DEP, MPB, TCS, or a combination of these chemicals showed higher levels of Firmicutes. California mice and canines exposed to higher amounts of BPA had higher relative abundances of *Bacteroides* species [25].

7.2.2.5 Ingestion of Microplastics

Microplastics are present everywhere. In addition to acting as long-distance carriers of possible pathogens, they have been demonstrated to cause deadly malnutrition, carry chemicals that can disrupt hormone balance and organismal biochemistry, and draw in environmental chemicals that may be momentarily and partially concealed like a group of soldiers with various weapons inside a Trojan horse [26].

Exposure to microplastics changed gut microbial beta diversity, resulting in microbial communities that were different from the controls that were not treated with microplastics. This was the most often seen effect. Additionally, notable changes in the relative abundances of several phyla – often Bacteroidetes, Firmicutes, and Proteobacteria – that are frequently found in gut microbiome investigations were noted [26].

In what ways might microplastics affect the gut microbiota? We believe that the gut microbiome may be altered by the ingestion of microplastics because it is associated with a variety of factors that are known to cause gut dysbiosis, including inflammation, malnutrition, the introduction of pathogens, endocrine disruptors, and environmental chemicals. consists of alterations of alpha and beta diversity, an increase in potential pathogenic bacteria, a decline in bacteria characteristic of a healthy gut microbial community, and the development of negative interactions between possible pathogens and vital gut microorganisms, which makeup the core microbiome. Because microplastics are more likely to carry plasmids on their surface than in open seas, they have been identified as possible hot sites where bacteria may horizontally transmit antibiotic resistance genes [26].

7.2.3 Role of Gut Microbiota in Various Body Processes

Various types of microbial community residing in the human GI track that plays an important role in maintaining overall health by taking part in various physiological functions, such as digestion, immunity, metabolism, and neurological processes [27]. The gut microbiota, which takes nutrients from host dietary components and maintain normal tissue turnover by shedding of epithelial cells. Shed epithelial cells is an organ by itself with significant functional plasticity and broad metabolic capacity. These features of the gut microbiome have been quickly moving focus from research from the diversity and abundance of the microbial members to the functional aspects [28]. This section gives a quick summary of the main role of the normal gut microbiota in various body processes **(Figure 7.2)**.

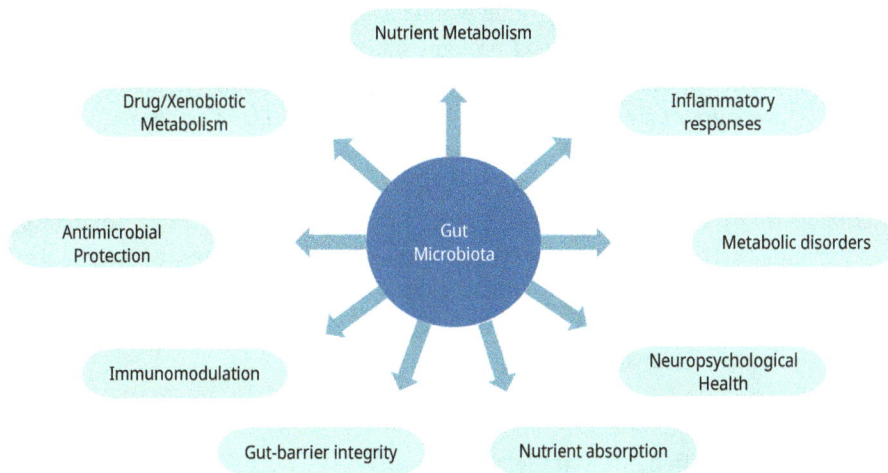

Figure 7.2: Role of gut microbiota in various physiological processes in human body.

7.2.3.1 Nutrient Metabolism

By improving the digestion and fermentation of dietary fibers, proteins, and peptides to produce energy and preserve gut health, the gut microbiota plays a critical role in nutrition metabolism. The metabolic activities of gut bacteria, particularly those belonging to the genera *Bacteroides*, *Roseburia,* and *Bifidobacterium*, are now being studied [5]. These bacteria were first identified for their function in storing undigested food. They ferment carbohydrates that evade proximal digestion to create short-chain fatty acids (SCFAs), such as butyrate, propionate, and acetate [18]. Not only do these SCFAs provide the host with energy, but they also control energy balance through G-protein-coupled receptors and affect hormone responses. Notably, butyrate helps to minimize the accumulation of toxic metabolic waste [29]. Furthermore, the biotransformation of dietary polyphenols, the production of vitamins K and B, and the metabolism of fats and proteins are all greatly influenced by the gut microbiota, which in turn affects general metabolic health [18, 30]. Maintaining a healthy microbiota is essential for efficient nutrition and general health because of the critical roles that gut bacteria and their metabolic products play in maximizing nutrient absorption and averting dietary-related diseases.

Along with the production of vital vitamins K and B and the biotransformation of dietary polyphenols, the gut microbiota has a major impact on the metabolism of fats and proteins. *Bacteroides thetaiotaomicron* and other gut bacteria specifically improve lipid digestion by increasing the synthesis of colipase, which facilitates lipid breakdown and improves fat absorption while fostering energy balance [31]. Different microbial enzymes break down food proteins into amino acids, which are then converted into bioactive compounds including histamine and gamma-aminobutyric acid (GABA), which support a number of physiological functions [32]. Furthermore, important vitamins, such as vitamin K and a number of B vitamins, which are essential for many metabolic processes, are synthesized by gut microorganisms. Additionally, the microbiota helps dietary polyphenols, like flavonoids, undergo biotransformation by transforming their glycosylated forms into their active metabolites, which the host may use [33]. For the maintenance of general health and metabolic equilibrium, the gut microbiota plays a complex role in lipid and protein metabolism, vitamin production, and polyphenol transformation.

7.2.3.2 Drug and Xenobiotic Metabolism

Throughout the human body, gut microbiota plays a very important role in various physiological processes, significantly affecting drug and xenobiotic metabolism. Drugs are among the many substances that have been demonstrated to be metabolized by these gut microbes. For example, p-cresol is a metabolite that is produced by gut bacteria; it can inhibit the liver's efficiency to metabolize acetaminophen by interfering with hepatic sulfotransferases [34]. Similarly, by metabolic change that results in inactivation, bacteria like *Eggerthella lenta* can affect the effectiveness of cardiac glycosides like digoxin [28]. Furthermore, the anticancer medication irinotecan can be deconjugated by microbial β-glucuronidase, which increases its toxicity by causing inflammation and diarrhea [35].

7.2.3.3 Antimicrobial Protection

The gut microbiota plays a crucial role in preserving antimicrobial protection by avoiding pathogenic overgrowth and regulating tolerance to advantageous commensals. In the large intestine, a two-tiered mucus layer made of mucin glycoproteins produced by goblet cells serves as a barrier. The outside layer supplies nutrition for commensals, while the denser inner layer keeps microorganisms from coming into contact [28]. Paneth cells generate antimicrobial proteins (AMPs), such as cathelicidins and defensins, in the small intestine, where the mucus layer is less efficient [36]. Microbial components trigger this production through pattern recognition receptors (PRRs). While bacterial metabolites like SCFAs encourage the development of antimicrobial peptides, the interaction of certain gut microbiota, particularly *Bacteroides thetaiotaomicron* and *Lactobacillus innocua*, increases AMP synthesis and activity [28]. Furthermore, gut microorganisms encourage intestinal dendritic cells to produce secretory IgA (sIgA), which covers the microbiota and inhibits translocation into the circulation, hence limiting systemic immune responses and preserving immunological homeostasis [37].

7.2.3.4 Immunomodulation

By interacting with both the innate and adaptive immune systems, especially through elements of the gut-associated lymphoid tissues (GALT) and different immune cell types, such as effector and regulatory T cells, IgA-producing plasma cells, innate lymphoid cells (ILCs), resident macrophages, and dendritic cells, the gut microbiota plays a critical role in the immunomodulation of the gut [38]. Changes in IgE^+ B-cell numbers and Th2 dominance in effector T-cell responses are two indicators that microbiota dysregulation might hinder the development of GALT [28]. Through processes that may include PRRs and metabolites like SCFAs, which epigenetically control the Foxp3 locus, microbiota affect the formation and function of $Foxp3^+$ regulatory T cells [39]. ILCs react to gut-derived cytokines in a way that is dependent on the microbiota, whereas dendritic cells use MyD88 signaling to induce sIgA synthesis by mucosal plasma cells. MyD88-dependent processes are used by resident macrophages in the lamina propria to generate cytokines such as IL-1β, which are essential for immunological responses [40]. Gut immunity is also influenced by other elements, such as intestinal alkaline phosphatase and the regulation of neutrophil recruitment, highlighting the intricate relationship between the immune system and gut bacteria [28].

7.2.3.5 Gut Barrier Integrity

The integrity of the gut barrier depends on the gut microbiota, as well as the immune system. The expression of proteins that preserve tight connections between epithelial cells and prevent dangerous chemicals from penetrating the gut lining is stimulated by microbial interactions [41]. Notably, it has been demonstrated that *Bacteroides thetaiotaomicron* promotes the expression of proteins required for desmosome maintenance, hence, improving epithelial health [42]. Microbial fermentation products, especially SCFAs, are essential for sustaining the gut mucosal barrier and meeting the energy requirements of intestinal epithelial cells [43].

7.2.3.6 Nutrient Absorption

The gut flora also affects how nutrients are absorbed. There are numerous mechanisms behind the intricate link between gut microbiota and bones, including the control of dietary nutrient absorption (e.g., Ca and P), the movement of microbial products (such lipopolysaccharide (LPS) and SCFAs) across the gut endothelium, and immune system modulation [44]. Through SCFAs, the gut microbiota has been shown to have a major influence on bone formation and development. Increased absorption of calcium has been generated when the gut microbiota ferments SCFAs [45]. Calcium is absorbed and deposited as calcium hydroxyapatite in teeth and bones, providing strength to tissues [18].

7.2.3.7 Neuropsychological Health

The gut-brain axis provides an example of how gut microbes might impact the central nervous system in relation to cognitive health. Because of this reciprocal link, gut microorganisms can interact with the brain to influence mood, behavior, and cognitive processes [18]. According to research, SCFAs help maintain the blood-brain barrier's integrity, which prevents potentially dangerous substances from entering the brain tissue [18]. Additionally, microbial metabolites can affect neurological functioning by releasing pro-inflammatory cytokines such as TNF-α, IL-6, MCP-1, and IL-8 [46].

2.3.8 Metabolic Disorders

Apart from their functions in cognitive and metabolic health, gut microbiota is also linked to the emergence of metabolic diseases as type 2 diabetes (T2D). Impaired insulin signaling and elevated inflammation linked to metabolic syndrome have been linked to alterations in the gut microbiota's composition, including imbalances in bacterial populations. Furthermore, because gut microbes regulate metabolism and digestion, they can affect body weight and fat distribution [18]. For example, research has demonstrated that recipients with metabolic problems can benefit from increased insulin sensitivity when lean people's microbiota is transferred [47].

7.2.3.9 Inflammatory Responses

By maintaining and competing with pathogenic species, they offer a natural defense against them. The microbes living in the gut can either cause or stop inflammation through their interactions with the immune system. Their association with anti-inflammatory mechanisms may involve inducing immune system regulatory cells to suppress inflammation [48]. However, some types of bacteria can encourage a "leaky gut," in which metabolites linked to the microorganisms escape the gut and enter the bloodstream, because they control the permeability of the intestines. Consequently, the body initiates an inflammatory response by producing cytokines and other mediators [49]. Likewise, the gut's epithelial cells transfer bacterial metabolites to immune cells, causing inflammation at the local and systemic levels. The continuation of this situation may result in subacute or chronic inflammation, which can lead to the development of disorders including IBD, diabetes, or cardiovascular disease [48].

Understanding the multiple relationships between gut microbiota and these processes can result in the novel therapeutic strategies aimed at harboring the microbiota to improve health outcomes and prevent diseases. Maintaining a diverse and balanced microbiome is important for supporting overall health and well-being, highlighting the importance of dietary and lifestyle interventions to promote a healthy gut environment.

7.2.4 Microbiome Dysbiosis

Gut dysbiosis, also known as gut microbial dysbiosis, is the term used to describe pathological changes in the quantity and composition of gut microorganisms that result in altered gut-immune and neuroimmune status [3]. The word dysbiosis was initially used more than a century ago by Nobel Prize winner Elie Metchnikoff to characterize disturbed symbiosis and to propose that introducing the good bacteria present in yogurt intestinal dysbiosis could improve human health [50]. Numerous definitions attempted to connect dysbiosis to illness, such as "any change in the composition of resident commensal communities relative to the community found in healthy individuals" [51]. More nuanced assessments were reached by scholars, though, when they examined the human gut microbiota from the perspective of contemporary ecological theory. According to these authors' radical suggestion, a dysbiotic microbiota configuration should meet the requirements of Koch's postulates for defining a disease-causing microbial agent, which are now expanded to include a disease-causing microbiome [51].

The billions of microorganisms that build the human gut microbiota are primarily bacterial and viral in origin and are thought to be nonpathogenic [52]. To prevent pathogen colonization and invasion, the microbiota works in concert with the host's defenses and immune system. It also carries out a vital metabolic function, serving as a source of vital vitamins and nutrients and facilitating the process of obtaining energy and nutrients from food, including amino acids and SCFAs. Since the host ultimately depends on the gut microbiota for several essential processes, the microbiota may enhance health [52]. In addition to promoting intestinal cell regeneration and the synthesis of mucin and other metabolites, including as bile acids, ethanol, acetaldehyde, acetate, and other SCFAs, gut bacteria also preserve the integrity of the intestinal epithelial barrier [3].

In their 2017 study, Levy et al. identified three distinct types of dysbiosis: "bloom of pathobionts, loss of commensals, or loss of diversity." In their 2015 study, Vangay et al. identified four types of dysbiosis: "loss of keystone taxa, loss of diversity, shifts in metabolic capacity, or blooms of pathogens" [51]. A species is considered a keystone if, in relation to its abundance, it has a disproportionately large impact on its natural environment. Numerous definitions attempted to connect dysbiosis to illness, such as "any change in the composition of resident commensal communities relative to the community found in healthy individuals." Here, the term "pathobiont" refers to commensal microbiota members that possess the capacity to induce pathology [51].

To sustain intestinal immunity and overall body homeostasis, the gut microbiota must maintain a delicate equilibrium; any disturbance of this balance could have disastrous pathophysiological repercussions. Dysbiosis is the general term for an imbalance in the gut flora [11]. Even though the process of characterizing a healthy microbiota is still in its early stages, the F/B ratio – a measure of the proportion of the microbial communities Firmicutes (F) and Bacteroidetes (B) – may be utilized as a diagnostic for pathological diseases, according to growing data [11]. More than 90% of the human gut microbiota is made up of the two major bacterial phyla, Firmicutes and Bacteroidetes, along with additional subdominant taxa, such as Proteobacteria, Actinobacteria, and Verrucomicrobia [53].

Advances in high-throughput sequencing have fueled the development of culture-independent techniques for characterizing complex microbial communities, which forms the foundation of contemporary microbiome research. Early research was centered on the topic of what a typical species constitutes the human gut microbiota looks like, as this technique first produced catalogs of microbial species names. Identification of core species shared by the human colonic microbiota is the first step in identifying a "healthy" microbial community [10]. Because the human colonic microbiota lacks core species, it is challenging to characterize gut homeostasis using microbiota composition analysis. Some speculated that a healthy human microbiome might not even exist as a result of this deadlock [10].

Host-specific factors like genetic background, health status (infections, inflammation), and lifestyle choices can promote dysbiosis. However, environmental factors like nutrition (high sugar, low fiber), xenobiotics (antibiotics, medications, and food additives), and cleanliness are more significant [54]. The process by which gut bacteria move from the intestinal lumen into extraintestinal locations is known as "leaky gut" or bacterial translocation, and it is brought on by several of these dysbiosis causes [2].

7.2.4.1 The Diversity Issue

Numerous definitions of dysbiosis have included loss of diversity as a criterion, implying a reduction in the microbiota's "health" and a microbial service to the host [51]. The gut microbiota is unquestionably characterized by a high species diversity. Increased collaboration within such a complex community is anticipated to improve overall metabolic performance; nevertheless, the host must trade this for a reduction in ecological stability, as a large number of species tends to destabilize the system [51].

Scientists warned against using the diversity argument naively. Bottle-fed babies, for instance, consistently displayed a more varied gut microbiota than breastfed babies, but this difference did not necessarily translate into bottle-fed babies being healthier than breastfed babies [51]. Low-diversity dysbiosis can be caused by a variety of causes that vary depending on the condition. Broad-spectrum antibiotic use is one documented factor contributing to low-diversity dysbiosis. People who experience recurrent *Clostridioides difficle* infection(CDI) frequently exhibit low-diversity dysbiosis-like microbiome conditions, and CDI is closely linked to exposure to antibiotics [55]. Supportive therapy, in the event of severe diarrheal illness, or more complicated procedures like liver transplantation can all be used to correct low-diversity dysbiosis. Fecal microbiota transplant (FMT) was initially used to treat severe pseudomembranous colitis in individuals with CDI from healthy donors in the 1950s [56].

7.2.4.2 Environmental and Physical Factors

The kind of delivery (vaginal or caesarean) has a significant impact on the species that initially colonizes the gut microbiome of newborns, and it is one of the most significant factors influencing the formation of the newborn gut microbiota. According to research, cesarean births can change the type and diversity of the first colonizing bacteria, which can have a detrimental effect on the development of the infant's gut microbiome and immune system. Infants born via caesarean section are also more susceptible to immune and allergy disorders, specifically IBD, asthma, arthritis, and immunological deficits. This might be the consequence of intestinal dysbiosis brought on by cesarean deliveries [57].

7.2.4.3 Antibiotics Use

Antibiotics have many advantages and are among the most often recommended drugs. Their usage has, however, also been connected to both immediate and long-term health risks, including an elevated risk of asthma and autoimmune illnesses. Antibiotics' disruption of the gut bacteria's normal assembly and decreased variety may be connected to the negative health effects. Following antibiotic usage, CDI is one of the most prevalent infections in all age groups, accounting for 15–25% of antibiotic-associated diarrhea and most cases of antibiotic-associated colitis [57].

7.2.4.4 Eubiosis

The phrase "dysbiosis" is a quite elusive concept, which is the opposite of the similarly poorly defined term "eubiosis." Beyond having a "balanced" microbiota or one that is present in healthy individuals, no exact definition of eubiosis has been provided. It's crucial for case-control studies that the eubiotic microbiota be clearly defined [51].

It would be sufficient to use microbiome data from a limited number of control subjects if the healthy eubiotic microbiota was clearly characterized both linguistically and by its microbial composition. This effect is suggested by the "Anna Karenina principle," which holds that the composition of the microbial community varies more among dysbiotics than among healthy individuals. Leo Tolstoy's book, which begins with the line, "All happy families look alike, each unhappy family is unhappy in its own way," is the source of this principle's odd moniker [51].

7.2.4.5 Future Approaches: Restoration of the Intestinal Microbiota Through Bacteriotherapy

Determining what a "healthy" microbiome is throughout life is a crucial prerequisite for bacteria-based therapy, or bacteriotherapy. This can be done differently for each individual and for the group. The variety of microbial genes (microbiome), the species and strain diversity in the GI tract, and their role in the GI tract from birth to death all require further investigation [52].

Probiotic-based therapeutic methods have been employed for centuries with varying degrees of effectiveness and has the more extreme and unsanitary method of fecal transplantation-based wholesale microbiota replacement procedures. Understanding their molecular mechanisms of action and the unique host characteristics that necessitate a customized approach will be essential to the development and application of these and other more sophisticated methods utilizing chemically defined bacterial products in the clinic. This will allow bacterial/probiotic therapies to reach their maximum potential in the management and treatment of human health [52].

7.3 Functional Foods and Their Influence on Gut Microbiome

7.3.1 Definition and Classification of Functional Foods

Functional foods confer health benefits beyond basic nourishment attributable to physiologically active components [58]. The European Commission's Concerted Action on Functional Food Science in Europe explains a functional food as one that "beneficially impacts one or more target functions in the body, in a way related to enhanced health and well-being and/or lessening of disease risk" [59] listed in Table 7.1.

Functional foods can be stratified as:

a. **Natural Functional Foods**: Foods which naturally contains bioactive components that are beneficial to health, for instance polyphenols rich berries or omega-3 fatty acids in fatty fish [60].
b. **Modified Conventional Foods**: Traditional foods improved or altered to confer added benefits:
    - Foods having additional beneficial components (e.g., vitamin D-fortified milk)
    - Foods having abridged detrimental components (e.g., low-sodium products)
    - Foods having augmented naturally occurring components (e.g., omega-3 fatty acid enriched eggs) [61].

c. **Synthesized Functional Constituents**: Recently advanced food ingredients that have discrete health benefits, involving novel prebiotic complexes and plant sterols for management of cholesterol [62].

d. **Processed Functional Foods**: Products articulated explicitly for functional benefits, such as fermented foods intended to deliver definite probiotic strains of microbes [63].

In the framework of gut microbiome accentuation, functional foods can directly impact microbial populations and events inside the GI tract, which denotes a promising tactic to anticipatory healthcare and management of disease [64].

7.3.2 Major Bioactive Components in Functional Foods

Some bioactive compounds, their mechanism of action, metabolic products, and their effects are listed in Table 7.2.

7.3.2.1 Probiotics

Probiotics are living microorganisms, which confer a health benefit on the host when administered in adequate amounts to the host [72]. The genera *Lactobacillus*, *Bifidobacterium*, *Streptococcus*, and some nonpathogenic strains of *Escherichia*, along with some species of yeast such as *Saccharomyces boulardii* are some of the most frequently used probiotic microbes [16].

Current research accentuates intricated interactions among probiotics, immunity of host, and the barrier function. Some *Lactobacillus* strains upregulate tight junction proteins, augmenting the integrity of epithelial barrier [83]. The function of dendritic cells is modulated by some specific strains of bacteria, which prompts the balance among pro-inflammatory and regulatory T-cell responses [84]. The effectiveness of probiotics is markedly strain-specific, with different strains displaying distinct physiological impacts even within the same species. For instance, while *Lactobacillus rhamnosus* GG may shine at preventing antibiotic-associated diarrhea, *L. rhamnosus* GR-1 exhibits particular effectiveness against urogenital infections [85].

7.3.2.2 Prebiotics

Prebiotics are "selectively fermented constituents resulting in particular variations in the composition and activity of the microbiota of GI tract, thus conferring advantage upon health of the host" [78]. In contrast to probiotics, prebiotics act as selective substrates encouraging the growth or commotion of indigenous beneficial microbes. Con-

Table 7.1: Classification and characteristics of functional foods.

Functional food type	Subdivision	Main examples	Chief active compounds	Health benefits	References
Natural functional foods	Omega-3-rich fish	Salmon, mackerel, and sardines	EPA and DHA	Abridged inflammation and enhanced metabolic health	[65, 66]
	Polyphenol-rich fruits	Blueberries, cranberries, and pomegranates	Anthocyanins, proanthocyanidins, and ellagitannins	Anti-inflammatory, neuroprotective, and cardiovascular health	[67, 68]
	Cruciferous vegetables	Broccoli, cauliflower, and kale	Glucosinolates and sulforaphane	Detoxification and cancer prevention	[69, 70]
Modified functional foods	Plant-based alternatives	Fermented soy products and enriched plant milks	Isoflavones, added cultures, and plant fibers	Hormone balance and protein adequacy	[71, 72]
	Fortified dairy products	Probiotic yogurt and kefir with added prebiotics	Live cultures + FOS/GOS	Enhanced digestive health and immune function	[73, 74]
	Enriched cereals	Whole grain cereals with added resistant starch	β-Glucan, resistant starch, and added probiotics	Metabolic health and satiety enhancement	[75, 76]
Processed functional foods	Synbiotic formulations	Commercial probiotic supplements with prebiotics	Multi-strain probiotics and selective substrates	Personalized microbiome modulation	[77, 78]
	Novel protein foods	Insect-based products and algae supplements	Novel proteins, chitin, and omega-3 from algae	Sustainable nutrition and novel health benefits	[79, 80]
	Fermented beverages	Kombucha, water kefir, and kvass	SCOBY cultures, organic acids, and bioactive peptides	Digestive health and antioxidant activity	[81, 82]

Table 7.2: Bioactive compounds in functional foods.

Bioactive compound	Main source	Mechanism of action	Target microbes	Metabolic products	Clinical outcomes	References
Probiotics						
Bifidobacterium longum	Infant formulas and fermented foods	Immune modulation and barrier function enhancement	Cross-feeding with other beneficial bacteria	SCFAs and immunomodulatory compounds	Improved atopic dermatitis in infants	[100, 101]
Lactobacillus rhamnosus GG	Fermented dairy and supplements	Competitive exclusion and antimicrobial peptide production	Pathogenic *E. coli* and *Salmonella* spp.	Lactic acid and bacteriocins	Reduced antibiotic-associated diarrhea	[102, 103]
Akkermansia muciniphila	Naturally present and emerging supplements	Mucin degradation and barrier strengthening	Supports mucus layer regeneration	Propionate, acetate, and mucin-derived peptides	Improved glucose metabolism and reduced inflammation	[104, 105]
Prebiotics						
Fructooligosaccharides (FOS)	Onions, bananas, and asparagus	Rapid fermentation in proximal colon	Preferential *Bifidobacterium* growth	Short-chain fatty acids and B-vitamins	Enhanced calcium absorption and improved lipid profile	[106, 107]
Human milk oligosaccharides (HMOs)	Human breast milk and synthetic production	Selective bacterial nutrition and pathogen blocking	Increases *Bifidobacterium infantis* and decline in pathogenic bacteria	Anti-adhesion compounds and immune factors	Reduced infections in infants and immune development	[108, 109]
Inulin	Chicory root, Jerusalem artichoke, and garlic	Selective fermentation in colon	Increases *Bifidobacterium* and increases *Lactobacillus*	Butyrate, acetate, and immunomodulatory metabolites	Improved mineral absorption and reduced inflammation	[78, 110]

Postbiotics

Lactate	Lactic acid fermentation products	pH reduction and competitive exclusion	Supports beneficial lactobacilli	Precursor to other SCFAs	Pathogen inhibition and metabolic benefits	[111, 112]
Butyrate	Fermentation of resistant starch and fiber	Histone deacetylase inhibition and energy for colonocytes	Supports butyrate-producing bacteria	Direct metabolic substrate	Improved barrier function and reduced inflammation	[71, 113]
Bacteriocins	Fermented foods and probiotic cultures	Direct antimicrobial activity	Targeted pathogen inhibition	Antimicrobial peptides	Reduced foodborne pathogens	[114, 115]

ventional prebiotics comprise galactooligosaccharides (GOSs), fructooligosaccharides (FOSs), and inulin. The notion has prolonged to comprise:

- **Human Milk Oligosaccharides (HMOs)**: These are the intricate glycans naturally profuse in breast milk of humans, which selectively promote the growth of *Bifidobacterium* species in the gut of infant [86].
- **Polyphenols**: These are generally found in plant-based foods such as berries, cocoa, and tea. Some polyphenolic compounds endure biotransformation by gut microbes to synthesize metabolites that have boosted bioactivity [33].
- **Xylooligosaccharides (XOSs)**: These are derived from agricultural by-products such as sugarcane bagasse and corn cobs. XOS exhibits strong bifidogenic effects at lesser doses than conventional prebiotics [87].

The dosage essential for prebiotic effectiveness fluctuates considerably among compounds, with conventional prebiotics such as FOS generally requiring 5–10 g/day [88].

7.3.2.3 Synbiotics

Synbiotics amalgamates probiotics and prebiotics to augment overall efficiency. The term differentiates among:

- **Complementary Synbiotics**: These are the amalgamations in which prebiotic supports general growth of advantageous gut microbes.
- **Synergistic Synbiotics**: These are the combinations where the prebiotic selectively supports the development of the accompanying probiotic microbes [89].

Topical advancements comprise "precision synbiotics," where definite strains of bacteria are combined with substrates, which selectively encourages their metabolic activities inside the intestinal environment of the host [90].

7.3.2.4 Postbiotics

Postbiotics may be defined as the formulations of inanimate microorganisms and/or their compounds that confers health benefits on to the host [91]. This set of functional foods comprises bacterial lysates, exuded metabolites, fragments of cell-wall, cell-free supernatants, and purified bioactive complexes synthesized during fermentation. Important postbiotic compounds are as follows:

- **SCFAs**: Mainly acetate, propionate, and butyrate, fabricated through microbial fermentation of dietary fiber. Butyrate functions as the prime source of energy for colonocytes and exhibits anti-inflammatory and barrier-enhancing characteristics [92].

- **Bacterial Exopolysaccharides (EPS)**: High molecular weight polymers emanated by several lactic acid bacteria which can regulate the mucosal immunity [93].
- **Bacteriocins**: Antimicrobial peptides that can constrain pathogenic microbes while often reprieving beneficial commensals [94].
- **Bacterial Enzymes**: Consisting bile salt hydrolases which facilitates reduction of cholesterol and β-galactosidases, which alleviates symptoms of lactose intolerance [95].

7.3.2.5 Other Bioactive Compounds

- **Dietary Polyphenols**: These are the compounds from plants, which cooperate bi-directionally with gut microbiome. Polyphenols are transformed into more bio-available metabolites by certain bacteria of the gut, and polyphenols can even selectively regulate the microbial communities [96].
- **Bioactive Peptides**: These are the gut microbial composition influencing protein fragments, which also exerts direct effects on epithelial cells of intestine [97].
- **Omega-3 Fatty Acids**: Beyond their direct anti-inflammatory influence, these compounds can impact composition of the gut microbiota often prompting the growth of beneficial microbes [98].
- **Amino Acids**: Definite amino acids such as glutamine and arginine act as nutrients for intestinal cells and effects bacterial metabolism in the gut [99].

7.3.3 Examples of Functional Foods

7.3.3.1 Fermented Foods

Fermented foods characterize one of the oldest types of functional foods, with consumption courting back thousands of years across various cultures. Yogurt remains the most widely researched fermented food. Beyond acquainting viable probiotic microbes, the fermentation procedure pre-digests lactose that makes yogurt apt for several lactose-intolerant individuals. Explicit *Lactobacillus* strains present in yogurt prompt the production of mucin, which augments the intestinal mucus layer and intensifies the barrier function [116].

Kefir, which is a fermented milk beverage consisting of a complex community of bacteria and yeasts, has exhibited that consistent consumption can upsurge the growth of beneficial microbes while plummeting opportunistic pathogens in the gut microbiome. The complex polysaccharide kefiran displays prebiotic characteristics and may contribute to immunomodulatory effects of kefir [117].

Designer fermented foods denote a noteworthy trend, where explicit strains of bacteria are selected based on their functional qualities rather than conventional fermentation competences alone [118].

7.3.3.2 Plant-Based Functional Foods

Whole grain offers complex sugars, resistant starch, and dietetic fiber that act as substrates for advantageous gut microbes. Barley and oats are predominantly renowned for their high β-glucan content, which exhibits prebiotic effects and is linked with amplified SCFA production. Studies show that substituting from refined grains to whole grains can encourage substantial shifts in gut microbiome within as less as 2 weeks [119]. Legumes provides an exclusive blend of resistant starch, dietary fibers, and bioactive components encompassing polyphenols and GOSs. These ingredients selectively encourage the growth of beneficial microbes in the lower GI tract [120].

Fruits and vegetables comprise of varied types of fiber and polyphenolic complexes that can selectively impact gut microbial populations. Apples deliver a mixture of soluble and insoluble fibers together with procyanidins, which upsurge *Bifidobacterium* populations. Berries impart anthocyanins and proanthocyanidins that experience wide-ranging biotransformation by gut microbes [121]. Nuts and seeds offer fiber and polyphenols, which impacts the gut microbiome. Walnuts have been shown to enhance butyrate-producing microbes and augment the diversity of microbes in human intercession studies [122].

7.3.3.3 Specialized Functional Foods

Probiotic-fortified foods apart from dairy now comprises cereals, juices, snack bars, and chocolate goods. Microencapsulation methods have extended the array of food matrices apt for delivery of probiotic [123]. Polyphenol-rich functional foods comprising specialty chocolates, teas, and fruit preparations, are being established with explicit attention to their gut microbiome regulating attributes. Dark chocolates, which are high in flavanols, upsurge the populations of *Bifidobacterium* and *Lactobacillus* while reducing potentially detrimental *Clostridium* species [124].

7.3.4 Mechanisms of Action

7.3.4.1 Prebiotic Effects and Selective Growth Promotion

The selective stimulation of advantageous microorganisms denotes the primary mechanism of many functional food constituents. Substrate specificity plays a critical role in defining which bacterial populations will profit from a certain prebiotic compound, depending on:

- **Molecular Structure**: The glycosidic linkages in oligosaccharides mainly govern which microbial species own the essential enzymes for their metabolism [125].
- **Degree of Polymerization**: The chain length of sugar polymers impacts their rate of fermentation and the intestinal site where metabolism largely happens [126].
- **Branching Patterns**: Highly forked fiber structures may necessitate complex enzymatic systems for degradation, encouraging microbial range [127].

7.3.4.2 Production of Bioactive Metabolites

SCFAs, mainly acetate, propionate, and butyrate, denote the key end-products of carbohydrate fermentation in the colon. Apart from serving as energy substrates for colonocytes, SCFAs works as signaling molecules through communication with G-protein coupled receptors, prompting diverse physiological retorts [92]. Diverse dietary fibers produce erratic SCFA profiles based on their chemical structure and the bacterial populations they assist. Resistant starch classically produces relatively higher amounts of butyrate compared to pectin, which inclines to produce more acetate [128].

Microbial conversion of phenolic compounds produces metabolites with transformed bioavailability and bioactivity. Ellagitannins from berries and pomegranates experience microbial translation to urolithins, which displays anti-inflammatory and barrier-protective influence in the colon [129].

7.3.4.3 Modulation of Intestinal Barrier Function

Functional food constituents can augment intestinal barrier integrity through several mechanisms:

- Direct interaction among probiotic microbes and intestinal epithelial cells can trigger the expression of tight junction proteins [130].
- SCFAs, principally butyrate, support barrier function by assisting as an energy source for colonocytes and by epigenetically modulating genes involved in formation of tight junction [71].
- Specific bioactive peptides from food proteins can interact directly with epithelial cells of intestine to augment the production and the secretion of mucin [97].

7.3.4.4 Immunomodulatory Effects

Explicit bacterial cell wall components encompassing peptidoglycan, EPS, and lipoteichoic acids can be identified by receptors of pattern recognition on immune cells, which regulates their activity [84]. Microbial metabolites encompassing SCFAs, impacts development and function of immune cell. Butyrate encourages the differentiation of regulatory T cells via inhibition of histone deacetylases, potentially diminishing extreme inflammatory responses [131]. Some bioactive components in functional foods can directly stimulate immune cells. Definite milk oligosaccharides cooperate with dendritic cells, controlling their production of cytokine [132].

7.3.4.5 Competitive Exclusion of Pathogens

Functional foods can improve colonization resistance against pathogens through numerous mechanisms:
-   The inhabitation of intestinal niches by advantageous microbes can physically hamper pathogen attachment to surfaces of epithelium [133].
-   The production of antimicrobial complexes by beneficial microbes signifies another vital mechanism of pathogen exclusion.
-   Competition for vital nutrients can limit pathogen advancement in the intestinal atmosphere. Specific probiotic strains yield siderophores that sequester iron, potentially curbing its accessibility to pathogens [134].

7.3.4.6 Modulation of Host Metabolism

SCFAs influence the homeostasis of energy and metabolism of glucose through several mechanisms. Propionate can instigate intestinal gluconeogenesis, significantly enhancing the glucose indulgence. SCFAs interrelate with G-protein coupled receptors (GPCRs) on enteroendocrine cells, instigating the secretion of hormones implicated in modulation of appetite and homeostasis of energy [131]. The metabolism of dietary choline and L-carnitine by explicit gut microorganisms produces trimethylamine, which is afterward oxidized to trimethylamine N-oxide, a complex allied with cardiovascular risk. Some functional food ingredients can regulate this metabolic pathway [135].

7.4 Functional Foods and Gut Microbiome in Disease Prevention

Functional foods used in the prevention of various diseases and the microbiome alterations are listed in Table 7.3.

7.4.1 Obesity disease prevention and Metabolic Syndrome

In comparison to lean individuals, obese people generally have lower microbial diversity, a lower Bacteroidetes to Firmicutes ratio, and an increase in potentially harmful microbes. Obesity and its associated metabolic disorders also show notable differences in microbial composition. These microbial variations contribute in the advancement of obesity through various mechanisms, including improved harvest of energy from food, different storage of fat, interrupted barrier function of intestine, and enduring low-grade inflammation [136].

Given these recognized networks among dysbiosis of gut microbiome and metabolic dysfunction, the functional foods arise as hopeful therapeutic gears for preventing and managing obesity and metabolic syndrome. By tactically controlling the composition of gut microbiome in the direction of a healthier contour, the ensuing functional food groups have exhibited certain effectiveness in addressing obesity and metabolic syndrome through targeted regulation of gut microbiome:

7.4.1.1 Polyphenol-Rich Foods

The use of cranberry extract increases the abundance of *Akkermansia muciniphila*, a bacterium that is commonly associated with inflammation, diabetes, and obesity [67]. Green tea polyphenols correlate with reduced adiposity and improved glucose tolerance by increasing *Bacteroides* and decreasing Firmicutes inhabitants [137]. After arriving in the colon largely unaltered, polyphenols undergo extensive biotransformation by resident microorganisms to produce bioactive metabolites with increased bioavailability and therapeutic effects [33].

7.4.1.2 Whole Grains and Dietary Fiber

An 8-week whole-grain diet amplified the profusion of useful bacteria encompassing *Lactobacillus, Bifidobacterium*, and *Faecalibacterium prausnitzii*, while concurrently reducing inflammatory markers and enhancing insulin sensitivity equated to a refined grain diet [75]. The metabolic profits of whole grains are mainly credited to

Table 7.3: Functional foods in the prevention of disease and regulation of the gut microbiome.

Types of diseases	Explicit conditions	Functional food intercessions	Microbiome alterations	Mechanism	Clinical evidence	References
Immune and infectious diseases	Antibiotic-associated diarrhea	*Saccharomyces boulardii* and multi-strain probiotics	Restoration of colonization resistance	Competitive exclusion and antimicrobial production	64% reduction in AAD incidence	[167, 168]
	Upper respiratory infections	*Lactobacillus casei* and fermented milk products	Enhanced immune cell populations and increase in IgA production	Strengthened immune response and reduced pathogen adhesion	27% reduction in infection incidence	[151, 169]
	Allergic diseases	Multi-strain probiotics during pregnancy/infancy	Early microbiome establishment, immune tolerance	Th1/Th2 balance and regulatory T-cell development	50% reduction in atopic dermatitis risk	[170, 171]
Metabolic disorders	Metabolic syndrome	Polyphenol-rich berry extracts	Increase in *Akkermansia muciniphila* and *Bifidobacterium*	Anti-inflammatory pathways and improved gut barrier	23% reduction in insulin resistance and improved lipid profile	[67, 172]
	Nonalcoholic fatty liver	Omega-3-enriched foods and fish oil	Increase in *Lactobacillus* and *Bifidobacterium*, and reduced Proteobacteria	Reduced hepatic inflammation and improved fat metabolism	25% reduction in liver fat content	[173, 174]
	Obesity and type 2 diabetes	Whole grain cereals and resistant starch	Increase in *Prevotella, and Faecalibacterium,* Decline in Firmicutes/Bacteroidetes ratio	Enhanced SCFA production, improved insulin sensitivity, and reduced inflammation	15% reduction in postprandial glucose and 8% weight loss	[75, 175]
Gastrointestinal disorders	Irritable bowel syndrome	Multi-strain probiotics and low-FODMAP + prebiotic reintroduction	Restored microbial diversity and increase in beneficial bacteria	Reduced visceral hypersensitivity and normalized gut-brain signaling	40% improvement in IBS symptom scores	[147, 176]

	Condition	Intervention	Microbiome change	Mechanism/effect	Outcome	
	Inflammatory bowel disease	Fermented dairy products and VSL#3 probiotic	Increase in *Lactobacillus* and *Bifidobacterium*, and decline in pathogenic bacteria	Immune modulation, barrier function improvement	65% remission maintenance in patients	[177, 178]
	Colorectal cancer prevention	Fermented soy products and cruciferous vegetables	Increased *Bacteroides* and reduced sulfate-reducing bacteria	Reduced secondary bile acids and enhanced detoxification	28% reduced CRC risk (prospective cohort)	[149, 179]
Neurological disorders	Cognitive decline and dementia	Polyphenol-rich foods (blueberries and green tea)	Increase in *Bifidobacterium* and *Akkermansia*	Reduced neuroinflammation and enhanced neuroprotection	Improved cognitive test scores in older adults	[159, 180]
	Depression and anxiety	Psychobiotic formulations and fermented foods	Increase in GABA-producing bacteria and serotonin precursors	Gut-brain axis modulation and neurotransmitter production	Significant reduction in depression scores	[162, 181]
	Autism spectrum disorders	Targeted probiotic interventions	Normalized microbiome diversity and reduced pathogenic bacteria	Improved gut-brain communication and reduced inflammation	Reduced GI symptoms and improved behavioral outcomes	[163, 164]

their high content of nondigestible sugars, which aid as substrates for microbial fermentation in the colon. The subsequent SCFAs wield pleiotropic effects on the metabolism of host [138].

7.4.1.3 Fermented Foods and Probiotics

A meta-analysis of 21 randomized controlled examination determined that supplementation with probiotic led to substantial decline in the body mass index, body weight, and percentage of fat in contrast to placebo, with *Lactobacillus gasseri* strains exhibiting the most distinct impacts [139]. Kefir administration to high-fat diet nourished mice prohibited weight increase, reestablished gut microbiome composition, abridged intestinal permeability, and alleviated hepatic steatosis [140].

7.4.1.4 Omega-3 Fatty Acids and Fish Oil

Omega-3 fatty acids' plasma levels were positively associated with microbial variety and the copiousness of SCFA-producing microbes in healthy adults [66]. Omega-3 polyunsaturated fatty acids modify the composition of bile acid and encourage the growth of useful microbes while at the same time suppressing potentially pathogenic microbes. These microbial alterations associate with enhancements in metabolic parameters [141].

7.4.2 Gastrointestinal Disorders

7.4.2.1 Inflammatory Bowel Disease

Patients with IBD characteristically displays diminished microbial assortment, reduction of anti-inflammatory microbes such as *Faecalibacterium prausnitzii*, and augmentation of pro-inflammatory Proteobacteria [142]. Diets rich in plant foods, exclusively those high in polyphenols and dietary fiber, were allied with deceased risk of IBD flares and ailment evolution [143]. Plant-based diets encourage SCFA manufacturing microbes that sustain integrity of colonic epithelium and control immune responses [43].

Curcumin from turmeric has exhibited effectiveness in upholding remission in ulcerative colitis patients through several mechanisms consisting NF-κB inhibition and microbiome regulation [144]. Pomegranate extract endorses the development of advantageous *Akkermansia* and *Parabacteroides* while restraining pro-inflammatory Proteobacteria [145].

7.4.2.2 Irritable Bowel Syndrome

Patients with IBS every so often show abridged microbial diversity, declined levels of *Lactobacillus* and *Bifidobacterium*, and amplified profusion of gas-producing microbes [146]. A meta-analysis of 53 randomized controlled experiments determined that explicit probiotic preparations, predominantly those comprising *Bifidobacterium infantis*, *Lactobacillus plantarum*, and *Saccharomyces boulardii*, substantially enhanced IBS symptoms [147]. Premeditated re-establishment of definite prebiotics and resistant starches succeeding symptom enhancement can help reinstate beneficial microorganisms while upholding symptomatic relief [148].

7.4.2.3 Colorectal Cancer

The anticarcinogenic impacts of yogurt and other fermented dairy goods comprises manufacturing of bioactive peptides with antiproliferative characteristics, competitive exclusion of pathogenic microbes, binding of carcinogens, and regulation of host immune retortions [149]. Dietary polyphenols from fruits, vegetables, whole grains, and tea displays chemopreventive potential through microbiome regulation. Ellagitannins from pomegranate, berries, and walnuts are metabolized by gut microbiome to generate urolithins with anti-inflammatory and anticarcinogenic characteristics [150].

7.4.3 Immune System and Infections

7.4.3.1 Respiratory Infections

Everyday consumption of a synbiotic preparation comprising *Lactobacillus rhamnosus* GG and GOSs decreased the prevalence and sternness of upper respiratory tract infections in healthy adults by 27% in relation to placebo [151]. Consistent ingestion of shiitake mushroom extracts rich in lentinan augments activity of natural killer cells and diminishes production of inflammatory cytokine, consequences partly accredited to enhancement of beneficial *Faecalibacterium* and *Roseburia* species in the gut environment [152].

7.4.3.2 Gastrointestinal Infections

Probiotic preparations, chiefly those comprising *Saccharomyces boulardii* and some *Lactobacillus* species, abridged the risk of antibiotic-associated diarrhea and *Clostridioides difficile* infection when administered simultaneously with antibiotics [153].

Cranberry proanthocyanidins impede the adhesion of enteropathogenic *E. coli* to epithelial cells of intestines while at the same time encouraging advantageous *Akkermansia* and *Bifidobacterium* species [154].

7.4.3.3 Allergic Diseases

Exposure to larger variety of fermented foods through early life was linked with condensed risk of allergic diseases, comprising atopic dermatitis and food allergies by the age of 6 years [155]. Fermented foods and probiotics stimulate allergic responses through advancement of oral tolerance, solidification of intestinal barrier function, and regulation of T-helper cell differentiation for favoring the regulatory phenotypes [156].

7.4.3.4 Autoimmune Diseases

Mediterranean diet, which is rich in fruits, vegetables, olive oil, fish, and fermented foods, imparts safety against numerous autoimmune diseases through microbiome regulation. The defensive effects are accredited to its high content of polyphenols, omega-3 fatty acids, and dietary fiber, which jointly endorse a microbiome profile considered by boosted production of SCFAs [157].

7.4.4 Neurological Disorders

7.4.4.1 Neurodegenerative Diseases

Alzheimer's disease patients display abridged microbial range and enhancement of pro-inflammatory microbes in contrast to healthy controls [158]. Everyday consumption of extract of blueberry for 3 months enhanced cognitive performance in grown-up adults with minor cognitive damage, correlating with amplified profusion of *Bifidobacterium* and *Akkermansia* species [159]. Resveratrol supplementation enhanced cognitive function in older adults with slight cognitive impairment, correlating with augmented copiousness of butyrate-producing microbes [160].

7.4.4.2 Mood Disorders and Anxiety

Butyrate-synthesizing *Faecalibacterium* and *Coprococcus* positively associate with higher eminence of life, whereas *Bacteroides* contrarywise associates with depression [161]. A meta-analysis determined that probiotic supplementation substantially diminished depressive symptoms equated to placebo in both the clinical and nonclinical populations. Multi-strain preparations encompassing both *Lactobacillus* and *Bifidobacterium* species exhibited superior effectiveness [162].

7.4.4.3 Autism Spectrum Disorders

Children with autism spectrum disorder (ASD) display abridged microbial variety, transformed Bacteroidetes to Firmicutes ratio, and augmentation of potentially pathogenic microbes in contrast to neurotypical children [163]. A 6-month supplementation with a probiotic blend comprising *Bifidobacterium*, *Lactobacillus*, and *Streptococcus* strains notably enhanced GI symptoms and decreased irritability and hyperactivity in children suffering from ASD [164].

7.4.4.4 Epilepsy and Seizure Disorders

Antibiotic instigated disruption of the gut microbiota decreased the effectiveness of ketogenic diet in governing seizures in the mouse models of epilepsy [165]. The ketogenic diet's anticonvulsant influences are somewhat interceded through gut microbiota fluctuations, predominantly amelioration of *Akkermansia muciniphila* and *Parabacteroides* species that yield GABA and other inhibitory neurotransmitters [166].

7.5 Recent Advances in Functional Food Research

7.5.1 Food Design for Gut Microbiome Health

In the context of gut microbiome modulation, food design refers to the planned formulation and processing to maximize nutrient and substrate delivery to the gut bacteria. This method seeks to improve the health and diversity of the microbiome by supplying a varied range of substrates, such as dietary fiber, resistant starch, proteins, lipids, and phytochemicals, which are not entirely accessible to humans but serve as nutrients for gut bacteria [182]. The role of the diet is very critical in shaping the composition and metabolic activity of the gut microbiota, which in turn significantly influence human health. The gut microbiota is a complex ecology that influences the host's immune system, metabolism, and overall well-being. Dietary changes can cause rapid changes in the gut microbiota, impacting up to 57% of its composition, while host genetics account for no more than 12% of these changes [183].

Dietary fibers and prebiotics both are important dietary components that can significantly affect the gut microbiome. Due to their functionality overlapping, it is difficult to differentiate them. Dietary fibers, carbohydrate polymers that are resistant to digestion, and absorption in the human small intestine but are fermented in the colon, come in soluble and in-soluble form, both affecting gut health differentially. Soluble fibers like pectin, readily ferment to produce SCFAs, on the other hand insoluble fibers such as cellulose, which has bulking effect and can promote regularity in bowel movement by preventing constipation [184, 185]. According to Quigly (2020), prebiotics are nondigestible food components that selectively stimulate the growth of beneficial gut microbiome, particularly *Lactobacillus* and *Bifidobacterium*, resulting in a more balanced microbiome [186]. This selective fermentation eventually leads to the production of SCFAs such as butyrate, acetate, and propionate, which enhances gut barrier integrity, modulate immune response, and regulate metabolic processes. To support a healthy and diversified gut microbiome, increase the generation of SCFAs, and promote general gut health, it is essential to consume both dietary fibers and prebiotics [187].

Probiotics and postbiotics are also becoming important components in the food design landscape for modifying the gut microbiome and treating metabolic disorders. Probiotics, which are described as beneficial live microorganisms, can influence the composition of the gut microbiota, with certain strains such as *Bifidobacterium, Lactobacillus* showing promise in restoring *Akkermansia muciniphila* and Rikenellaceae abundances while decreasing Lactobacillaceae. This regulation is critical since gut microbiota dysbiosis is frequently related with metabolic disorders such as obesity and T2D [188]. Furthermore, probiotics can reduce the growth of dangerous bacteria, improve gut barrier function, and boost immune function, making them useful tools for controlling metabolic health. Postbiotics on the other hand, are nonliving microbial compounds that provide health benefits, making them a safer and more stable option for food design [189]. Postbiotics, which modulate gut microbial metabolites and improve intestinal barrier integrity, can be used to treat metabolic disorders [188]. Although clinical strategies for appropriate synbiotic formulae and the health effects of postbiotics deserve additional investigation, the promising function of both probiotics and postbiotics in addressing gut microbiota to ameliorate metabolic disorders [190, 191].

Synbiotics are a mixture of prebiotics and probiotics intended to work in concert to improve the gut microbiota and eventually metabolic health. By giving beneficial bacteria introduced by probiotics a selective substrate, the prebiotic, this synergistic strategy seeks to increase their survival and activity and promote a more resilient and balanced gut ecology. The potential of combining probiotics and prebiotics for better gut health outcomes is highlighted by Li et al.'s [188] observation that synbiotics containing *Bifidobacterium lactis, Lactobacillus rhamnosus*, and oligofructose-enriched inulin may be able to regulate the intestinal microenvironment more successfully.

Fruits, vegetables, and whole grain products contain phytochemicals and polyphenols, which play an important role in modulating human gut microbiome. These compounds can act as prebiotics by enhancing the growth of beneficial bacteria like

Bifidobacteriaceae, like dietary fibers they stimulate the production of SCFAs like butyrate and propionate, which are good for gut health [191, 192]. Phytochemicals with anti-inflammatory and antioxidant qualities, such as polyphenols and carotenoids, can improve the gut microbiome by lowering inflammation and oxidative stress [193]. In order to preserve the integrity of the immune system and the operation of the gut barrier, they can also interact with the gut mucosa, limiting its degradation and promoting the colonization of the beneficial bacteria. Furthermore, certain phytochemicals may affect the equilibrium of the gut microbiome by influencing bacterial communication through interactions with quorum sensing processes [193, 194]. The metabolism of phytochemicals in the gut can lead to the formation of bioactive molecules, such as equol from daidzein and enterolactone from secoisolariciresinol, which contribute to general health benefits. Despite these connections, further in vivo research is needed to completely understand phytochemicals' impact on the human gut flora [191, 192].

7.5.2 Personalized Nutrition

The idea that dietary guidelines for the general public should be further tailored for particular consumer segments or subgroups in order to promote individual health management and disease prevention is not new; it was first proposed in the 1970s [195]. People differ in their dietary requirements depending on their age or specific physiological state; for instance, pregnant women and infants have distinct demands. Patients with allergies or long-term conditions like diabetes, dyslipoproteinemia, or liver disease also need to follow particular diets. The selection of dietary components to effectively maintain optimal health is influenced by the social, physical, and economic environments in addition to physiological status [196]. Dietary interventions are tailored to each participant based on their unique characteristics. The more specialized the intervention, the more complex and possibly costly it will be to gather, examine, and respond to those participant traits [197].

What conceptual underpinnings support personalized nutrition? Customizing dietary advice, products, or services will be more beneficial than more general methods, according to the theory behind personalized nutrition. Biological evidences of distinct reactions to meals and nutrients based on either phenotypic or genotypic characteristics can serve as the basis for personalization. Evaluation of present behavior, preferences, obstacles, and goals and subsequent administration of therapies encourage and facilitate each individual's ability to modify their eating habits [195].

A novel approach to providing dietary guidance has arisen as a result of the capacity to identify the hundreds of stable genetic variations that exist in the human population. This approach is known as personalized nutrition. The idea that people differ in how they respond to diet and how they metabolize nutrients is encapsulated in personalized nutrition, which is similar to the ideas of pharmacogenomics and personalized medicine. Important aspects of the personalized nutrition approach include the fact that disease risk, not treatment, differs from person to person and that genetic and nutritional factors combine to influence risk [198].

The fast development of high-throughput techniques for evaluating genome-wide genetic variants and patterns of gene expression has been facilitated by the sequencing of the human genome. Individuals in the human population can now have their expression of the vast number of stable genetic variants, including single-nucleotide polymorphisms (SNPs), characterized. Studying genetic influences on disease risk has changed dramatically as a result of the subsequent sharp decline in sequencing costs and the advancement of tools that evaluate a large number of SNPs, such as genome-wide association studies [198].

The Human Genome Project, the International HapMap consortium, and the Personal Genome Project are examples of large research collaborations that have been started to collect data on genetic variation with the ultimate goals of establishing a connection between genetic variation and the risk of human disease and advancing the development of personalized medicine [198].

7.5.2.1 Controlling and Preventing Metabolic Syndrome

The scientific community widely concurs that nutrigenetics alone will not be the foundation of precision nutrition in the future, as was previously suggested [199]. It is obvious that while creating customized or personalized diets, variables other than genetics must also be considered. One of the primary objectives of precision nutrition is to ensure that customized dietary recommendations are effective in anticipating how each individual will react to dietary intakes. connected to genetic or dietary factors, such as lifestyle factors like metabolomics, gut microbiomics, or physical activity habits, are also becoming important contributions that should be taken into account in the field of precision nutrition [200].

More precisely, precision nutrition aims to create more dynamic and thorough dietary recommendations based on variables that change and interact with one another during an individual's life. In light of this, precision nutrition strategies incorporate not just genetics but also dietary practices, eating behavior, physical activity, the microbiota, and the metabolome [200]. Recent published studies on gene-environment interactions have provided valuable insights into the role of macronutrient intake in the relationship of genetic markers with body composition, fat mass accumulation, and metabolic health in relation to obesity and metabolic syndrome. This has significant implications for precision nutrition since the findings of these research, which concentrated on macronutrient intake, allow for effective diet customization depending on a person's genetic composition [200].

In light of this, the American Society of Nutrition recently named a better understanding of inter-individual variability in diet response as one of the six nutrition science research priorities to address in order to meet the upcoming challenges in population health management.

7.5.2.2 Personalized Nutrition Based on Microbiota

Numerous microbial species that coevolved within the host and are known to have a significant impact on both health and disease makeup the rich and complex ecosystem that is the human gut microbiome [201]. Dependent on host genotype and environmental circumstances, the composition of the gut microbiota varies from person to person and during development. The development and preservation of microbial diversity in the gut have been linked to early microbial exposure, nutrition, age, location, and antibiotic exposure. Despite being subjected to external stressors on a continual basis, the gut microbiota can recover from disruptions like pathogen infection or antibiotic treatment. Another name for this ability to regenerate itself is the resilience phenomenon [201].

Pathogenesis, progression, and treatment of diseases, from neurological disorders to metabolic disorders, are all influenced by the microbiota [202]. Using tailored nutrition to alter host microbiota interactions is a novel therapeutic approach for illness prevention and control. The composition and function of the gut microbiota are shaped beginning in infancy when the person is colonized by bacteria from their caregivers and the environment. This process has a significant impact on the microbiota's composition in maturity [202].

In order to build a diet that would produce positive results, personalized nutrition approaches seek to uncover important microbiome characteristics that forecast the body's reaction to specific food components. Determining how the host, microbiome, and dietary exposures interact to shape dietary responses is the primary issue in maximizing the promise of microbiome-informed personalized nutrition [202]. Major risk factors for noncommunicable diseases (NCDs), such as T2D, CVDs, and numerous malignancies, include poor diet and inactivity. In order to lower the burden of NCDs, the majority of population strategies have relied on "one-size-fits-all" public health guidelines, such as "eat at least five portions of fruit and vegetables daily." The need for more effective prevention is highlighted by the fact that the worldwide burden of NCDs is still increasing [197].

The gut microbiota supplies the necessary resources for the fermentation of indigestible materials, such as dietary fibers and endogenous intestinal mucus. This fermentation promotes the development of specialist microbes that generate SCFAs and gases. The main SCFAs generated are butyrate, propionate, acetate, and propionate [203].

The first strategy involves identifying certain microbial signatures and the metabolic characteristics that go along with them. These signatures might be as basic as the presence or lack of particular species, genes, or enterotypes in the microbiome, or they can be as complicated as a variety of traits. Finding meals that are good for all types of microbiomes and for the intended results follows next once the population has been stratified [202].

The lack of consideration for interindividual variations in dietary responses and gut microbiota profile may be the cause of the nutritional therapies' unsatisfactory effectiveness in managing microbiota-associated disorders to date. The addition of already-processed NDCs, which appear to be less successful than unprocessed alternatives and/or techniques combining slowly and rapidly fermentable NDCs, is another drawback [201]. The two separate weight-loss tactics suggested for either the *Bacteroides* enterotype or the *Prevotella* enterotype, given their varying reactions to NDCs, are examples of how the idea of enterotypes opened the door for more attempts at subject-specific therapy [201].

Future developments in microbiome-based customized nutrition will involve the application of advanced statistical techniques and next-generation "omics" tools, which are widely used in research, to clinical settings [201]. The identification of high-level clusters, like enterotypes, that are intended to categorize people according to the composition of their gut microbiota might be viewed as a step toward customized approaches from the standpoint of personalized microbiomes [201].

To go beyond the existing stratification techniques, a more thorough understanding of the variables influencing each person's gut microbiota makeup would be necessary to develop individualized recommendations and interventions. A deeper comprehension of the mechanisms governing stability and resilience will be essential to applying precision nutrition to enhance the gut ecosystem's recovery following disruptions [196]. Along with taking into account the ideal daily intakes, this information will enable the identification of which dietary components, prebiotics, or probiotics are likely to be the most helpful for a particular patient [201].

7.5.2.3 Dietary Responses That Are Altered by Metabotypes and Genotypes

The study of how genetic composition influences a person's vulnerability to health and disease (nutrigenetics) has gained a lot of attention in the last 10 years. This has led to a better understanding of the basic molecular and metabolic processes that are influenced by diet (nutrigenomics) [201]. Twin studies provide the best evidence for the genetic influence on food responses. When it comes to their adaptability to chronic overfeeding, weight growth, and fat distribution, monozygotic twins have demonstrated a high degree of intra-pair resemblance [201], as well as strongly associated variations in body weight and lipoprotein metabolism while alternating between low- and high-fat diets, even though the twins' baseline levels of physical activity varied greatly [204]. As a result of an individual risk-benefit analysis of dietary components, nutrigenetics and nutrigenomics have already opened up new avenues for integrating genetic testing with dietary recommendations in order to further reduce diseases linked to poor nutrition [198].

7.5.2.4 Precision Nutrition

Personalized nutrition's conventional idea is to modify a person's diet to suit their needs and tastes. Precision nutrition can finally help reduce and prevent disease by using genetic information to anticipate whether or not an individual would respond to particular nutritional patterns. This is made possible by the development of high-throughput technologies. The foundation of personalized nutrition is the idea that a person's DNA sequence may influence how much a certain food or nutrient quantity may change their risk of developing a disease [205].

Three levels of precision nutrition can be distinguished: (1) personalized nutrition that incorporates phenotypic data about an individual's current nutritional status (such as anthropometry, biochemical, and metabolic analysis, physical activity, and others); (2) genotype-directed nutrition based on uncommon or common gene variation; and (3) conventional nutrition based on general guidelines for population groups by age, gender, and social determinants. The ultimate objective is to integrate these information sources so that medical professionals such as dietitians, doctors, pharmacists, and genetic counselors have enough knowledge of nutrigenetics and nutrigenomics to determine the best course of action for achieving precision nutrition, which incorporates social, environmental, and metabolic factors in addition to phenotypical and genotypical issues [205].

The International Society of Nutrigenetics/Nutrigenomics states that three levels of discussion should be used to discuss the future of precision nutrition: genetic-directed nutrition based on rare genetic variants that have high penetrance and influence how people react to specific foods; individual approaches presented from a deep and refined phenotyping; and stratification of conventional nutritional guidelines into population subgroups by age, gender, and other social determinants [200].

7.5.2.5 Nutrition Therapy

Supportive care methods for critical illnesses must include nutrition therapy. The early acute phase, the immediate post-acute phase, and the recovery period are the three stages of critical illness. Amino acids are recruited as a substrate for acute-phase protein and immune-system products during the hypercatabolic state that dominates the acute phase. Additionally, the immunological dysregulation, the subsequent dysbiosis, and the quick breakdown of the intestinal barrier all contribute to the inflammatory response [199].

Overall results also tend to support evidence from observational studies and clinical trials that people with markers of higher risk for diabetes, prediabetes, or insulin resistance have lower risk when they consume less calories, carbohydrates, or saturated fat and/or more fiber or protein than their peers [206].

It is acknowledged that "one-size-fits-all" and "set and forget" nutritional strategies fall short in addressing the intricate immunological, hormonal, and metabolic alterations that accompany critical illness. Understanding these pathways and how they affect nutrition metabolism is crucial for doctors. Nutrition Therapy Guidelines for Critical Illness, the dietary management of critically ill patients can now be guided by four worldwide clinical practice recommendations: (1) ASPEN/SCCM (2016); (2) Canadian Clinical Practice Guidelines; (3) ESICM clinical practice guidelines (2017); and (4) ESPEN [207].

Main or auxiliary therapy for diseases ranging from neurological disorders and cancer to metabolic and immune diseases of the gut; prophylaxis for diseases for which a person is more susceptible due to genetics or lifestyle; and performance enhancement and the accomplishment of various physiological goals, as required, for instance, in sports, are all areas of the future of personalized nutrition [202]. A future vision of a personalized nutrition is depicted in this scenario, where either (a) people will pay commercial businesses for advice based on genetic data, or (b) medical professionals, nutritionists, or dietitians consider advanced genetic data when making dietary recommendations to minimize disease risk and maximize health information [198].

7.5.3 Emerging Trends and Technological Innovation

Development and management of functional foods, especially those that target gut health in humans, have seen a dramatic change as a result of the food industry's use of cutting-edge technologies and innovative approaches.

7.5.3.1 Personalized Nutrition

On the basis of an individual's unique characteristics, including genetics, metabolic responses, and microbiome composition, personalized nutrition involves tailoring dietary plans to optimize health and well-being [208]. Recent developments in bioinformatics and artificial intelligence (AI) are transforming the way that nutrition is customized. Large data sets can be analyzed by AI system to find trends linking particular food ingredients to particular health outcomes. For example, dietitian might suggest customized probiotic strains for the best gut health by knowing how particular probiotics interact with various gut microbial profiles. Furthermore, wearable technology improves adherence to specific suggestion by providing real time feedback on nutritional choices. This customized strategy guarantees that functional meals are more successful in promoting health objectives including better digestion, increased immunity, and weight management, opening the door for astute dietary innovations [209].

7.5.3.2 Next-Generation Probiotics

Next-generation probiotics (NGPs), which boost the conventional range of probiotic use, are changing the probiotic environment. Live microorganisms with health-promoting properties, prebiotics, and postbiotics microbial metabolites that promote health can all be included in NGPs. The goal of the study is to pinpoint particular strains that potentially offer particular health advantages, like those that improve the function of the gut barrier, alter immunological responses, or generate advantageous SCFAs that support gut cells. To increase effectiveness and retention in the stomach, new technologies are being investigated, such as genetically modified probiotics. This novel strategy improves probiotics' therapeutic benefits and makes it possible to use them in common functional meals like yogurt and smoothies, which will increase consumer acceptance and dietary adherence [210].

7.5.3.3 Fortified and Functional Foods

Fortified foods containing probiotics, prebiotics, vitamins, minerals, and other bioactive components are increasingly popular among health-conscious consumers [211]. The current trend is not just to give nutrients, but also to incorporate features that promote intestinal health. Kefir and kombucha, for example, are high in probiotics, as well as other nutrients that promote a healthy microbiome [212]. Advances in processing technology, like as microencapsulation, enable to maintain the viability of these additional probiotics during manufacture and storage. Furthermore, consumers are becoming more aware of the importance of functional foods in preventing chronic diseases associated with gut health, resulting in increased demand for foods such as fiber-enriched cereals or snacks, which can improve gut microbiota composition and general health [209].

7.5.3.4 Innovative Processing Technologies

Innovative processing procedures are crucial for preserving the bioactive components of functional foods while also assuring food safety. Nonthermal processing techniques such as high-pressure processing, cold plasma, and ultrasonic technologies are gaining popularity. These procedures eradicate hazardous pathogens and spoilage organisms while preserving sensitive nutrients and bioactive components that would otherwise be damaged during traditional cooking or pasteurization [213]. The application of these technologies has the potential to result in the development of new types of functional foods that retain their health-promoting features, providing customers with improved options that promote gut health without compromising safety or nutritional value [214].

7.5.3.5 Sustainability and Alternative Proteins

There is a notable trend toward plant-based and cultured meat products as alternative proteins due to growing concerns about environmental sustainability. These developments offer nutritional profiles that support gut health in addition to addressing moral issues regarding animal cruelty. For example, a lot of plant-based diets are inherently richer in fiber, which is necessary to support good gut flora. Furthermore, dairy substitutes and cultured meat are being created to replicate the nutritional characteristics of conventional animal products. As these goods develop, they have the potential to create new markets for functional meals that promote both individual and global health, in addition to reducing the environmental impact of animal agriculture [209, 215, 216].

7.5.3.6 Biotechnological Applications

There are new opportunities to improve gut health – thanks to the application of biotechnology in the creation of functional food ingredients. Using particular microbial strains that are known to provide health benefits to ferment food products is an example of a biotechnological breakthrough. Through this process, new bioactive compounds that can promote gut microbiota balance are also produced, in addition to increasing the availability of functional components like vitamins and antioxidants. For instance, nutritional supplements that improve the gut microbiome may be developed as a result of the extraction of advantageous metabolites from specific gut bacteria. Biotechnology holds great promise for developing highly functional foods that are suited to certain dietary requirements and medical situations, putting these advancements at the forefront of managing gut health [209, 217].

7.6 Future Perspectives and Challenges

7.6.1 Potential Breakthroughs

7.6.1.1 Advanced Microbiome Analysis and Modeling

Technological developments in microbiome analysis, involving next-generation sequencing, bioinformatics, and metabolomics are reforming our knowledge of diet-microbiome interactions [203]. Recently advanced ingestible sensors and noninvasive monitoring tactics can offer dynamic visions into how functional foods impacts the gut environment over time [218].

7.6.1.2 Precision Fermentation and Synthetic Biology

Advances in precision fermentation permit production of very specific bioactive complexes through microbial fermentation methods [219]. Engineered probiotics proficient of sensing inflammation of gut and anti-inflammatory compounds production in response exemplify the potential for "smart" probiotics that retort dynamically to conditions of the gut [220, 221].

7.6.1.3 Innovative Delivery Systems

Nanoencapsulation methods can shield sensitive complexes through the harsh environments of the upper GI tract and attain meticulous release in the colon [222]. 3D food printing can generate complex food structures with accurate spatial dispersal of bioactive components, permitting customized functional foods personalized to distinct microbiome profiles [223, 224].

7.6.1.4 Integration with Digital Health Platforms

Mobile applications that tails gut symptoms, dietary consumption, and added health parameters can deliver treasured data for optimizing functional food consumption grounded on responses of individuals [225]. AI algorithms can mix various data sources to create personalized endorsements for consumption of functional food [226, 202].

7.6.2 Challenges in Functional Food Development

7.6.2.1 Scientific and Technical Challenges

Inter-individual erraticism in composition and function of microbiome influences how people answer to dietary intercessions [227]. Polyphenols are expansively metabolized by gut microbes, creating metabolites with dissimilar bioactivities equated to the parent compounds, necessitating extra elucidation of these metabolic lanes [157]. Founding clear cause-effect associations among specific functional food ingredients and health results arbitrated by gut microbiota amendments remains challenging [228].

7.6.2.2 Regulatory Challenges

The European Food Safety Authority in the European Union inflicts rigorous standards for approving health claims associated to gut microbiome regulation [229]. The FDA differentiates among 'structure/function claims' and 'health claims' with the latter necessitating significant scientific contract built on the weight of scientific indication [230]. Regulatory frameworks may not keep pace through scientific developments in microbiome studies [63].

7.6.2.3 Commercial and Consumer Challenges

Several bioactive complexes that positively influence gut microbiome can confer undesirable flavors or textures to food goods [231]. Certifying sustainable manufacture of functional food constituents poses challenges in meeting snowballing global demand [232].

7.6.3 Conclusion

The domain of functional foods for gut microbiota supervision stands at a thrilling intersection, with substantial contests to overcome but also unparalleled prospects for invention and effect. The future will probably be molded by interdisciplinary partnership, bringing together proficiency from, nutrition, food science, microbiology, engineering, medicine, and data science.

Regulatory outlines will necessitate to progress to lodge evolving scientific understanding while safeguarding consumer well-being and averting misleading prerogatives. Consumer education characterizes another serious factor of future advancement, enabling informed selections and proper use of functional foods for maintenance of gut health and disease prevention. As our understanding of the gut microbiome endures to extend, functional foods deliver incredible potential to contribute to the improvement of public well-being on a worldwide scale.

Conflicts of interest: The authors do not declare any conflicts of interest.

References

[1] Rinninella E, Raoul P, Cintoni M, Franceschi F, Miggiano GA, Gasbarrini A, Mele MC. What is the healthy gut microbiota composition? A changing ecosystem across age, environment, diet, and diseases. Microorganisms. 2019 Jan 10;7(1):14.

[2] Kigerl KA, Hall JC, Wang L, Mo X, Yu Z, Popovich PG. Gut dysbiosis impairs recovery after spinal cord injury. Journal of Experimental Medicine. 2016 Nov 14;213(12):2603–2620.

[3] Chidambaram SB, Rathipriya AG, Mahalakshmi AM, Sharma S, Hediyal TA, Ray B, Sunanda T, Rungratanawanich W, Kashyap RS, Qoronfleh MW, Essa MM. The influence of gut dysbiosis in the pathogenesis and management of ischemic stroke. Cells. 2022 Apr 6;11(7):1239.

[4] Sencio V, Machado MG, Trottein F. The lung–gut axis during viral respiratory infections: The impact of gut dysbiosis on secondary disease outcomes. Mucosal Immunology. 2021 Mar 1;14(2):296–304.

[5] Rowland I, Gibson G, Heinken A, Scott K, Swann J, Thiele I, Tuohy K. Gut microbiota functions: Metabolism of nutrients and other food components. European Journal of Nutrition. 2018 Feb;57:1–24.

[6] Hou K, Wu ZX, Chen XY, Wang JQ, Zhang D, Xiao C, Zhu D, Koya JB, Wei L, Li J, Chen ZS. Microbiota in health and diseases. Signal Transduction and Targeted Therapy. 2022 Apr 23;7(1):135.

[7] Greenblum S, Turnbaugh PJ, Borenstein E. Metagenomic systems biology of the human gut microbiome reveals topological shifts associated with obesity and inflammatory bowel disease. Proceedings of the National Academy of Sciences. 2012 Jan 10;109(2):594–599.

[8] Marchesi JR, Adams DH, Fava F, Hermes GD, Hirschfield GM, Hold G, Quraishi MN, Kinross J, Smidt H, Tuohy KM, Thomas LV. The gut microbiota and host health: A new clinical frontier. Gut. 2016 Feb 1;65(2):330–339.

[9] Li J, Zhao F, Wang Y, Chen J, Tao J, Tian G, Wu S, Liu W, Cui Q, Geng B, Zhang W. Gut microbiota dysbiosis contributes to the development of hypertension. Microbiome. 2017 Dec;5:1–9.

[10] Winter SE, Bäumler AJ. Gut dysbiosis: Ecological causes and causative effects on human disease. Proceedings of the National Academy of Sciences. 2023 Dec 12;120(50):e2316579120.

[11] Yang T, Santisteban MM, Rodriguez V, Li E, Ahmari N, Carvajal JM, Zadeh M, Gong M, Qi Y, Zubcevic J, Sahay B. Gut dysbiosis is linked to hypertension. hypertension. 2015 Jun;65(6):1331–1340.

[12] Langley-Evans SC. Nutrition in early life and the programming of adult disease: A review. Journal of Human Nutrition and Dietetics. 2015 Jan;28:1–4.

[13] Petroni ML, Brodosi L, Marchignoli F, Sasdelli AS, Caraceni P, Marchesini G, Ravaioli F. Nutrition in patients with type 2 diabetes: Present knowledge and remaining challenges. Nutrients. 2021 Aug 10;13(8):2748.

[14] Radu F, Potcovaru CG, Salmen T, Filip PV, Pop C, Fierbințeanu-Braticievici C. The link between NAFLD and metabolic syndrome. Diagnostics. 2023 Feb 7;13(4):614.

[15] Milani C, Duranti S, Bottacini F, Casey E, Turroni F, Mahony J, Belzer C, Delgado Palacio S, Arboleya Montes S, Mancabelli L, Lugli GA. The first microbial colonizers of the human gut: Composition, activities, and health implications of the infant gut microbiota. Microbiology and Molecular Biology Reviews. 2017 Dec;81(4):10–128.

[16] Hasan N, Yang H. Factors affecting the composition of the gut microbiota, and its modulation. PeerJ. 2019 Aug 16;7:e7502.

[17] Rinninella E, Cintoni M, Raoul P, Lopetuso LR, Scaldaferri F, Pulcini G, Miggiano GA, Gasbarrini A, Mele MC. Food components and dietary habits: Keys for a healthy gut microbiota composition. Nutrients. 2019 Oct 7;11(10):2393.

[18] Gomaa EZ. Human gut microbiota/microbiome in health and diseases: A review. Antonie Van Leeuwenhoek. 2020 Dec;113(12):2019–2040.

[19] Deo PN, Deshmukh R. Oral microbiome: Unveiling the fundamentals. Journal of Oral and Maxillofacial Pathology. 2019 Jan 1;23(1):122–128.

[20] Ruan W, Engevik MA, Spinler JK, Versalovic J. Healthy human gastrointestinal microbiome: Composition and function after a decade of exploration. Digestive Diseases and Sciences. 2020 Mar;65:695–705.

[21] Li X, Liu Y, Yang X, Li C, Song Z. The oral microbiota: Community composition, influencing factors, pathogenesis, and interventions. Frontiers in Microbiology. 2022 Apr 29;13:895537.

[22] Ozbey G, Sproston E, Hanafiah A. Helicobacter pylori infection and gastric microbiota. Euroasian Journal of Hepato-gastroenterology. 2020 Jan;10(1):36.

[23] Yadav M, Verma MK, Chauhan NS. A review of metabolic potential of human gut microbiome in human nutrition. Archives of Microbiology. 2018 Mar;200:203–217.

[24] Fröhlich EE, Farzi A, Mayerhofer R, Reichmann F, Jačan A, Wagner B, Zinser E, Bordag N, Magnes C, Fröhlich E, Kashofer K. Cognitive impairment by antibiotic-induced gut dysbiosis: Analysis of gut microbiota-brain communication. Brain, Behavior, and Immunity. 2016 Aug 1;56:140–155.

[25] Rosenfeld CS. Gut dysbiosis in animals due to environmental chemical exposures. Frontiers in Cellular and Infection Microbiology. 2017 Sep 8;7:396.

[26] Fackelmann G, Sommer S. Microplastics and the gut microbiome: How chronically exposed species may suffer from gut dysbiosis. Marine Pollution Bulletin. 2019 Jun 1;143:193–203.

[27] Cresci GA, Izzo K. Gut microbiome. In: Adult Short Bowel Syndrome. 2019 Jan 1. pp. 45–54, Academic Press.

[28] Jandhyala SM, Talukdar R, Subramanyam C, Vuyyuru H, Sasikala M, Reddy DN. Role of the normal gut microbiota. World Journal of Gastroenterology: WJG. 2015 Aug 7;21(29):8787.

[29] Mirzaei R, Dehkhodaie E, Bouzari B, Rahimi M, Gholestani A, Hosseini-Fard SR, Keyvani H, Teimoori A, Karampoor S. Dual role of microbiota-derived short-chain fatty acids on host and pathogen. Biomedicine & Pharmacotherapy. 2022 Jan 1;145:112352.

[30] Wang X, Qi Y, Zheng H. Dietary polyphenol, gut microbiota, and health benefits. Antioxidants. 2022 Jun 20;11(6):1212.

[31] Wang X, Zhang B, Jiang R. Microbiome interplays in the gut-liver axis: Implications for liver cancer pathogenesis and therapeutic insights. Frontiers in Cellular and Infection Microbiology. 2025 Jan 28;15:1467197.

[32] Otaru N, Ye K, Mujezinovic D, Berchtold L, Constancias F, Cornejo FA, Krzystek A, De Wouters T, Braegger C, Lacroix C, Pugin B. GABA production by human intestinal Bacteroides spp.: Prevalence, regulation, and role in acid stress tolerance. Frontiers in Microbiology. 2021 Apr 15;12:656895.

[33] Cardona F, Andrés-Lacueva C, Tulipani S, Tinahones FJ, Queipo-Ortuño MI. Benefits of polyphenols on gut microbiota and implications in human health. The Journal of Nutritional Biochemistry. 2013 Aug 1;24(8):1415–1422.

[34] Wilson ID, Nicholson JK. Gut microbiome interactions with drug metabolism, efficacy, and toxicity. Translational Research. 2017 Jan 1;179:204–222.

[35] Mahdy MS, Azmy AF, Dishisha T, Mohamed WR, Ahmed KA, Hassan A, Aidy SE, El-Gendy AO. Irinotecan-gut microbiota interactions and the capability of probiotics to mitigate Irinotecan-associated toxicity. BMC Microbiology. 2023 Mar 2;23(1):53.

[36] Wang SL, Shao BZ, Zhao SB, Fang J, Gu L, Miao CY, Li ZS, Bai Y. Impact of paneth cell autophagy on inflammatory bowel disease. Frontiers in Immunology. 2018 Apr 5;9:693.

[37] Pickard JM, Zeng MY, Caruso R, Núñez G. Gut microbiota: Role in pathogen colonization, immune responses, and inflammatory disease. Immunological Reviews. 2017 Sep;279(1):70–89.

[38] Bemark M, Pitcher MJ, Dionisi C, Spencer J. Gut-associated lymphoid tissue: A microbiota-driven hub of B cell immunity. Trends in Immunology. 2024 Mar 1;45(3):211–223.

[39] Shao T, Hsu R, Rafizadeh DL, Wang L, Bowlus CL, Kumar N, Mishra J, Timilsina S, Ridgway WM, Gershwin ME, Ansari AA. The gut ecosystem and immune tolerance. Journal of Autoimmunity. 2023 Dec 1;141:103114.

[40] Kinnebrew MA, Pamer EG. Innate immune signaling in defense against intestinal microbes. Immunological Reviews. 2012 Jan;245(1):113–131.

[41] Barbara G, Barbaro MR, Fuschi D, Palombo M, Falangone F, Cremon C, Marasco G, Stanghellini V. Inflammatory and microbiota-related regulation of the intestinal epithelial barrier. Frontiers in Nutrition. 2021 Sep 13;8:718356.

[42] Gul L, Modos D, Fonseca S, Madgwick M, Thomas JP, Sudhakar P, Booth C, Stentz R, Carding SR, Korcsmaros T. Extracellular vesicles produced by the human commensal gut bacterium Bacteroides thetaiotaomicron affect host immune pathways in a cell-type specific manner that are altered in inflammatory bowel disease. Journal of Extracellular Vesicles. 2022 Jan;11(1):e12189.

[43] Parada Venegas D, De la Fuente MK, Landskron G, González MJ, Quera R, Dijkstra G, Harmsen HJ, Faber KN, Hermoso MA. Short chain fatty acids (SCFAs)-mediated gut epithelial and immune regulation and its relevance for inflammatory bowel diseases. Frontiers in Immunology. 2019 Mar 11;10:277.

[44] Medina-Gomez C. Bone and the gut microbiome: A new dimension. Journal of Laboratory and Precision Medicine. 2018 Nov 23;3:96, doi: 10.21037/jlpm.2018.1

[45] Wallimann A, Magrath W, Thompson K, Moriarty TF, Richards RG, Akdis CA, O'Mahony L, Hernandez CJ. Gut microbial-derived short-chain fatty acids and bone: A potential role in fracture healing. European Cells & Materials. 2021 Apr 21;41:454.

[46] Adak A, Khan MR. An insight into gut microbiota and its functionalities. Cellular and Molecular Life Sciences. 2019 Feb 15;76:473–493.

[47] Den Besten G, Bleeker A, Gerding A, van Eunen K, Havinga R, van Dijk TH, Oosterveer MH, Jonker JW, Groen AK, Reijngoud DJ, Bakker BM. Short-chain fatty acids protect against high-fat diet–induced obesity via a PPARγ-dependent switch from lipogenesis to fat oxidation. Diabetes. 2015 Jul 1;64(7):2398–2408.

[48] Al Bander Z, Nitert MD, Mousa A, Naderpoor N. The gut microbiota and inflammation: An overview. International Journal of Environmental Research and Public Health. 2020 Oct;17(20):7618.

[49] Brandsma E, Kloosterhuis NJ, Koster M, Dekker DC, Gijbels MJ, Van Der Velden S, Ríos-Morales M, Van Faassen MJ, Loreti MG, De Bruin A, Fu J. A proinflammatory gut microbiota increases systemic inflammation and accelerates atherosclerosis. Circulation Research. 2019 Jan 4;124(1):94–100.

[50] Zeng MY, Inohara N, Nuñez G. Mechanisms of inflammation-driven bacterial dysbiosis in the gut. Mucosal Immunology. 2017 Jan 1;10(1):18–26.

[51] Brüssow H. Problems with the concept of gut microbiota dysbiosis. Microbial Biotechnology. 2020 Mar;13(2):423–434.

[52] Carding S, Verbeke K, Vipond DT, Corfe BM, Owen LJ. Dysbiosis of the gut microbiota in disease. Microbial Ecology in Health and Disease. 2015 Dec 1;26(1):26191.

[53] Magne F, Gotteland M, Gauthier L, Zazueta A, Pesoa S, Navarrete P, Balamurugan R. The firmicutes/bacteroidetes ratio: A relevant marker of gut dysbiosis in obese patients?. Nutrients. 2020 May 19;12(5):1474.

[54] Hrncir T. Gut microbiota dysbiosis: Triggers, consequences, diagnostic and therapeutic options. Microorganisms. 2022 Mar 7;10(3):578.

[55] Martinez KB, Leone V, Chang EB. Western diets, gut dysbiosis, and metabolic diseases: Are they linked?. Gut Microbes. 2017 Mar 4;8(2):130–142.

[56] Kriss M, Hazleton KZ, Nusbacher NM, Martin CG, Lozupone CA. Low diversity gut microbiota dysbiosis: Drivers, functional implications and recovery. Current Opinion in Microbiology. 2018 Aug 1;44:34–40.

[57] Parkin K, Christophersen CT, Verhasselt V, Cooper MN, Martino D. Risk factors for gut dysbiosis in early life. Microorganisms. 2021 Sep 30;9(10):2066.

[58] Shahidi F. Functional foods: Their role in health promotion and disease prevention. Journal of Food Science. 2004 Jun;69(5):R146-9.

[59] Martins ZE, Pinho O, Ferreira IM. Food industry by-products used as functional ingredients of bakery products. Trends in Food Science & Technology. 2017 Sep 1;67:106–128.

[60] Danneskiold-Samsøe NB, Barros HD, Santos R, Bicas JL, Cazarin CB, Madsen L, Kristiansen K, Pastore GM, Brix S, Júnior MR. Interplay between food and gut microbiota in health and disease. Food Research International. 2019 Jan 1;115:23–31.

[61] Bigliardi B, Galati F. Innovation trends in the food industry: The case of functional foods. Trends in Food Science & Technology. 2013 Jun 1;31(2):118–129.

[62] Jäger R, Zaragoza J, Purpura M, Iametti S, Marengo M, Tinsley GM, Anzalone AJ, Oliver JM, Fiore W, Biffi A, Urbina S. Probiotic administration increases amino acid absorption from plant protein: A placebo-controlled, randomized, double-blind, multicenter, crossover study. Probiotics and Antimicrobial Proteins. 2020 Dec;12:1330–1339.

[63] Marco ML, Sanders ME, Gänzle M, Arrieta MC, Cotter PD, De Vuyst L, Hill C, Holzapfel W, Lebeer S, Merenstein D, Reid G. The International Scientific Association for Probiotics and Prebiotics (ISAPP) consensus statement on fermented foods. Nature Reviews Gastroenterology & Hepatology. 2021 Mar;18(3):196–208.

[64] Turroni F, van Sinderen D, Ventura M. Bifidobacteria: From ecology to genomics. Front Biosci. 2009 Jan 1;14(4673):84.

[65] Watson H, Mitra S, Croden FC, Taylor M, Wood HM, Perry SL, Spencer JA, Quirke P, Toogood GJ, Lawton CL, Dye L. A randomised trial of the effect of omega-3 polyunsaturated fatty acid supplements on the human intestinal microbiota. Gut. 2018 Nov 1;67(11):1974–1983.

[66] Menni C, Zierer J, Pallister T, Jackson MA, Long T, Mohney RP, Steves CJ, Spector TD, Valdes AM. Omega-3 fatty acids correlate with gut microbiome diversity and production of N-carbamylglutamate in middle aged and elderly women. Scientific Reports. 2017 Sep 11;7(1):11079.

[67] Anhê FF, Roy D, Pilon G, Dudonné S, Matamoros S, Varin TV, Garofalo C, Moine Q, Desjardins Y, Levy E, Marette A. A polyphenol-rich cranberry extract protects from diet-induced obesity, insulin resistance and intestinal inflammation in association with increased Akkermansia spp. population in the gut microbiota of mice. Gut. 2015 Jun 1;64(6):872–883.

[68] Rodriguez-Mateos A, Istas G, Boschek L, Feliciano RP, Mills CE, Boby C, Gomez-Alonso S, Milenkovic D, Heiss C. Circulating anthocyanin metabolites mediate vascular benefits of blueberries: Insights from randomized controlled trials, metabolomics, and nutrigenomics. The Journals of Gerontology: Series A. 2019 Jun 18;74(7):967–976.

[69] Fahey JW, Holtzclaw WD, Wehage SL, Wade KL, Stephenson KK, Talalay P. Sulforaphane bioavailability from glucoraphanin-rich broccoli: Control by active endogenous myrosinase. PloS One. 2015 Nov 2;10(11):e0140963.

[70] Traka MH, Saha S, Huseby S, Kopriva S, Walley PG, Barker GC, Moore J, Mero G, Van den Bosch F, Constant H, Kelly L. Genetic regulation of glucoraphanin accumulation in Beneforté® broccoli. New Phytologist. 2013 Jun;198(4):1085–1095.

[71] Liu H, Wang J, He T, Becker S, Zhang G, Li D, Ma X. Butyrate: A double-edged sword for health?. Advances in Nutrition. 2018 Jan 1;9(1):21–29.

[72] Singh BP, Rateb ME, Rodriguez-Couto S, Polizeli MD, Li WJ. Microbial secondary metabolites: Recent developments and technological challenges. Frontiers in Microbiology. 2019 Apr 26;10:914.

[73] Hill C, Guarner F, Reid G, Gibson GR, Merenstein DJ, Pot B, Morelli L, Canani RB, Flint HJ, Salminen S, Calder PC. The International Scientific Association for Probiotics and Prebiotics consensus statement on the scope and appropriate use of the term probiotic. Nature Reviews Gastroenterology & Hepatology. 2014 Aug;11(8):506–514.

[74] Ailioaie LM, Litscher G. Probiotics, photobiomodulation, and disease management: Controversies and challenges. International Journal of Molecular Sciences. 2021 May 6;22(9):4942.

[75] Roager HM, Vogt JK, Kristensen M, Hansen LB, Ibrügger S, Mærkedahl RB, Bahl MI, Lind MV, Nielsen RL, Frøkiær H, Gøbel RJ. Whole grain-rich diet reduces body weight and systemic low-grade inflammation without inducing major changes of the gut microbiome: A randomised cross-over trial. Gut. 2019 Jan 1;68(1):83–93.

[76] Martínez I, Stegen JC, Maldonado-Gómez MX, Eren AM, Siba PM, Greenhill AR, Walter J. The gut microbiota of rural papua new guineans: Composition, diversity patterns, and ecological processes. Cell Reports. 2015 Apr 28;11(4):527–538.

[77] Swanson KS, Gibson GR, Hutkins R, Reimer RA, Reid G, Verbeke K, Scott KP, Holscher HD, Azad MB, Delzenne NM, Sanders ME. The International Scientific Association for Probiotics and Prebiotics (ISAPP) consensus statement on the definition and scope of synbiotics. Nature Reviews Gastroenterology & Hepatology. 2020 Nov;17(11):687–701.

[78] Gibson GR, Hutkins R, Sanders ME, Prescott SL, Reimer RA, Salminen SJ, Scott K, Stanton C, Swanson KS, Cani PD, Verbeke K. Expert consensus document: The International Scientific Association for Probiotics and Prebiotics (ISAPP) consensus statement on the definition and scope of prebiotics. Nature Reviews Gastroenterology & Hepatology. 2017 Aug;14(8):491–502.

[79] Huis AV, Itterbeeck JV, Klunder H, Mertens E, Halloran A, Muir G, Vantomme P. Edible Insects: Future Prospects for Food and Feed Security.

[80] Puljić L, Banožić M, Kajić N, Vasilj V, Habschied K, Mastanjević K. Advancements in Research on Alternative Protein Sources and Their Application in Food Products: A Systematic Review. Processes. 2025 Jan 3;13(1):108.

[81] Jayabalan R, Malbaša RV, Lončar ES, Vitas JS, Sathishkumar M. A review on kombucha tea – Microbiology, composition, fermentation, beneficial effects, toxicity, and tea fungus. Comprehensive Reviews in Food Science and Food Safety. 2014 Jul;13(4):538–550.

[82] Papadopoulou D, Chrysikopoulou V, Rampaouni A, Tsoupras A. Antioxidant and anti-inflammatory properties of water kefir microbiota and its bioactive metabolites for health-promoting bio-functional products and applications. AIMS Microbiology. 2024 Sep 5;10(4):756.

[83] Gao X, Liu J, Li L, Liu W, Sun M. A brief review of nutraceutical ingredients in gastrointestinal disorders: Evidence and suggestions. International Journal of Molecular Sciences. 2020 Mar 6;21 (5):1822.

[84] Castro-Bravo N, Wells JM, Margolles A, Ruas-Madiedo P. Interactions of surface exopolysaccharides from Bifidobacterium and Lactobacillus within the intestinal environment. Frontiers in Microbiology. 2018 Oct 11;9:2426.

[85] La Fata G, Weber P, Mohajeri MH. Probiotics and the gut immune system: Indirect regulation. Probiotics and Antimicrobial Proteins. 2018 Mar;10:11–21.

[86] Thongaram T, Hoeflinger JL, Chow J, Miller MJ. Human milk oligosaccharide consumption by probiotic and human-associated bifidobacteria and lactobacilli. Journal of Dairy Science. 2017 Oct 1;100(10):7825–7833.

[87] Patel S, Goyal A. Functional oligosaccharides: Production, properties and applications. World Journal of Microbiology and Biotechnology. 2011 May;27:1119–1128.

[88] Davani-Davari D, Negahdaripour M, Karimzadeh I, Seifan M, Mohkam M, Masoumi SJ, Berenjian A, Ghasemi Y. Prebiotics: Definition, types, sources, mechanisms, and clinical applications. Foods. 2019 Mar 9;8(3):92.

[89] Swanson KS, Gibson GR, Hutkins R, Reimer RA, Reid G, Verbeke K, Scott KP, Holscher HD, Azad MB, Delzenne NM, Sanders ME. The International Scientific Association for Probiotics and Prebiotics (ISAPP) consensus statement on the definition and scope of synbiotics. Nature Reviews Gastroenterology & Hepatology. 2020 Nov;17(11):687–701.

[90] Cunningham M, Azcarate-Peril MA, Barnard A, Benoit V, Grimaldi R, Guyonnet D, Holscher HD, Hunter K, Manurung S, Obis D, Petrova MI. Shaping the future of probiotics and prebiotics. Trends in Microbiology. 2021 Aug 1;29(8):667–685.

[91] Salminen S, Collado MC, Endo A, Hill C, Lebeer S, Quigley EM, Sanders ME, Shamir R, Swann JR, Szajewska H, Vinderola G. The International Scientific Association of Probiotics and Prebiotics (ISAPP) consensus statement on the definition and scope of postbiotics. Nature Reviews Gastroenterology & Hepatology. 2021 Sep;18(9):649–667.

[92] Silva YP, Bernardi A, Frozza RL. The role of short-chain fatty acids from gut microbiota in gut-brain communication. Frontiers in Endocrinology. 2020 Jan 31;11:508738.

[93] Lynch KM, Zannini E, Coffey A, Arendt EK. Lactic acid bacteria exopolysaccharides in foods and beverages: Isolation, properties, characterization, and health benefits. Annual Review of Food Science and Technology. 2018 Mar 25;9(1):155–176.

[94] Devi SM, Kurrey NK, Halami PM. In vitro anti-inflammatory activity among probiotic Lactobacillus species isolated from fermented foods. Journal of Functional Foods. 2018 Aug 1;47:19–27.

[95] O'Flaherty S, Briner Crawley A, Theriot CM, Barrangou R. The Lactobacillus bile salt hydrolase repertoire reveals niche-specific adaptation. MSphere. 2018 Jun 27;3(3):10–128.

[96] Wang S, Huang XF, Zhang P, Wang H, Zhang Q, Yu S, Yu Y. Chronic rhein treatment improves recognition memory in high-fat diet-induced obese male mice. The Journal of Nutritional Biochemistry. 2016 Oct 1;36:42–50.

[97] Martínez-Maqueda D, Miralles B, De Pascual-Teresa S, Reverón I, Muñoz R, Recio I. Food-derived peptides stimulate mucin secretion and gene expression in intestinal cells. Journal of Agricultural and Food Chemistry. 2012 Sep 5;60(35):8600–8605.

[98] Costantini L, Molinari R, Farinon B, Merendino N. Impact of omega-3 fatty acids on the gut microbiota. International Journal of Molecular Sciences. 2017 Dec 7;18(12):2645.

[99] Cruzat V, Macedo Rogero M, Noel Keane K, Curi R, Newsholme P. Glutamine: Metabolism and immune function, supplementation and clinical translation. Nutrients. 2018 Nov;10(11):1564.

[100] Fiocchi A, Pawankar R, Cuello-Garcia C, Ahn K, Al-Hammadi S, Agarwal A, Beyer K, Burks W, Canonica GW, Ebisawa M, Gandhi S. World allergy organization-McMaster University guidelines for allergic disease prevention (GLAD-P): Probiotics. World Allergy Organization Journal. 2015 Dec;8:1–3.

[101] Ancuceanu R, Anghel AI, Hovaneț MV, Ciobanu AM, Lascu BE, Dinu M. Antioxidant activity of essential oils from Pinaceae species. Antioxidants. 2024 Feb 26;13(3):286.

[102] Szajewska H, Kołodziej M. Systematic review with meta-analysis: Lactobacillus rhamnosus GG in the prevention of antibiotic-associated diarrhoea in children and adults. Alimentary Pharmacology & Therapeutics. 2015 Nov;42(10):1149–1157.

[103] Hojsak I, Pavić AM, Kos T, Dumančić J, Kolaček S. Bifidobacterium animalis subsp. lactis in prevention of common infections in healthy children attending day care centers–randomized, double blind, placebo-controlled study. Clinical Nutrition. 2016 Jun 1;35(3):587–591.

[104] Plovier H, Everard A, Druart C, Depommier C, Van Hul M, Geurts L, Chilloux J, Ottman N, Duparc T, Lichtenstein L, Myridakis A. A purified membrane protein from Akkermansia muciniphila or the pasteurized bacterium improves metabolism in obese and diabetic mice. Nature Medicine. 2017 Jan;23(1):107–113.

[105] Depommier C, Everard A, Druart C, Plovier H, Van Hul M, Vieira-Silva S, Falony G, Raes J, Maiter D, Delzenne NM, de Barsy M. Supplementation with Akkermansia muciniphila in overweight and obese human volunteers: A proof-of-concept exploratory study. Nature Medicine. 2019 Jul;25 (7):1096–1103.

[106] Wilson B, Whelan K. Prebiotic inulin-type fructans and galacto-oligosaccharides: Definition, specificity, function, and application in gastrointestinal disorders. Journal of Gastroenterology and Hepatology. 2017 Mar;32:64–68.

[107] Sabater-Molina M, Larqué E, Torrella F, Zamora S. Dietary fructooligosaccharides and potential benefits on health. Journal of Physiology and Biochemistry. 2009 Sep;65:315–328.

[108] Bode L. Human milk oligosaccharides in the prevention of necrotizing enterocolitis: A journey from in vitro and in vivo models to mother-infant cohort studies. Frontiers in Pediatrics. 2018 Dec 4;6:385.

[109] Goehring KC, Kennedy AD, Prieto PA, Buck RH. Direct evidence for the presence of human milk oligosaccharides in the circulation of breastfed infants. PloS One. 2014 Jul 7;9(7):e101692.

[110] Yasukawa Z, Inoue R, Ozeki M, Okubo T, Takagi T, Honda A, Naito Y. Effect of repeated consumption of partially hydrolyzed guar gum on fecal characteristics and gut microbiota: A randomized, double-blind, placebo-controlled, and parallel-group clinical trial. Nutrients. 2019 Sep 10;11(9):2170.

[111] Pessione E. Lactic acid bacteria contribution to gut microbiota complexity: Lights and shadows. Frontiers in Cellular and Infection Microbiology. 2012 Jun 22;2:86.

[112] Mortensen PB, Clausen MR. Short-chain fatty acids in the human colon: Relation to gastrointestinal health and disease. Scandinavian Journal of Gastroenterology. 1996 Jan 1;31(sup216):132–148.

[113] Ríos-Covián D, Ruas-Madiedo P, Margolles A, Gueimonde M, De Los Reyes-gavilán CG, Salazar N. Intestinal short chain fatty acids and their link with diet and human health. Frontiers in Microbiology. 2016 Feb 17;7:185.

[114] Cotter PD, Ross RP, Hill C. Bacteriocins – A viable alternative to antibiotics?. Nature Reviews Microbiology. 2013 Feb;11(2):95–105.

[115] Yang SC, Lin CH, Sung CT, Fang JY. Antibacterial activities of bacteriocins: Application in foods and pharmaceuticals. Frontiers in Microbiology. 2014 May 26;5:241.

[116] Marco ML, Heeney D, Binda S, Cifelli CJ, Cotter PD, Foligné B, Gänzle M, Kort R, Pasin G, Pihlanto A, Smid EJ. Health benefits of fermented foods: Microbiota and beyond. Current Opinion in Biotechnology. 2017 Apr 1;44:94–102.

[117] Bourrie BC, Willing BP, Cotter PD. The microbiota and health promoting characteristics of the fermented beverage kefir. Frontiers in Microbiology. 2016 May 4;7:196946.

[118] Kok CR, Hutkins R. Yogurt and other fermented foods as sources of health-promoting bacteria. Nutrition Reviews. 2018 Dec 1;76:(Supplement_1):4–15

[119] Vitaglione P, Mennella I, Ferracane R, Rivellese AA, Giacco R, Ercolini D, Gibbons SM, La Storia A, Gilbert JA, Jonnalagadda S, Thielecke F. Whole-grain wheat consumption reduces inflammation in a randomized controlled trial on overweight and obese subjects with unhealthy dietary and lifestyle behaviors: Role of polyphenols bound to cereal dietary fiber. The American Journal of Clinical Nutrition. 2015 Feb 1;101(2):251–261.

[120] Foyer CH, Lam HM, Nguyen HT, Siddique KH, Varshney RK, Colmer TD, Cowling W, Bramley H, Mori TA, Hodgson JM, Cooper JW. Neglecting legumes has compromised human health and sustainable food production. Nature Plants. 2016 Aug 2;2(8):1–0.

[121] Williamson G, Clifford MN. Role of the small intestine, colon and microbiota in determining the metabolic fate of polyphenols. Biochemical Pharmacology. 2017 Sep 1;139:24–39.

[122] Lamuel-Raventos RM, Onge MP. Prebiotic nut compounds and human microbiota. Critical Reviews in Food Science and Nutrition. 2017 Sep 22;57(14):3154–3163.

[123] Min M, Bunt CR, Mason SL, Bennett GN, Hussain MA. Effect of non-dairy food matrices on the survival of probiotic bacteria during storage. Microorganisms. 2017 Aug 1;5(3):43.

[124] Bhandarkar NS, Brown L, Panchal SK. Chlorogenic acid attenuates high-carbohydrate, high-fat diet–induced cardiovascular, liver, and metabolic changes in rats. Nutrition Research. 2019 Feb 1;62:78–88.

[125] Manach C, Scalbert A, Morand C, Rémésy C, Jiménez L. Polyphenols: Food sources and bioavailability. The American Journal of Clinical Nutrition. 2004 May 1;79(5):727–747.

[126] Singh RK, Chang HW, Yan DI, Lee KM, Ucmak D, Wong K, Abrouk M, Farahnik B, Nakamura M, Zhu TH, Bhutani T. Influence of diet on the gut microbiome and implications for human health. Journal of Translational Medicine. 2017 Dec;15(16):1–7.

[127] Cockburn DW, Koropatkin NM. Polysaccharide degradation by the intestinal microbiota and its influence on human health and disease. Journal of Molecular Biology. 2016 Aug 14;428 (1):3230–3252.

[128] Baxter NT, Schmidt AW, Venkataraman A, Kim KS, Waldron C, Schmidt TM. Dynamics of human gut microbiota and short-chain fatty acids in response to dietary interventions with three fermentable fibers. MBio. 2019 Feb 26;10(1):10–128.

[129] Selma MV, Beltrán D, Luna MC, Romo-Vaquero M, García-Villalba R, Mira A, Espín JC, Tomás-Barberán FA. Isolation of human intestinal bacteria capable of producing the bioactive metabolite isourolithin a from ellagic acid. Frontiers in Microbiology. 2017 Aug 7;8(13):1521.

[130] Rose EC, Odle J, Blikslager AT, Ziegler AL. Probiotics, prebiotics and epithelial tight junctions: A promising approach to modulate intestinal barrier function. International Journal of Molecular Sciences. 2021 Jun 23;22(13):6729.

[131] Dalile B, Van Oudenhove L, Vervliet B, Verbeke K. The role of short-chain fatty acids in microbiota–gut–brain communication. Nature Reviews Gastroenterology & Hepatology. 2019 Aug;16(8):461–478.

[132] Perdijk O, Van Splunter M, Savelkoul HF, Brugman S, Van Neerven RJ. Cow's milk and immune function in the respiratory tract: Potential mechanisms. Frontiers in Immunology. 2018 Feb 12;9:143.

[133] Khanna S, Pardi DS, Kelly CR, Kraft CS, Dhere T, Henn MR, Lombardo MJ, Vulic M, Ohsumi T, Winkler J, Pindar C. A novel microbiome therapeutic increases gut microbial diversity and prevents recurrent Clostridium difficile infection. The Journal of Infectious Diseases. 2016 Jul 15;214 (2):173–181.

[134] Kortman GA, Mulder ML, Richters TJ, Shanmugam NK, Trebicka E, Boekhorst J, Timmerman HM, Roelofs R, Wiegerinck ET, Laarakkers CM, Swinkels DW. Low dietary iron intake restrains the intestinal inflammatory response and pathology of enteric infection by food-borne bacterial pathogens. European Journal of Immunology. 2015 Sep;45(9):2553–2567.

[135] DiNicolantonio JJ, McCarty M, OKeefe J. Association of moderately elevated trimethylamine N-oxide with cardiovascular risk: Is TMAO serving as a marker for hepatic insulin resistance. Open Heart. 2019 Feb 1;6(1):e000890.

[136] Vallianou N, Stratigou T, Christodoulatos GS, Dalamaga M. Understanding the role of the gut microbiome and microbial metabolites in obesity and obesity-associated metabolic disorders: Current evidence and perspectives. Current Obesity Reports. 2019 Sep 15;8:317–332.

[137] Chen F, Wen Q, Jiang J, Li HL, Tan YF, Li YH, Zeng NK. Could the gut microbiota reconcile the oral bioavailability conundrum of traditional herbs?. Journal of Ethnopharmacology. 2016 Feb 17;179:253–264.

[138] Canfora EE, Meex RC, Venema K, Blaak EE. Gut microbial metabolites in obesity, NAFLD and T2DM. Nature Reviews Endocrinology. 2019 May;15(5):261–273.

[139] Borgeraas H, Johnson LK, Skattebu J, Hertel JK, Hjelmesaeth J. Effects of probiotics on body weight, body mass index, fat mass and fat percentage in subjects with overweight or obesity: A systematic review and meta-analysis of randomized controlled trials. Obesity Reviews. 2018 Feb;19(2):219–232.

[140] Kim DH, Kim H, Jeong D, Kang IB, Chon JW, Kim HS, Song KY, Seo KH. Kefir alleviates obesity and hepatic steatosis in high-fat diet-fed mice by modulation of gut microbiota and mycobiota:

Targeted and untargeted community analysis with correlation of biomarkers. The Journal of Nutritional Biochemistry. 2017 Jun 1;44:35–43.

[141] Bidu C, Escoula Q, Bellenger S, Spor A, Galan M, Geissler A, Bouchot A, Dardevet D, Morio B, Cani PD, Lagrost L. The transplantation of ω3 PUFA-altered gut microbiota of FAT-1 mice to wild-type littermates prevents obesity and associated metabolic disorders. Diabetes. 2018 Aug 1;67 (8):1512–1523.

[142] Ni J, Wu GD, Albenberg L, Tomov VT. Gut microbiota and IBD: Causation or correlation?. Nature Reviews Gastroenterology & Hepatology. 2017 Oct;14(10):573–584.

[143] Limketkai BN, Iheozor-Ejiofor Z, Gjuladin-Hellon T, Parian A, Matarese LE, Bracewell K, MacDonald JK, Gordon M, Mullin GE. Dietary interventions for induction and maintenance of remission in inflammatory bowel disease. Cochrane Database of Systematic Reviews. 2019;(2).

[144] Lang A, Salomon N, Wu JC, Kopylov U, Lahat A, Har-Noy O, Ching JY, Cheong PK, Avidan B, Gamus D, Kaimakliotis I. Curcumin in combination with mesalamine induces remission in patients with mild-to-moderate ulcerative colitis in a randomized controlled trial. Clinical Gastroenterology and Hepatology. 2015 Aug 1;13(8):1444–1449.

[145] Larrosa M, González-Sarrías A, Yáñez-Gascón MJ, Selma MV, Azorín-Ortuño M, Toti S, Tomás-Barberán F, Dolara P, Espín JC. Anti-inflammatory properties of a pomegranate extract and its metabolite urolithin-A in a colitis rat model and the effect of colon inflammation on phenolic metabolism. The Journal of Nutritional Biochemistry. 2010 Aug 1;21(8):717–725.

[146] Dale HF, Rasmussen SH, Asiller ÖÖ, Lied GA. Probiotics in irritable bowel syndrome: An up-to-date systematic review. Nutrients. 2019 Sep 2;11(9):2048.

[147] Ford AC, Harris LA, Lacy BE, Quigley EM, Moayyedi P. Systematic review with meta-analysis: The efficacy of prebiotics, probiotics, synbiotics and antibiotics in irritable bowel syndrome. Alimentary Pharmacology & Therapeutics. 2018 Nov;48(10):1044–1060.

[148] Whelan K, Martin LD, Staudacher HM, Lomer M. The low FODMAP diet in the management of irritable bowel syndrome: An evidence-based review of FODMAP restriction, reintroduction and personalisation in clinical practice. Journal of Human Nutrition and Dietetics. 2018 Apr;31 (2):239–255.

[149] Song M, Garrett WS, Chan AT. Nutrients, foods, and colorectal cancer prevention. Gastroenterology. 2015 May 1;148(6):1244–1260.

[150] Tomás-Barberán FA, González-Sarrías A, García-Villalba R, Núñez-Sánchez MA, Selma MV, García-Conesa MT, Espín JC. Urolithins, the rescue of "old" metabolites to understand a "new" concept: Metabotypes as a nexus among phenolic metabolism, microbiota dysbiosis, and host health status. Molecular Nutrition & Food Research. 2017 Jan;61(1):1500901.

[151] Tapiovaara L, Lehtoranta L, Poussa T, Mäkivuokko H, Korpela R, Pitkäranta A. Absence of adverse events in healthy individuals using probiotics–analysis of six randomised studies by one study group. Beneficial Microbes. 2016 Mar 11;7(2):161–170.

[152] Dai X, Stanilka JM, Rowe CA, Esteves EA, Nieves JC, Spaiser SJ, Christman MC, Langkamp-Henken B, Percival SS. Consuming Lentinula edodes (Shiitake) mushrooms daily improves human immunity: A randomized dietary intervention in healthy young adults. Journal of the American College of Nutrition. 2015 Nov 2;34(6):478–487.

[153] McFarland LV, Surawicz CM, Greenberg RN, Fekety R, Elmer GW, Moyer KA, Melcher SA, Bowen KE, Cox JL, Noorani Z, Harrington G. A randomized placebo-controlled trial of Saccharomyces boulardii in combination with standard antibiotics for Clostridium difficile disease. Jama. 1994 Jun 22;271 (24):1913–1918.

[154] Rodríguez-Pérez C, Quirantes-Piné R, Uberos J, Jiménez-Sánchez C, Peña A, Segura-Carretero A. Antibacterial activity of isolated phenolic compounds from cranberry (Vaccinium macrocarpon) against Escherichia coli. Food & Function. 2016;7(3):1564–1573.

[155] Roduit C, Frei R, Ferstl R, Loeliger S, Westermann P, Rhyner C, Schiavi E, Barcik W, Rodriguez-Perez N, Wawrzyniak M, Chassard C. High levels of butyrate and propionate in early life are associated with protection against atopy. Allergy. 2019 Apr;74(4):799–809.

[156] Suez J, Zmora N, Segal E, Elinav E. The pros, cons, and many unknowns of probiotics. Nature Medicine. 2019 May;25(5):716–729.

[157] Gutiérrez-Díaz I, Fernández-Navarro T, Sánchez B, Margolles A, González S. Mediterranean diet and faecal microbiota: A transversal study. Food & Function. 2016;7(5):2347–2356.

[158] Vogt NM, Kerby RL, Dill-Mcfarland KA, Harding SJ, Merluzzi AP, Johnson SC, Carlsson CM, Asthana S, Zetterberg H, Blennow K, Bendlin BB. Gut microbiome alterations in Alzheimer's disease. Scientific Reports. 2017 Oct 19;7(1):13537.

[159] Krikorian R, Shidler MD, Nash TA, Kalt W, Vinqvist-Tymchuk MR, Shukitt-Hale B, Joseph JA. Blueberry supplementation improves memory in older adults. Journal of Agricultural and Food Chemistry. 2010 Apr 14;58(7):3996–4000.

[160] Zhu F, Du B, Xu B. Anti-inflammatory effects of phytochemicals from fruits, vegetables, and food legumes: A review. Critical Reviews in Food Science and Nutrition. 2018 May 24;58(8):1260–1270.

[161] Valles-Colomer M, Falony G, Darzi Y, Tigchelaar EF, Wang J, Tito RY, Schiweck C, Kurilshikov A, Joossens M, Wijmenga C, Claes S. The neuroactive potential of the human gut microbiota in quality of life and depression. Nature Microbiology. 2019 Apr;4(4):623–632.

[162] Liu RT, Walsh RF, Sheehan AE. Prebiotics and probiotics for depression and anxiety: A systematic review and meta-analysis of controlled clinical trials. Neuroscience & Biobehavioral Reviews. 2019 Jul 1;102:13–23.

[163] Kang DW, Adams JB, Coleman DM, Pollard EL, Maldonado J, McDonough-Means S, Caporaso JG, Krajmalnik-Brown R. Long-term benefit of Microbiota Transfer Therapy on autism symptoms and gut microbiota. Scientific Reports. 2019 Apr 9;9(1):5821.

[164] Santocchi E, Guiducci L, Fulceri F, Billeci L, Buzzigoli E, Apicella F, Calderoni S, Grossi E, Morales MA, Muratori F. Gut to brain interaction in Autism Spectrum Disorders: A randomized controlled trial on the role of probiotics on clinical, biochemical and neurophysiological parameters. BMC Psychiatry. 2016 Dec;16:1–6.

[165] Olson CA, Vuong HE, Yano JM, Liang QY, Nusbaum DJ, Hsiao EY. The gut microbiota mediates the anti-seizure effects of the ketogenic diet. Cell. 2018 Jun 14;173(7):1728–1741.

[166] Dahlin M, Prast-Nielsen S. The gut microbiome and epilepsy. EBioMedicine. 2019 Jun 1;44:741–746.

[167] McFarland LV, Huang Y, Wang L, Malfertheiner P. Systematic review and meta-analysis: Multi-strain probiotics as adjunct therapy for Helicobacter pylori eradication and prevention of adverse events. United European Gastroenterology Journal. 2016 Aug;4(4):546–561.

[168] Goldenberg JZ, Yap C, Lytvyn L, Lo CK, Beardsley J, Mertz D, Johnston BC. Probiotics for the prevention of Clostridium difficile-associated diarrhea in adults and children. Cochrane Database of Systematic Reviews. 2017;(12).

[169] King S, Glanville J, Sanders ME, Fitzgerald A, Varley D. Effectiveness of probiotics on the duration of illness in healthy children and adults who develop common acute respiratory infectious conditions: A systematic review and meta-analysis. British Journal of Nutrition. 2014 Jul;112(1):41–54.

[170] Rautava S, Luoto R, Salminen S, Isolauri E. Microbial contact during pregnancy, intestinal colonization and human disease. Nature Reviews Gastroenterology & Hepatology. 2012 Oct;9(10):565–576.

[171] Kalliomäki M, Salminen S, Arvilommi H, Kero P, Koskinen P, Isolauri E. Probiotics in primary prevention of atopic disease: A randomised placebo-controlled trial. The Lancet. 2001 Apr 7;357(9262):1076–1079.

[172] Gu Y, Wang X, Li J, Zhang Y, Zhong H, Liu R, Zhang D, Feng Q, Xie X, Hong J, Ren H. Analyses of gut microbiota and plasma bile acids enable stratification of patients for antidiabetic treatment. Nature Communications. 2017 Nov 27;8(1):1785.

[173] Parker HM, Johnson NA, Burdon CA, Cohn JS, O'Connor HT, George J. Omega-3 supplementation and non-alcoholic fatty liver disease: A systematic review and meta-analysis. Journal of Hepatology. 2012 Apr 1;56(4):944–951.

[174] Scorletti E, Bhatia L, McCormick KG, Clough GF, Nash K, Hodson L, Moyses HE, Calder PC, Byrne CD, Welcome Study Investigators, Sheron N. Effects of purified eicosapentaenoic and docosahexaenoic acids in nonalcoholic fatty liver disease: Results from the Welcome* study. Hepatology. 2014 Oct;60 (4):1211–1221.

[175] Korem T, Zeevi D, Zmora N, Weissbrod O, Bar N, Lotan-Pompan M, Avnit-Sagi T, Kosower N, Malka G, Rein M, Suez J. Bread affects clinical parameters and induces gut microbiome-associated personal glycemic responses. Cell Metabolism. 2017 Jun 6;25(6):1243–1253.

[176] Staudacher HM, Lomer MC, Farquharson FM, Louis P, Fava F, Franciosi E, Scholz M, Tuohy KM, Lindsay JO, Irving PM, Whelan K. A diet low in FODMAPs reduces symptoms in patients with irritable bowel syndrome and a probiotic restores bifidobacterium species: A randomized controlled trial. Gastroenterology. 2017 Oct 1;153(4):936–947.

[177] Fedorak RN, Feagan BG, Hotte N, Leddin D, Dieleman LA, Petrunia DM, Enns R, Bitton A, Chiba N, Paré P, Rostom A. The probiotic VSL# 3 has anti-inflammatory effects and could reduce endoscopic recurrence after surgery for Crohn's disease. Clinical Gastroenterology and Hepatology. 2015 May 1;13(5):928–935.

[178] Derwa Y, Gracie DJ, Hamlin PJ, Ford AC. Systematic review with meta-analysis: The efficacy of probiotics in inflammatory bowel disease. Alimentary Pharmacology & Therapeutics. 2017 Aug;46 (4):389–400.

[179] Bamia C, Lagiou P, Buckland G, Grioni S, Agnoli C, Taylor AJ, Dahm CC, Overvad K, Olsen A, Tjønneland A, Cottet V. Mediterranean diet and colorectal cancer risk: Results from a European cohort. European Journal of Epidemiology. 2013 Apr;28:317–328.

[180] Miller MG, Shukitt-Hale B. Berry fruit enhances beneficial signaling in the brain. Journal of Agricultural and Food Chemistry. 2012 Jun 13;60(23):5709–5715.

[181] Chao L, Liu C, Sutthawongwadee S, Li Y, Lv W, Chen W, Yu L, Zhou J, Guo A, Li Z, Guo S. Effects of probiotics on depressive or anxiety variables in healthy participants under stress conditions or with a depressive or anxiety diagnosis: A meta-analysis of randomized controlled trials. Frontiers in Neurology. 2020 May 22;11:421.

[182] Ercolini D, Fogliano V. Food design to feed the human gut microbiota. Journal of Agricultural and Food Chemistry. 2018 Mar 22;66(15):3754–3758.

[183] Zhang N, Ju Z, Zuo T. Time for food: The impact of diet on gut microbiota and human health. Nutrition. 2018 Jul 1;51:80–85.

[184] Stribling P, Ibrahim F. Dietary fibre definition revisited-The case of low molecular weight carbohydrates. Clinical Nutrition ESPEN. 2023 Jun 1;55:340–356.

[185] Holscher HD. Dietary fiber and prebiotics and the gastrointestinal microbiota. Gut Microbes. 2017 Mar 4;8(2):172–184.

[186] Quigly EM. Nutraceuticals as modulators of gut microbiota: Role in therapy. British Journal of Pharmacology. 2020 Mar;177(6):1351–1362.

[187] Nogal A, Valdes AM, Menni C. The role of short-chain fatty acids in the interplay between gut microbiota and diet in cardio-metabolic health. Gut Microbes. 2021 Jan 1;13(1):1897212.

[188] Li HY, Zhou DD, Gan RY, Huang SY, Zhao CN, Shang AO, Xu XY, Li HB. Effects and mechanisms of probiotics, prebiotics, synbiotics, and postbiotics on metabolic diseases targeting gut microbiota: A narrative review. Nutrients. 2021 Sep 15;13(9):3211.

[189] Ji J, Jin W, Liu SJ, Jiao Z, Li X. Probiotics, prebiotics, and postbiotics in health and disease. MedComm. 2023 Dec;4(6):e420.

[190] Liu C, Ma N, Feng Y, Zhou M, Li H, Zhang X, Ma X. From probiotics to postbiotics: Concepts and applications. Animal Research and One Health. 2023 Aug;1(1):92–114.

[191] Plamada D, Vodnar DC. Polyphenols – Gut microbiota interrelationship: A transition to a new generation of prebiotics. Nutrients. 2021 Dec 28;14(1):137.

[192] Dingeo G, Brito A, Samouda H, Iddir M, La Frano MR, Bohn T. Phytochemicals as modifiers of gut microbial communities. Food & Function. 2020;11(10):8444–8471.

[193] Santhiravel S, Bekhit AE, Mendis E, Jacobs JL, Dunshea FR, Rajapakse N, Ponnampalam EN. The impact of plant phytochemicals on the gut microbiota of humans for a balanced life. International Journal of Molecular Sciences. 2022 Jul 23;23(15):8124.

[194] Kumar Singh A, Cabral C, Kumar R, Ganguly R, Kumar Rana H, Gupta A, Rosaria Lauro M, Carbone C, Reis F, Pandey AK. Beneficial effects of dietary polyphenols on gut microbiota and strategies to improve delivery efficiency. Nutrients. 2019 Sep 13;11(9):2216.

[195] Ordovas JM, Ferguson LR, Tai ES, Mathers JC. Personalised nutrition and health. bmj. 2018 Jun 13;361.

[196] De Roos B. Personalised nutrition: Ready for practice?. Proceedings of the Nutrition Society. 2013 Feb;72(1):48–52.

[197] Celis-Morales C, Livingstone KM, Marsaux CF, Macready AL, Fallaize R, O'Donovan CB, Woolhead C, Forster H, Walsh MC, Navas-Carretero S, San-Cristobal R. Effect of personalized nutrition on health-related behaviour change: Evidence from the Food4Me European randomized controlled trial. International Journal of Epidemiology. 2017 Apr 1;46(2):578–588.

[198] Hesketh J. Personalised nutrition: How far has nutrigenomics progressed?. European Journal of Clinical Nutrition. 2013 May;67(5):430–435.

[199] Martindale R, Patel JJ, Taylor B, Arabi YM, Warren M, McClave SA. Nutrition therapy in critically ill patients with coronavirus disease 2019. Journal of Parenteral and Enteral Nutrition. 2020 Sep;44 (7):1174–1184.

[200] De Toro-martín J, Arsenault BJ, Després JP, Vohl MC. Precision nutrition: A review of personalized nutritional approaches for the prevention and management of metabolic syndrome. Nutrients. 2017 Aug 22;9(8):913.

[201] Fassarella M, Blaak EE, Penders J, Nauta A, Smidt H, Zoetendal EG. Gut microbiome stability and resilience: Elucidating the response to perturbations in order to modulate gut health. Gut. 2021 Mar 1;70(3):595–605.

[202] Kolodziejczyk AA, Zheng D, Elinav E. Diet–microbiota interactions and personalized nutrition. Nature Reviews Microbiology. 2019 Dec;17(12):742–753.

[203] Valdes AM, Walter J, Segal E, Spector TD. Role of the gut microbiota in nutrition and health. Bmj. 2018 Jun 13;361.

[204] Chen TC, Clark J, Riddles MK, Mohadjer LK, Fakhouri TH. National Health and Nutrition Examination Survey. 2015– 2018, sample design and estimation procedures.

[205] Ferguson LR, De Caterina R, Görman U, Allayee H, Kohlmeier M, Prasad C, Choi MS, Curi R, De Luis DA, Gil Á, Kang JX. Guide and position of the international society of nutrigenetics/nutrigenomics on personalised nutrition: Part 1-fields of precision nutrition. Lifestyle Genomics. 2016 May 12;9 (1):12–27.

[206] Evert AB, Dennison M, Gardner CD, Garvey WT, Lau KH, MacLeod J, Mitri J, Pereira RF, Rawlings K, Robinson S, Saslow L. Nutrition therapy for adults with diabetes or prediabetes: A consensus report. Diabetes Care. 2019 Apr 15;42(5):731.

[207] Lambell KJ, Tatucu-Babet OA, Chapple LA, Gantner D, Ridley EJ. Nutrition therapy in critical illness: A review of the literature for clinicians. Critical Care. 2020 Dec;24:1–1.

[208] Lagoumintzis G, Patrinos GP. Triangulating nutrigenomics, metabolomics and microbiomics toward personalized nutrition and healthy living. Human Genomics. 2023 Dec 8;17(1):109.

[209] Hassoun A, Bekhit AE, Jambrak AR, Regenstein JM, Chemat F, Morton JD, Gudjónsdóttir M, Carpena M, Prieto MA, Varela P, Arshad RN. The fourth industrial revolution in the food industry – Part II: Emerging food trends. Critical Reviews in Food Science and Nutrition. 2024 Jan 13;64(2):407–437.

[210] Abouelela ME, Helmy YA. Next-generation probiotics as novel therapeutics for improving human health: Current trends and future perspectives. Microorganisms. 2024 Feb 20;12(3):430.

[211] Obayomi OV, Olaniran AF, Owa SO. Unveiling the role of functional foods with emphasis on prebiotics and probiotics in human health: A review. Journal of Functional Foods. 2024 Aug 1;119:106337.

[212] Leeuwendaal NK, Stanton C, O'toole PW, Beresford TP. Fermented foods, health and the gut microbiome. Nutrients. 2022 Apr 6;14(7):1527.

[213] Allai FM, Azad ZA, Mir NA, Gul K. Recent advances in non-thermal processing technologies for enhancing shelf life and improving food safety. Applied Food Research. 2023 Jun 1;3(1):100258.

[214] Ballini A, Charitos IA, Cantore S, Topi S, Bottalico L, Santacroce L. About functional foods: The probiotics and prebiotics state of art. Antibiotics. 2023 Mar 23;12(4):635.

[215] Jafarzadeh S, Qazanfarzadeh Z, Majzoobi M, Sheiband S, Oladzadabbasabad N, Esmaeili Y, Barrow CJ, Timms W. Alternative proteins; A path to sustainable diets and environment. Current Research in Food Science. 2024 Oct;10:100882.

[216] Bryant CJ. Plant-based animal product alternatives are healthier and more environmentally sustainable than animal products. Future Foods. 2022 Dec 1;6:100174.

[217] Abbaspour N. Fermentation's pivotal role in shaping the future of plant-based foods: An integrative review of fermentation processes and their impact on sensory and health benefits. Applied Food Research. 2024 Aug;9:100468.

[218] Mimee M, Nadeau P, Hayward A, Carim S, Flanagan S, Jerger L, Collins J, McDonnell S, Swartwout R, Citorik RJ, Bulović V. An ingestible bacterial-electronic system to monitor gastrointestinal health. Science. 2018 May 25;360(6391):9158.

[219] Johnson M, Zaretskaya I, Raytselis Y, Merezhuk Y, McGinnis S, Madden TL. NCBI BLAST: A better web interface. Nucleic Acids Research. 2008 Apr 24;36(suppl_2):W5–9.

[220] Riglar DT, Giessen TW, Baym M, Kerns SJ, Niederhuber MJ, Bronson RT, Kotula JW, Gerber GK, Way JC, Silver PA. Engineered bacteria can function in the mammalian gut long-term as live diagnostics of inflammation. Nature Biotechnology. 2017 Jul;35(7):653–658.

[221] Mimee M, Citorik RJ, Lu TK. Microbiome therapeutics – Advances and challenges. Advanced Drug Delivery Reviews. 2016 Oct 1;105:44–54.

[222] Díaz-Montes E. Wall materials for encapsulating bioactive compounds via spray-drying: A review. Polymers. 2023 Jun 12;15(12):2659.

[223] Sun J, Zhou W, Huang D, Yan L. 3D food printing: Perspectives. Polymers for Food Applications. 2018 Aug;10:725–755.

[224] Liu Z, Zhang M, Bhandari B, Wang Y. 3D printing: Printing precision and application in food sector. Trends in Food Science & Technology. 2017 Nov 1;69:83–94.

[225] Zeevi D, Korem T, Zmora N, Israeli D, Rothschild D, Weinberger A, Ben-Yacov O, Lador D, Avnit-Sagi T, Lotan-Pompan M, Suez J. Personalized nutrition by prediction of glycemic responses. Cell. 2015 Nov 19;163(5):1079–1094.

[226] Berry SE, Valdes AM, Drew DA, Asnicar F, Mazidi M, Wolf J, Capdevila J, Hadjigeorgiou G, Davies R, Al Khatib H, Bonnett C. Human postprandial responses to food and potential for precision nutrition. Nature Medicine. 2020 Jun 1;26(6):964–973.

[227] Zmora N, Suez J, Elinav E. You are what you eat: Diet, health and the gut microbiota. Nature Reviews Gastroenterology & Hepatology. 2019 Jan;16(1):35–56.

[228] Fan Y, Pedersen O. Gut microbiota in human metabolic health and disease. Nature Reviews Microbiology. 2021 Jan;19(1):55–71.

[229] Binda S, Hill C, Johansen E, Obis D, Pot B, Sanders ME, Tremblay A, Ouwehand AC. Criteria to qualify microorganisms as "probiotic" in foods and dietary supplements. Frontiers in Microbiology. 2020 Jul 24;11:1662.

[230] Binns N. Probiotics, Prebiotics and the Gut Microbiota. 2013.

[231] Granato D, Barba FJ, Bursać KD, Lorenzo JM, Cruz AG, Putnik P. Functional foods: Product development, technological trends, efficacy testing, and safety. Annual Review of Food Science and Technology. 2020 Mar 25;11(1):93–118.

[232] Willett W, Rockström J, Loken B, Springmann M, Lang T, Vermeulen S, Garnett T, Tilman D, DeClerck F, Wood A, Jonell M. Food in the Anthropocene: The EAT–Lancet Commission on healthy diets from sustainable food systems. The Lancet. 2019 Feb 2;393(10170):447–492.

Susmitha B., Kalidoss A.*, Ashwitha Kodaparthi, Yogananth N.

Chapter 8
Fermented Foods: Sources of Probiotics and Prebiotics

Abstract: Fermented foods serve as rich sources of probiotics and prebiotics, both essential for maintaining a healthy gut microbiome. Probiotics are live beneficial microorganisms – predominantly bacteria such as *Lactobacillus* and *Bifidobacterium*, or yeast – that provide health benefits when consumed in adequate amounts. These probiotics occur naturally in fermented foods such as yoghurt and sauerkraut. Prebiotics, primarily nondigestible fibers found in foods like garlic, onions, and bananas, fuel the growth of beneficial gut bacteria. When probiotics and prebiotics combine, they form synbiotics, which work together to enhance gut health more effectively than either component alone. This chapter explores the origin and fermentation process of foods, such as idli, dosa, dhokla, pickles, enduri pitha, hawaijar, khorisa, handia, gundruk, naan, bhatura, and kulcha, through examining their probiotic profiles and health benefits. By understanding how probiotics, prebiotics, and synbiotics function in fermented foods, this chapter provides insights into their contributions to gut health and overall well-being.

Keywords: probiotics, prebiotics, synbiotics, lactic acid bacteria (LAB), gut microbiome

8.1 Introduction

Fermentation, an age-old technique, has been employed across various cultures to produce and preserve food, yielding a diverse array of fermented products that are integral to regional cuisines and possess distinct nutritional profiles. This process involves the enzymatic breakdown of food components by microorganisms, leading to

*Corresponding author: Kalidoss A., Department of Microbiology, SRM Arts and Science College, Kattankulathur, Chengalpattu 603203, Tamil Nadu, India, e-mail: kalidassgene@gmail.com
Susmitha B., Department of Microbiology, St. Pious X Degree and PG College for Women (Autonomous), Nacharam, Hyderabad 500076, Telangana, India, e-mail: kalidassgene@gmail.com
Ashwitha Kodaparthi, Department of Microbiology, MNR PG College, Kukatpally, Hyderabad 500085, Telangana, India
Yogananth N., Department of Biotechnology, Mohamed Sathak College of Arts and Science, Sholinganallur, Chennai 600119, Tamil Nadu, India

https://doi.org/10.1515/9783112205150-008

the production of beneficial compounds such as organic acids, alcohols, and gases. These transformations contribute to the characteristic of fermented foods such as distinct flavors, textures, and aromas.

The human gastrointestinal tract hosts a complex community of microorganisms, collectively known as the gut microbiota, which influences various physiological functions, including digestion, immune response, and even mental health. Maintaining a balanced gut microbiota is essential, and dietary interventions, particularly the inclusion of fermented foods, have been shown to positively impact its composition and function.

Fermented foods are rich in probiotics, live microorganisms that confer health benefits to the host when administered in adequate amounts. Strains such as *Lactobacillus* and *Bifidobacterium* are commonly found in fermented products like yogurt, kefir, and kimchi. These probiotics can enhance gut barrier function, modulate immune responses, and inhibit the growth of pathogenic bacteria. In addition to probiotics, fermented foods often contain prebiotics – nondigestible food components that selectively stimulate the growth or activity of beneficial microorganisms in the gut. Prebiotics, found in foods like garlic, onions, and bananas, serve as nourishment for probiotics, facilitating their beneficial effects. The synergistic relationship between probiotics and prebiotics is termed "synbiotics." Consuming foods that contain both components can enhance the survival and activity of beneficial microbes, thereby promoting a healthy gut microbiota (Figure 8.1).

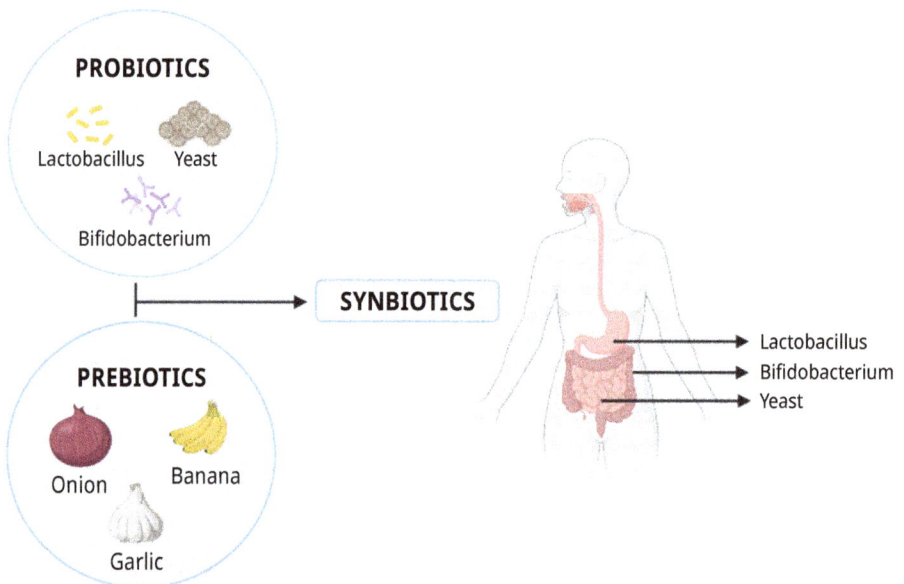

Figure 8.1: The synergistic relationship between probiotics and prebiotics.

This chapter explores the origin, microbial content, and health benefits of various fermented foods, both indigenous and nonindigenous. Understanding the roles of probiotics, prebiotics, and synbiotics in fermented foods provides valuable insights into their contributions to gut health and overall well-being [1, 2].

8.2 Fermented Foods

Fermented foods have long been an integral part of traditional diets across various cultures, providing not only distinctive flavors and textures but also significant nutritional and health benefits. These foods result from controlled microbial fermentation, where beneficial bacteria, yeasts, and molds act on raw ingredients, often derived from plant or animal sources. This process enhances the digestibility, safety, and shelf life of the foods while imparting unique sensory qualities. Many fermented foods are made from ingredients, such as cereals, grains, legumes, vegetables, and tubers, which naturally supply dietary fiber and function as prebiotics. Lactic acid bacteria (LAB) and other microorganisms involved in fermentation are essential for breaking down complex compounds, enhancing the nutritional profile, and serving as natural preservatives due to their antimicrobial properties.

Indigenous Indian fermented foods, such as idli, dosa, dhokla, and enduri pitha, exemplify cereal- and legume-based fermentation, enhancing protein digestibility while providing a soft, palatable texture. Fermented vegetable products, including pickles, hawaijar, khorisa, and gundruk, serve as flavorful condiments and side dishes, offering probiotic benefits and extended shelf stability. Additionally, fermented beverages like handia, made from rice, highlight the diversity of indigenous fermentation practices. Breads such as naan, bhatura, and kulcha depend on natural or added microbial cultures to achieve their distinctive softness, chewiness, and subtle tang. Collectively, all these traditional fermented foods are steeped in cultural heritage and serve as a source of dietary biotics, promoting gut health, enhancing nutrition, and supporting food security [3] (Figure 8.2).

8.2.1 Idli

8.2.1.1 Origin and Raw Materials Used

Idli, which originated from South India, is a delicious fermented breakfast dish [3] known for its fluffy texture and slightly sour taste – both results of the fermentation process and the raw materials used are urad dal, rice, and fenugreek [4]. In the preparation of certain fermented foods, fenugreek seeds (methi) are added in small quantities to enhance the fermentation process and improve both flavor and texture [4–7].

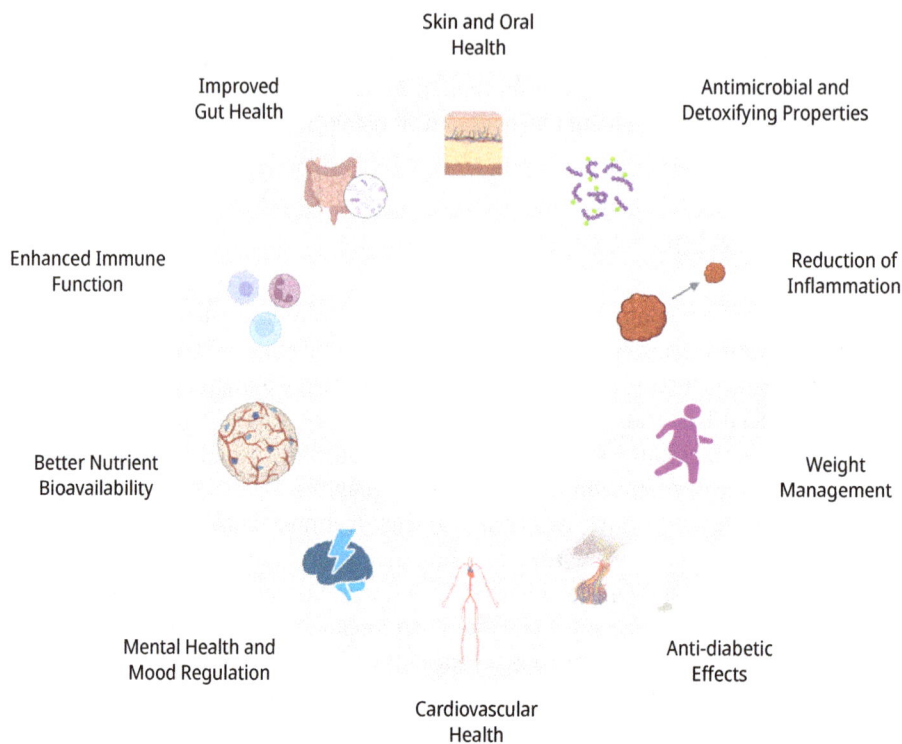

Figure 8.2: Potent health benefits of various fermented foods.

8.2.1.2 Probiotic Microflora Associated with Fermentation

The microorganisms that ferment idli batter are primarily probiotics. These include LAB, such as *Lactobacillus delbrueckii, Leuconostoc mesenteroides, Streptococcus faecalis*, and *Pediococcus cerevisiae*. Yeasts also play a role, including *Saccharomyces cerevisiae, Torulopsis candida*, and *Trichosporon pullulans* [8, 9].

8.2.1.3 Process of Idli Making

Fermentation of idli batter takes 18–20 h. During this time, microorganisms such as LAB, yeasts, and other beneficial bacteria like *L. delbrueckii* and *Enterococcus faecalis* break down the ingredients. This process produces carbon dioxide (which leavens the batter) and lactic acid (which increases the acidity) [10–13]:

	Black gram dal	**White rice (polished rice)**
Wash and soak	4–6 h	4–6 h
Finely/coarsely grind	In a mortar	In a mortar
	Combine slurries into a thick batter and mix well	
	Add salt for seasoning (~1% w/w of the total batter weight)	
	Incubate overnight in a warm place (30–32 °C)	
	Pour batter into small cups in idli cooker	
	Steam for 10 min	

8.2.1.4 Probiotic Health Benefits of Idli

Due to its fermentation process, idli contains beneficial microorganisms that support digestive health, improve nutrient absorption, and enhance the intestinal microflora. It helps in maintaining a healthy gut by reducing intestinal pH and promoting better intestinal functioning. Idli can restore the natural gut flora after antibiotic treatment and plays a role in preventing and treating acute diarrhea caused by rotaviruses. It is also known to increase lactose tolerance and improve lactose digestion. Nutritionally, idli is a low-glycemic-index food, making it suitable for individuals managing diabetes. It provides a steady release of energy, supports weight management, and contributes to heart health by aiding in cholesterol reduction. When consumed regularly as part of a balanced diet, idli serves as a complete, light, and nutritious meal with significant probiotic benefits [14].

8.2.2 Dosa

8.2.2.1 Origin and Raw materials

Dosa originated in South India and stands as a popular fermented food. The fermentation process plays a crucial role in creating the characteristic soft and fluffy batter. The primary raw materials include urad dal, rice, and fenugreek [15].

8.2.2.2 Predominant Probiotic Microflora of Dosa

The microorganisms that ferment dosa batter consist primarily of probiotics. These include LAB and yeasts, such as *Saccharomyces cerevisiae*, *T. candida*, *Trichosporon beigelli*, *Trichosporon pullulans*, and *Debaromyces hansenii*. Other LAB species present are *L. delbrueckii*, *Lactobacillus fermentum*, *Bacillus amyloliquefaciens*, *E. faecalis*, *Leuconostoc mesenteroides*, *S. faecalis*, and *P. cerevisiae* [16, 17].

8.2.2.3 Process of Dosa Making

Dosa batter fermentation typically requires 18–20 h. During this period, microorganisms naturally present in rice and black gram – including LAB, yeasts, and other LAB species like *L. delbrueckii* and *E. faecalis* – break down the ingredients. This process produces carbon dioxide, which leavens the batter, and lactic acid, which increases the batter's acidity [18]:

	Black gram dhal	White rice (polished rice)
Wash and soak	4–6 h	4–6 h
Finely/coarsely grind	In a mortar	In a mortar
Combine slurries into a thick batter and mix well		
Add salt for seasoning (~1% w/w of the total batter weight)		
Incubate overnight in a warm place (30–32 °C)		
Pour batter onto a pan, make a crepe, and add ghee/oil		
Heat for 2 min		

8.2.2.4 Probiotic Health Benefits of Dosa

Dosa provides multiple health benefits, especially related to digestive health. Its probiotic activity aids in easy digestion and enhances nutrient absorption. Being nutrient-rich, dosa offers a complete nutritional profile with a steady release of energy, making it suitable for individuals managing diabetes due to its low glycemic index. It also helps increase lactose tolerance and digestion, positively influences the intestinal microflora, reduces intestinal pH, and improves overall intestinal function, thereby supporting a healthy immune system [19].

8.2.3 Dhokla

8.2.3.1 Origin and Raw Materials

Dhokla is a popular fermented Indian snack that originates from the state of Gujarat. It is a steamed cake made from a fermented batter of rice and chickpea flour, resulting in a spongy texture. The primary ingredients are rice and chickpea flour, and the batter also typically consists of a mixture of rice with the pulse (Bengal gram). However, there are variants with in which the Bengal gram is replaced by chickpeas, pigeon peas, or urad beans [20].

8.2.3.2 Probiotic Microflora Associated with Fermentation

The microorganisms involved in the fermentation of dhokla are considered probiotics. This process primarily relies on LAB, including *L. fermentum*, various *Lactobacillus* species, *Leuconostoc mesenteroides*, and *Enterococcus* species, as well as yeasts such as *Pichia silvicola*, which contribute to its distinctive flavor and spongy texture [21, 22].

8.2.3.3 Process of Dhokla Making

	Bengal gram (1 part)	White polished rice (three parts)
Wash and soak	3 h	3 h
Grind	Finely in a mortar	Coarsely in a mortar
	Combine slurries into a thin batter and mix well	
	Add appropriately one part curd	
	Incubate overnight at a warm place (30–32 °C)	
	Add green chillies and table salt	
	Add batter on greasy plates and steam in suitable pan	
	Cut into squares and season with coriander leaves, green chillies, mustard, and asafoetida	

8.2.3.4 Health Benefits

Dhokla, a traditional fermented food from India, offers numerous health benefits attributed to its ingredients and preparation method. The fermentation process enhances the bioavailability of nutrients such as B-vitamins, iron, and zinc by reducing anti-nutritional factors like phytic acid. It also improves the digestibility of proteins and carbohydrates, making dhokla a suitable option for individuals with sensitive digestion. The LAB introduced through curd contribute probiotic-like effects, supporting gut health and microbial balance. Bengal gram, a primary ingredient, is high in dietary fiber and plant-based protein, which promote satiety, aid in digestion, and help manage cholesterol levels. Additionally, when combined with rice, the overall glycemic response of the dish is moderated, offering sustained energy release, which is beneficial for individuals managing blood sugar levels. Since dhokla is steamed rather than fried, it is lower in fat and thus more heart-healthy compared to many other Indian snacks. Furthermore, its traditional formulation is naturally gluten-free, making it appropriate for those with gluten intolerance or celiac disease. These attributes position dhokla not only as a culturally significant dish but also as a functional food with considerable nutritional advantages [23–25].

8.2.4 Pickles

8.2.4.1 Origin

Archaeological and textual evidence indicates that the practice of pickling dates back to approximately 2030 BCE in Mesopotamia, particularly in the Tigris Valley. Early preservation methods involved soaking various foods in a brine solution (brine is essentially saltwater, with a high concentration of dissolved salt), enabling them to be stored for extended periods without spoilage. This ancient technique not only preserved seasonal produce but also enhanced flavor and nutritional value, laying the groundwork for pickling traditions that spread across cultures over millennia [26, 27].

8.2.4.2 Raw Materials Used for Fermented Pickles

A wide variety of fruits and vegetables serve as raw materials in the preparation of fermented pickles. Commonly used produce includes cucumbers, mangoes, carrots, and green chilies, which act as substrates for microbial fermentation. Vinegar is often added to impart a tangy flavor and inhibit the growth of spoilage-causing microorganisms. Salt plays a critical role in drawing out moisture through osmosis, creating conditions that favor beneficial bacteria while suppressing harmful microbes. Sugar, in addition to enhancing sweetness, contributes to preservation by reducing water activity. Various spices, such as red chili powder, turmeric, and coriander powder, are incorporated to add flavor, color, and aroma. Water is used to create the brine solution necessary for anaerobic fermentation. While oil is not always essential, certain traditional pickle recipes include mustard oil or sesame oil to enhance flavor and further improve preservation [28].

8.2.4.3 Probiotic Microflora Associated with Fermentation

The fermentation of pickles is facilitated by probiotic microorganisms, primarily LAB, along with certain yeasts and molds. These microbes play a crucial role in both preserving the food and developing its characteristic flavor and texture. LAB initiate fermentation by producing lactic acid, which lowers the pH and creates an environment that is hostile to spoilage organisms. Key species involved include *Lactobacillus brevis*, *L. fermentum*, *Leuconostoc mesenteroides*, *Lactobacillus plantarum*, *Pediococcus pentosaceus*, and *P. cerevisiae*. These organisms are classified as probiotics, meaning they offer health benefits when consumed in adequate amounts. They should not be confused with prebiotics, which are nondigestible food components that promote the growth of beneficial microbes [29–31].

8.2.4.4 Process of Fermented Pickle Making

The preparation of fermented pickles [32] involves a series of steps designed to encourage natural fermentation by LAB while preserving the texture and flavor of the
raw vegetables. There are two commonly used approaches: the traditional spiced vegetable pickle and the brine-based fermented vegetable.

8.2.4.4.1 Steps for Vegetable Pickle Making
1. **Selection and preparation** – Choose fresh vegetables; peel and wash.
2. **Salting** – Add salt to the vegetables and store them in a container for 1–2 days.
   This step initiates the release of water and creates a favorable environment
   for LAB.
3. **Spicing** – Mix in traditional spices (such as turmeric, chilli powder, and fenugreek) along with a small amount of water.
4. **Sun-drying** – Allow the mixture to dry under the sun for 3–5 days to enhance
   flavor development and to reduce moisture.
5. **Packaging** – Pack the fermented product in sterilized containers for storage.

8.2.4.4.2 Brine-Based Fermented Pickles
In this method, washed and cut vegetables are submerged in a saltwater brine solution (typically 2–8% salt) and allowed to ferment naturally at ambient temperature
for several days to weeks. The anaerobic environment supports the growth of LAB,
which acidifies the medium and preserves the vegetables.

8.2.4.5 Health Benefits

Fermented pickles are not only a means of food preservation but also offer functional
health benefits, largely due to the presence of probiotic LAB. These beneficial microbes have been associated with enhanced resistance to gastrointestinal infections,
prevention of urogenital tract infections, and modulation of immune responses. Research studies have also shown potential roles in the suppression of certain cancers,
improvement of digestion, and reduction of serum cholesterol levels. Regular consumption of fermented pickles may contribute to improved gut health, better nutrient
absorption, and enhanced immune function. Furthermore, pickled vegetables retain
many of their original vitamins and antioxidants, making them a valuable addition to
a balanced diet [33–35].

8.2.5 Enduri Pitha

8.2.5.1 Origin and Raw Materials Used

Enduri pitha is a traditional steamed delicacy that originates from Odisha, India. It consists of a pancake made from a fermented batter of rice and urad dal (black gram), which is then wrapped in turmeric leaves before being steamed. The dish is typically filled with a sweet mixture of grated coconut, jaggery, and aromatic spices, such as cardamom and clove, which impart a unique flavor and aroma. The primary ingredients used to make enduri pitha are rice and urad dal, both of which undergo soaking and fermentation prior to preparation [30, 36].

8.2.5.2 Probiotic Microorganisms

The fermentation process in enduri pitha involves a consortium of probiotic microorganisms, primarily LAB, which include *L. fermentum* and *Leuconostoc mesenteroides*. These bacteria contribute to acidification and flavor development. Additionally, various yeast species, including *Saccharomyces* species, *Pichia silvicola*, *Torulopsis*, and *Candida*, play a significant role in fermentation by aiding in leavening and enhancing the complexity of the aroma and taste [37].

8.2.5.3 Process of Enduri Pitha Making

The preparation of enduri pitha involves a series of steps starting with the soaking of raw materials and ending with steaming the filled batter wrapped in turmeric leaves:

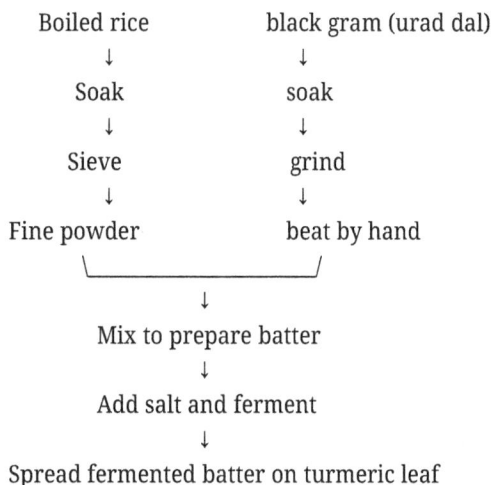

<pre>
        Boiled rice   black gram (urad dal)
            ↓   ↓
          Soak   soak
            ↓   ↓
          Sieve   grind
            ↓   ↓
      Fine powder   beat by hand
            _____/
                        ↓
               Mix to prepare batter
                        ↓
              Add salt and ferment
                        ↓
      Spread fermented batter on turmeric leaf
</pre>

$\downarrow$

Add filling (coconut + dahi channa + sugar)

$\downarrow$

Fold the turmeric leaf

$\downarrow$

Steam until cooked

$\downarrow$

Enduri pitha ready to serve

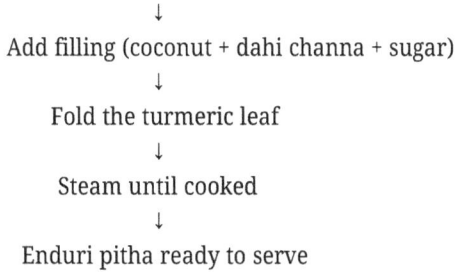

8.2.5.4 Health Benefits of Enduri Pitha

Enduri pitha has several reported health benefits attributed to both its ingredients and fermentation process. It is known to help boost the immune system and acts as a protective agent against various seasonal infections. The presence of turmeric leaves contributes anti-inflammatory and antimicrobial properties, while the LAB generated during fermentation aid digestion and enhance gut health. Together, these factors make enduri pitha not only a traditional delicacy but also a functional food with potential health-promoting effects [38–41].

8.2.6 Hawaijar

8.2.6.1 Origin

Hawaijar is a traditional fermented soybean product that originates from Manipur, a state in the Northeast region of India. It is particularly associated with the Meitei community, who prepare it as a household delicacy known for its distinctive pungent flavor, soft texture, and high nutritional value. Hawaijar plays a significant role in local diets and is customarily consumed as a protein-rich accompaniment to rice-based meals [42, 43].

8.2.6.2 Raw Materials Used

The primary raw material used for hawaijar preparation is medium- or small-sized soybean seeds. During fermentation, the wrapped soybeans are enclosed in fresh leaves of *Ficus hispida* or banana leaves, or alternatively in cotton cloth. The fermentation is carried out in a bamboo basket or earthen pot, which facilitates aeration and temperature regulation, both critical for microbial activity and flavor development [44].

8.2.6.3 Probiotics Associated with the Fermentation of Hawaijar

Hawaijar is a naturally fermented food involving a variety of probiotic microorganisms, particularly species of the genus *Bacillus*. The most dominant microorganism identified in the fermentation process is *Bacillus subtilis*, which contributes to the product's characteristic ammoniacal odor, texture softening, and enhanced protein digestibility. Other *Bacillus* species such as *Bacillus licheniformis* and *Bacillus cereus* may also be present. In some studies, *Staphylococcus aureus* and *Staphylococcus sciuri* have been detected, although their presence is generally considered incidental and their safety profile should be monitored [45].

8.2.6.4 Process of Hawaijar Making

The traditional preparation of hawaijar involves several steps of cleaning, boiling, and fermenting soybeans, typically over a 3- to 4-day period:

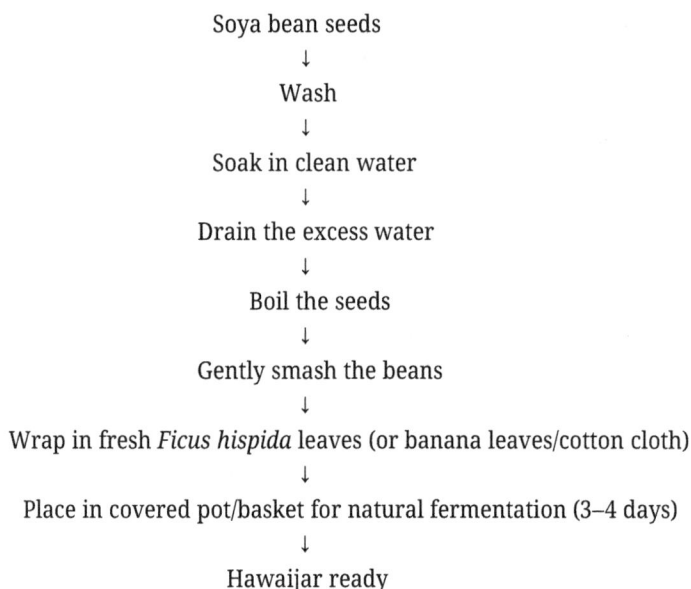

Soya bean seeds

↓

Wash

↓

Soak in clean water

↓

Drain the excess water

↓

Boil the seeds

↓

Gently smash the beans

↓

Wrap in fresh *Ficus hispida* leaves (or banana leaves/cotton cloth)

↓

Place in covered pot/basket for natural fermentation (3–4 days)

↓

Hawaijar ready

This traditional low-cost process relies on ambient temperature and natural microbial flora to achieve fermentation, making it a sustainable method of protein enhancement in rural and semi-urban communities.

8.2.6.5 Health Benefits

Hawaijar is widely recognized as a low-cost, protein-rich food, making it a crucial dietary component for economically disadvantaged populations. Beyond its macronutrient content, Hawaijar is believed to exhibit functional health benefits due to the bioactive compounds produced during fermentation. These include potential anti-osteoporotic, anticarcinogenic, and hypocholesterolemic properties, which are attributed to the metabolic activity of probiotic *Bacillus* species. Such properties align Hawaijar with the growing category of functional fermented foods that support both nutrition and preventive health [46, 47].

8.2.7 Khorisa

8.2.7.1 Origin

Khorisa is a traditional fermented food from Assam, a state in Northeast India, prepared from young bamboo shoots. It is a significant component of Assamese cuisine and is often consumed as a pickle or side dish. The fermentation process imparts a characteristic sour flavor and enhances the shelf life of the bamboo shoots, which are typically preserved for up to 2 years [48, 49].

8.2.7.2 Raw Materials Used

The primary raw material used in the preparation of khorisa is young bamboo shoots, which are harvested during the monsoon season. In its pickled form, khorisa may also include edible oil, salt, and green or red chilies and preserved in closed containers for up to 2 years. The fermentation takes place in sealed containers, traditionally earthen pots, using naturally occurring microorganisms. The inclusion of *Garcinia pedunculata Roxb.* (a souring agent) helps preserve and acidify the bamboo shoots.

8.2.7.3 Probiotic Microorganisms Associated with the Fermentation of Khorisa

The fermentation of khorisa is mediated primarily by LAB, which are known for their probiotic properties. Common LAB identified in khorisa include *L. plantarum*, *Lactobacillus pentosus*, *Lactobacillus paracasei*, *L. brevis*, and *Lactobacillus collinoides*. These microorganisms play a crucial role in acidification, flavor development, and pathogen inhibition, making khorisa a naturally preserved, functional food [50, 51].

8.2.7.4 Process of Making Khorisa

The preparation of khorisa is a straightforward traditional method involving minimal ingredients and relying on natural fermentation:

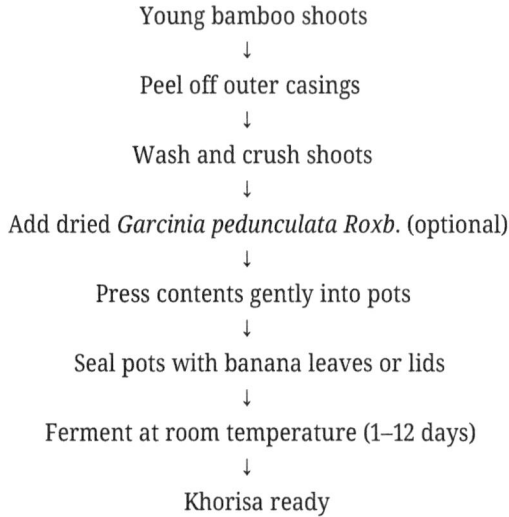

Young bamboo shoots
↓
Peel off outer casings
↓
Wash and crush shoots
↓
Add dried *Garcinia pedunculata Roxb.* (optional)
↓
Press contents gently into pots
↓
Seal pots with banana leaves or lids
↓
Ferment at room temperature (1–12 days)
↓
Khorisa ready

8.2.7.5 Health Benefits of Khorisa

Khorisa offers numerous health benefits due to both the nutritional properties of bamboo shoots and the presence of probiotic microorganisms. Bamboo shoots are high in dietary fiber, which promotes healthy digestion through stimulation of intestinal peristalsis. Additionally, studies have shown that bamboo shoots have antioxidant and anti-inflammatory effects, which may contribute to the prevention of cardiovascular diseases. They are also linked to anti-obesity, antiapoptotic, and cholesterol-lowering activities. Furthermore, bamboo lignin has been found to protect neurons from oxidative stress, suggesting a potential role in neuroprotection [52–55].

8.2.8 Handia

8.2.8.1 Origin

Handia is a traditional fermented rice-based beverage consumed widely by tribal communities in the eastern states of India, including Odisha, Jharkhand, Bihar, and West Bengal. It is culturally significant and is often prepared during festivals, rituals, and community gatherings. The beverage is mildly alcoholic and is known for its refreshing and digestive properties. The beverage is primarily made using parboiled rice and a unique starter culture known as ranu or bakhar. Ranu is a fermenting agent typically formed into small balls composed of powdered rice mixed with various indigenous herbs, which initiate and facilitate the fermentation process [56].

8.2.8.2 Microorganisms Associated with Fermentation

The fermentation of handia is a result of the natural activity of yeasts and LAB. The yeasts primarily involved include *Saccharomyces cerevisiae* and *Hanseniaspora guilliermondii*, which contribute to ethanol production and aroma development. In addition, several LAB such as *Pediococcus lolli*, *Brevibacillus agri*, and *Leuconostoc mesenteroides* play roles in acidification, flavor enhancement, and microbial safety.

8.2.8.3 Process of Making Handia

The traditional preparation of handia involves using a starter culture known as *ranu* or *bakhar*, which contains a mix of wild microorganisms adhered to dried herbal tablets:

Traditional Handia Preparation

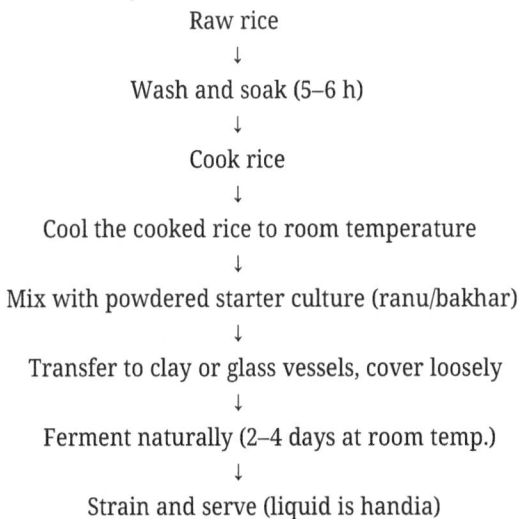

Raw rice
↓
Wash and soak (5–6 h)
↓
Cook rice
↓
Cool the cooked rice to room temperature
↓
Mix with powdered starter culture (ranu/bakhar)
↓
Transfer to clay or glass vessels, cover loosely
↓
Ferment naturally (2–4 days at room temp.)
↓
Strain and serve (liquid is handia)

8.2.8.4 Health Benefits

Handia is considered a functional fermented food with multiple health-promoting attributes. It contains probiotic microorganisms that support gut health, aid digestion, and enhance the immune response. Traditionally, it is used to reduce stress, promote hydration, and support metabolic activity. Studies have highlighted its potential antioxidant, antidiabetic, anti-obesity, and anti-inflammatory properties, making it more than just a cultural beverage [57–59].

8.2.9 Gundruk

8.2.9.1 Origin and Raw Materials Used

Gundruk is a traditional fermented leafy vegetable product that originated in Nepal and is widely consumed in neighboring regions of Sikkim, Darjeeling, Assam, and Arunachal Pradesh. It is prepared using leafy greens, such as mustard, radish, and cauliflower leaves, which are abundant during the harvest season. These greens are preserved through natural lactic acid fermentation, allowing communities to store and consume them during off-seasons [60].

8.2.9.2 Microorganisms Associated with Fermentation

The microbial profile of gundruk is dominated by LAB, which contribute significantly to flavor, preservation, and health benefits. Dominant LAB species include *L. brevis, L. plantarum, L. fermentum, Pediococcus pentosaceus, Pediococcus acidilactici,* and *Leuconostoc fallax*. Other microbial associates include *Bacillus pantothenticus, Bacillus pumilus,* and some wild yeasts that contribute to texture and mild acidity [61].

8.2.9.3 Process of Making Gundruk

The preparation of gundruk follows a natural fermentation method that utilizes household techniques and ambient microbial flora:

Fresh mustard, radish, or cauliflower leaves

↓

Wilt at room temperature (1–2 days)

↓

Shred and mildly crush

↓

Soak in warm water (optional)

↓

Pack tightly into earthen/plastic container or pot

↓

Ferment at room temperature (7–10 days)

↓

Sun-dry for 3–4 days

↓

Gundruk ready

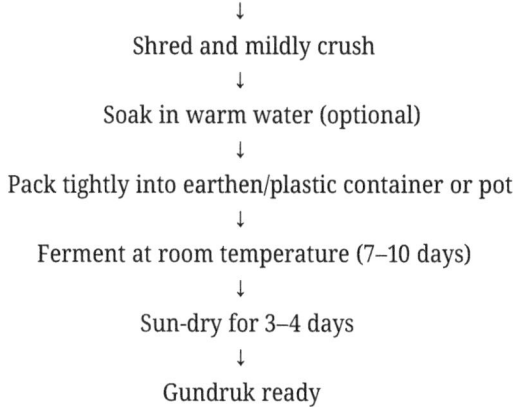

This process improves shelf life, imparts a sour taste, and develops probiotic properties beneficial to health.

8.2.9.4 Health Benefits of Gundruk

Gundruk offers a wide range of nutritional and therapeutic benefits, largely attributed to the activity of LAB during fermentation. These bacteria produce bioactive compounds such as conjugated linoleic acids (CLAs), exopolysaccharides, and bacteriocins, which possess multiple functional properties. Specifically, CLAs help lower blood pressure, exopolysaccharides exhibit prebiotic activity, promoting gut health, bacteriocins have antimicrobial effects, enhancing food safety, sphingolipids possess anticarcinogenic and antimicrobial properties. And other bioactive peptides offer antioxidant, anti-inflammatory, anti-allergenic, and opioid antagonist effects. As a result, gundruk demonstrates potential in supporting cardiovascular health, metabolic regulation, and immune function, and is often more nutritious than its unfermented counterpart [62–64].

8.2.10 Naan, Bhatura, and Kulcha

8.2.10.1 Origin and Raw Materials Used

Naan, bhatura, and kulcha are traditional fermented flatbreads that originated in Punjab, a region in the Indian subcontinent known for its rich culinary heritage. These flatbreads are staples in North Indian cuisine and are commonly served with lentils, curries, and pickles. The main raw material used is refined wheat flour (maida) and other ingredients such as milk, yoghurt, salt, and oil are typically added to enrich texture and flavor [65].

8.2.10.2 Microorganisms Associated with Fermentation

The leavening and fermentation of dough for naan, bhatura, and kulcha are mediated primarily by yeasts and LAB. The most prominent yeast is *Saccharomyces cerevisiae*, which produces carbon dioxide and ethanol, leading to dough expansion and development of soft, spongy texture. LAB such as *Lactobacillus* species, *Pediococcus* species, and *Leuconostoc* species also contribute by producing lactic acid, which imparts a subtle sour flavor and aids in partial preservation through pH reduction [66].

8.2.10.3 Process of Making Naan, Bhatura, and Kulcha

While each of these breads has unique characteristics, their preparation follows a similar base process with variation in fermentation time, cooking method, and final texture:

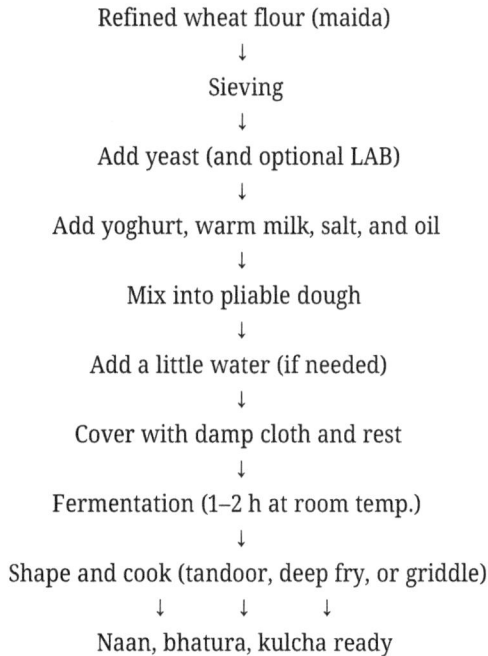

Refined wheat flour (maida)
↓
Sieving
↓
Add yeast (and optional LAB)
↓
Add yoghurt, warm milk, salt, and oil
↓
Mix into pliable dough
↓
Add a little water (if needed)
↓
Cover with damp cloth and rest
↓
Fermentation (1–2 h at room temp.)
↓
Shape and cook (tandoor, deep fry, or griddle)
↓   ↓   ↓
Naan, bhatura, kulcha ready

Naan is traditionally cooked in a tandoor (clay oven) and may be brushed with ghee or butter. Bhatura is deep-fried and typically puffed, served with chole (spicy chickpeas). Kulcha is usually pan-cooked or baked, often stuffed with potato or paneer fillings.

8.2.10.4 Health Benefits

Despite being made from refined flour, fermented breads such as naan and kulcha offer notable health benefits due to the fermentation process. Fermentation helps break down complex carbohydrates, making these breads easier to digest. The presence of LAB contributes to maintaining a healthy gut microbiome and may enhance metabolic functions. Additionally, fermented breads can lower the glycemic response, which is beneficial in managing and potentially preventing type 2 diabetes. When consumed in moderation, these foods may also reduce inflammation, lower the risk of gallstones, and support cardiovascular health. Emerging research further suggests that fermented whole wheat variants may aid in weight management, support hormonal balance, and potentially reduce the risk of conditions such as breast cancer [67–69] (Table 8.1).

Table 8.1: Summary of different fermented foods' microflora and health benefits.

Fermented food	Origin/ region	Raw materials used	Microflora – probiotics/ prebiotics /	Health benefits
Idli	South India	Rice and black gram	*Lactobacillus* species and *Leuconostoc* species	Improves digestion and rich in probiotics
Dosa	South India	Rice and black gram	*Lactobacillus* species and *Streptococcus* species	Gut health, light, and easy to digest
Dhokla	Gujarat	Gram flour and curd	*Lactococcus lactis* and *Lactobacillus* species	Promotes gut flora and protein-rich
Pickles	Pan India	Vegetables, salt, spices, and oil	LAB and yeast	Antioxidant-rich and improves gut flora
Enduri pitha	Odisha	Rice flour, jaggery, and coconut	*Lactobacillus* and *Leuconostoc*	Digestive support and mild probiotic activity
Hawaijar	Manipur	Soybeans	*Bacillus subtilis*	Protein-rich and antioxidant potential
Khorisa	Assam	Bamboo shoots	*Lactobacillus plantarum* and *Weissella* species	Fiber-rich and supports gut bacteria
Handia	Jharkhand and Odisha	Rice and herbal starter	Yeast and LAB	Probiotic-rich and energy booster
Gundruk	Nepal and Sikkim	Mustard leaves	*Lactobacillus plantarum*	Enhances gut flora and fiber-rich
Naan, bhatura, and kulcha	North India	Wheat flour, curd, and yeast	Yeast and *Lactobacillus* species	Improves gut function and energy source

Fermentation is a traditional Indian culinary practice, providing a remarkable intersection of nutrition, flavor, and cultural heritage. The fermented foods that are central to Indian households and regional identities – ranging from breakfast staples like idli and dosa to ceremonial dishes such as enduri pitha, as well as everyday items like pickles are of great health benefitted fermented foods. Products such as dhokla and handia demonstrate the diverse applications of fermentation across various textures and ingredients – from steamed cakes to fermented beverages. Gundruk and hawaijar exemplify the innovative methods by which fermentation preserves and enhances the nutritional value of vegetables and legumes. Fermented bamboo shoots like khorisa, along with leavened breads, such as bhatura, naan, and kulcha, highlight the wide range of substrates and fermentation techniques that characterize the pluralism of Indian cuisine.

There is a wide array of indigenous fermented foods such as *ziang sang, sinki, goyang, ngari, hentak, tungtap, kinema, hawaiijar, tungrumbai,* and *ankhone/bekang/ peruyyan,* each deeply rooted in local traditions and ecological knowledge. These products reflect the region's unique microbial heritage and traditional food preservation techniques. While nonindigenous fermented foods like *kefir, tempeh, natto, kombucha, miso, kimchi, sauerkraut,* and *yoghurt* were also remained on the richness and diversity of indigenous practices. Together, all the fermented foods highlight the cultural, nutritional, and scientific benefit of fermentation across the world.

8.3 Conclusion and Future Prospects

Fermented foods have long been integral to human diets, originally valued for their ability to extend the shelf life of seasonal foods. The consumption of fermented foods provides a holistic approach to gut health, offering a natural means to support the microbiome without the need for supplements. These foods serve as rich sources of probiotics, prebiotics, synbiotics, and postbiotics, each playing a distinct role in maintaining a balanced and healthy gut microbiome. Incorporating a variety of these foods into the diet can lead to improved digestion, enhanced immune function, and a reduced risk of certain chronic diseases.

As the scientific community continues to explore the complexities of the gut microbiome and its impact on health, fermented foods remain a valuable dietary component. Their multifaceted benefits highlight the importance of traditional food practices in modern nutrition. In conclusion, fermented foods are not merely dietary supplements; they are functional foods that offer significant health benefits. Embracing these foods can contribute to a balanced and thriving gut microbiome, laying the foundation for long-term health and vitality.

Conflicts of interest: The authors do not declare any conflicts of interest.

References

[1]   Leeuwendaal NK, Stanton C, O'toole PW, Beresford TP. Fermented foods, health and the gut microbiome. Nutrients. 2022;14(7):1527.

[2]   Salminen S, Collado MC, Endo A, Hill C, Lebeer S, Quigley EM, Sanders ME, Shamir R, Swann JR, Szajewska H, Vinderola G. The International Scientific Association of Probiotics and Prebiotics (ISAPP) consensus statement on the definition and scope of postbiotics. Nature Reviews Gastroenterology & Hepatology. 2021;18(9):649–667.

[3]   Tamang JP, Cotter PD, Endo A, Han NS, Kort R, Liu SQ, Mayo B, Westerik N, Hutkins R. Fermented foods in a global age: East meets West. Comprehensive Reviews in Food Science and Food Safety. 2020;19(1):184–217.

[4]   Paul Ross R, Morgan S, Hill C. Preservation and fermentation: Past, present and future International Journal of Food Microbiology. 2002;5.79(1–2):3–16.

[5]   Kumari A, Siripuram S. (2024). Fenugreek seeds based functional foods: A review. International Journal of Agriculture and Food Science, 6(1), Part A, 552.

[6]   Ramakrishnan CV. Terminal Report of American PL 480-project Nr GF-IN-491. Study of Indian Fermented Foods from Legumes and Production of Similar Fermented Foods from U.S. Soybean. 1979, Baroda: Biochemistry Dept. Baroda University.

[7]   Alu'datt MH, Rababah T, Al-ali S, Tranchant CC, Gammoh S, Alrosan M, Kubow S, Tan TC, Ghatasheh S. Current perspectives on fenugreek bioactive compounds and their potential impact on human health: A review of recent insights into functional foods and other high value applications. Journal of Food Science. 2024;89(4):1835–1864.

[8]   Soni S, Sandhu D, Vilkhu K, Kamra N. Microbiological studies on Dosa fermentation. Food Microbiology. 1986;3(1):45–53.

[9]   Mukherjee SK, Albury MN, Pederson CS, Vanveen AG, Steinkraus KH. Role of Leuconostoc mesenteroides in leavening the batter of idli, a fermented food of India. Applied Microbiology. 1965;13:227–3.

[10] Reddy N, Sathe S, Pierson M, Salunkhe D. Idli, an Indian fermented food: A review. Journal of Food Quality. 1982;5(2):89–101.

[11] Batra LR, Millner PD. Some Asian fermented foods and beverages, and associated fungi. Mycologia. 1974;66:942–948.

[12] Steinkraus K. Handbook of Indigenous Fermented Foods, Revised and Expanded. 2018, CRC press.

[13] Murthy C, Rao P. Thermal diffusivity of idli batter. Journal of Food Engineering. 1997;33(3):299–304.

[14] Nisha P, Ananthanarayan L, Singhal RS. Effect of stabilizers on stabilization of idli (traditional South Indian food) batter during storage. Food Hydrocolloids. 2005;19(2):179–186.

[15] Prado FC, Parada JL, Pandey A, Soccol CR. Trends in non-dairy probiotic beverages. Food Research International. 2008;41(2):111–123.

[16] Farnworth ERT. Handbook of Fermented Functional Foods. 2008, CRC press.

[17] Palanisamy BD, Rajendran V, Sathyaseelan S, Bhat R, Venkatesan BP. Enhancement of nutritional value of finger millet-based food (Indian dosa) by co-fermentation with horse gram flour. International Journal of Food Science and Nutrition. 2012;63:5–15.

[18] Iyer BK, Singhal RS, Ananthanarayan L. Characterization and in vitro probiotic evaluation of lactic acid bacteria isolated from Idli batter. Journal of Food Science and Technology. 2013;50:1114–1121.

[19] Ramalingam S, Kandasamy S, Bahuguna A, Kim M. Nutritional and Health Benefits of Idli and Dosa. In *Fermented Food Products*. 2019, pp. 181–194, CRC Press.

[20] Sreeja V, Prajapati JB. Ethn. Fermented Foods Beverages India Sci. *Hist. Cult.* 2020, pp. 157–187, *Springer, Singapore*.

[21] Tsafrakidou P, Michaelidou AM, G. Biliaderis C. Fermented cereal-based products: Nutritional aspects, possible impact on gut microbiota and health implications. Foods. 2020;9(6):734.

[22] Roopashri AN, Savitha J, Divyashree MS, Mamatha BS, Rani KU, Kumar A. Indian traditional fermented foods: The role of lactic acid bacteria. In: Lactobacillus-A Multifunctional Genus. 2023, IntechOpen.

[23] Battacharya S, Bhat KK. Steady shear rheology of rice blackgram suspensions and suitability of rheological models. Journal of Food Engineering. 1997;32:241–250.

[24] Vandana, Verma C, Sabharwal PK. Traditional foods: The inheritance for good health. Emerging Technologies in Food Science: Focus on the Developing World. 2020;135–138.

[25] Sharma S, Gautam P, Joshi S, Dobhal A, Anand J, Kumar S. Nutritional and functional profiling of major millets and its processed food products: A review. Environment Conservation Journal. 2024;25 (4):1180–1190.

[26] Ray M, Ghosh K, Singh S, Mondal KC. Folk to functional: An explorative overview of rice-based fermented foods and beverages in India: A review. Journal of Ethnic Foods. 2016;3:5–18.

[27] Toussaint-Samat M. A History of Food. 2009, John Wiley & Sons.

[28] Behera SS, El Sheikha AF, Hammami R, Kumar A. Traditionally fermented pickles: How the microbial diversity associated with their nutritional and health benefits?. Journal of Functional Foods. 2020;70:103971.

[29] Chakraborty R, Roy S. Exploration of the diversity and associated health benefits of traditional pickles from the Himalayan and adjacent hilly regions of Indian subcontinent. Journal of Food Science and Technology. 2018;55:1599–1613.

[30] http://www.fao.org/3/a-au116e.pdf .

[31] Marco ML, Heeney D, Binda S, Cifelli CJ, Cotter PD, Foligné B, Gänzle M, Kort R, Pasin G, Pihlanto A, Smid EJ. Health benefits of fermented foods: Microbiota and beyond. Current Opinion in Biotechnology. 2017;44:94–102.

[32] Ghnimi S, Guizani N. Vegetable fermentation and pickling. In: Handbook of Vegetables and Vegetable Processing. 2018, pp. 407–427.

[33] Guizani N. Vegetable fermentation and pickling. In: Handbook of Vegetables and Vegetable Processing. 2011, pp. 351–367.

[34] Zhou M, Zheng X, Zhu H, Li L, Zhang L, Liu M, Liu Z, Peng M, Wang C, Li Q, Li D. Effect of Lactobacillus plantarum enriched with organic/inorganic selenium on the quality and microbial communities of fermented pickles. Food Chemistry. 2021;365:130495.

[35] Tan X, Cui F, Wang D, Lv X, Li X, Li J. Fermented vegetables: Health benefits, defects, and current technological solutions. Foods. 2023;13(1):38.

[36] Nout MJR. Fermented foods and food safety. Food Research International. 1994;27:291–298.

[37] Ghosh K, Mondal SP, Mondal KC. Ethnic fermented foods and beverages of West Bengal and Odisha. In: Ethnic Fermented Foods and Beverages of India: Science History and Culture. 2020, pp. 647–685.

[38] Ray RC, Swain MR. Indigenous fermented foods and beverages of Odisha, India: An overview. In: Indigenous Fermented Foods of South Asia. 2013, pp. 1–6.

[39] Roy A, Moktan B, Sarkar PK. Traditional technology in preparing legume-based fermented foods of Orissa. Indian Journal of Traditional Knowledge. 2007;6:12–16.

[40] Kumari B, Vibhute P, Solanki H. "PITHA" the heritage food of Odisha during the traditional festival: A scientific studies on its nutraceutical and nutritional analysis. 2022, 1–11

[41] Tewari S, Shinde DB, Solanke GM, Maske Sachin V. Anti-oxidant activities of different types of indigenous pithas of Odisha. ISSN- 2394-5125 2020;7(18)

[42] Jeyaram K, Mohendro Singh W, Premarani T, Ranjita Devi A, Selina Chanu K, Talukdar NC, Rohinikumar Singh M. Molecular identification of dominant microflora associated with 'Hawaijar' – A traditional fermented soybean (Glycine max (L.)) food of Manipur, India. International Journal of Food Microbiology. 20 Mar 2008;122(3):259–268.

[43] Sarmah M, Chakraborty D, Haldar M, Chetia LR, Seal S. Probiotic microorganisms in fermented food of North Eastern States of India. Northeast Journal of Contemporary Research. Dec 2020;7(1):8–14.

[44] Sonar NR, Halami PM. Journal of Food Science and Technology. 2014 Dec;51(12):4143–4148.

[45] Das A, Deka S. Mini Review: Fermented foods and beverages of the North-East India. International Food Research Journal. 2012;19.

[46] Kabui KK, Rawson A, Athmaselvi KA. Selected fermented foods of Manipur, India: Traditional preparation methods, nutritional profile, and health benefits. Food Chemistry Advances. 2024;100864.

[47] Kavitake D, Kandasamy S, Devi PB, Shetty PH. New Innovations in Fermentation Biotechnology of Traditional Foods of the Indian Sub-Continent. In Functional Foods and Biotechnology. 2020, pp. 77–100, CRC Press.

[48] Devi P, Suresh Kumar P. Traditional, Ethnic and Fermented Foods of Different Tribes of Manipur. Indian Journal of Traditional Knowledge. 2012;11(1):70–77.

[49] Keishing S, Banu AT. Hawaijar–A Fermented Soya of Manipur, India. IOSR-JESTFT. 2013;4(2):29–33.

[50] Sharma N, Barooah M. Microbiology of khorisa, its proximate composition and probiotic potential of lactic acid bacteria present in Khorisa, a traditional fermented Bamboo shoot product of Assam. Indian Journal of Natural Products and Resources. Mar 2017;8(1):78–88.

[51] Nirmala C, Bisht MS, Sheena H. Nutritional properties of bamboo shoots: Potential and prospects for utilization as a health food. Comprehensive Reviews in Food Science and Food Safety. 2011;10:153–169.

[52] Bajwa HK, Santosh O, Koul A, Bisht MS, Nirmala C. Phytomodulatory effects of fresh and processed shoots of an edible bamboo Dendrocalamus hamiltonii Nees & Arn. Ex Munro on antioxidant defense system in mouse liver. Journal of Food Measurement and Characterization. 2019;13:3250–3256.

[53] Li X, Guo J, Ji K, Zhang P. Bamboo shoot fiber prevents obesity in mice by modulating the gut microbiota. Scientific Reports. 2016;6(1):32953.

[54] Joshi B, Indira A, Oinam S, Koul A, Chongtham N. Fermented bamboo shoots: A potential source of nutritional and health supplements. In: Bamboo Science and Technology. 2023, pp. 201–236, Singapore: Springer Nature Singapore.

[55] Singh M. A Review on Traditional Ecological Knowledge of Indigenous Communities of Northeast India. In: Learning 'From'and 'With'the Locals: Traditional Knowledge Systems for Environmental Sustainability in the Himalayas. 2024, pp. 259–292.

[56] Roy A, Khanra K, Bhattacharya C, Mishra A, Bhattacharyya N. Bakhar-Handia fermentation: General analysis and a correlation between traditional claims and scientific evidences. Advances in Bioresearch. 2012;3(3):28–32.

[57] Singh KA, Singh S, Yadav SS, Yadav S, Patra F, Duary RK. Traditional and Ethnic Fermented Beverages: Functional and Health Benefit Prospects. In: Traditional Foods: The Reinvented Superfoods. 2024, pp. 299–343, Cham: Springer.

[58] Hembrom S, Lal S, Kumar S, 2022. The Microbiology and Traditional State of Fermented Beverage Handia: A Review.

[59] Ignat MV, Salanță LC, Pop OL, Pop CR, Tofană M, Mudura E, Coldea TE, Borşa A, Pasqualone A. Current functionality and potential improvements of non-alcoholic fermented cereal beverages. Foods. 2020;9(8):1031.

[60] Kc S. Probiotic potential bacteria of indigenous fermented food (GUNDRUK) from local market of Sunsari, Nepal (Doctoral dissertation, A Dissertation Submitted to the Department of Microbiology. In: Partial Fulfillment of the Requirements for the Award of Degree of Master of Science in Microbiology (Environment and Public Health)). 2022, Dharan, Nepal: Central Campus of Technology, Tribhuvan University.

[61] Singh B, Mal G, Kalra RS, Marotta F. Traditional foods as sources of probiotics. In: Probiotics as Live Biotherapeutics for Veterinary and Human Health, Volume 1: Functional Feed and Industrial Applications. 2024, pp. 33–63, Cham: Springer Nature Switzerland.

[62] Tamang B, Tamang JP. In situ fermentation dynamics during production of gundruk and khalpi, ethnic fermented vegetable products of the Himalayas. Indian Journal of Microbiology. 2010;50:93–98.

[63] Tamang JP, Tamang B, Schillinger U, Franz CM, Gores M, Holzapfel WH. Identification of predominant lactic acid bacteria isolated from traditionally fermented vegetable products of the Eastern Himalayas. International Journal of Food Microbiology. 2005;105(3):347–356.

[64] Tamang JP. ed. Ethnic Fermented Foods and Beverages of India: Science History and Culture. 2020, Springer Nature.

[65] Sahoo M, Aradwad P, Sanwal N, Sahu JK, Kumar V, Naik SN. Fermented foods in health and disease prevention. In Microbes in the Food Industry. 2023, pp. 39–85.

[66] Bhalla TC, Savitri. Yeasts and traditional fermented foods and beverages. In: Yeast Diversity in Human Welfare. 2017, pp. 53–82.

[67] Sathe GB, Mandal S. Fermented products of India and its implication: A review. Asian Journal of Dairy and Food Research. 2016;35(1):1–9.

[68] Sharma P, Nagar R, Kaur R, Comparative Study of the Nutritive Values and Acceptability of Different Types of Bhatura in India.

[69] Parimala KR, Sudha ML. Wheat-based traditional flat breads of India. Critical Reviews in Food Science and Nutrition. 2015;55(1):67–81.

R. Nithyatharani*, R. Manikandan, M. Vinoth, R. Malathi

Chapter 9
Yeast as Potent Probiotics: Future Perspective

Abstract: The use of probiotics has traditionally been dominated by bacterial strains, particularly lactic acid bacteria; however, recent research has illuminated the significant potential of yeasts, especially *Saccharomyces cerevisiae* and *Saccharomyces boulardii*, as effective probiotic agents. Yeasts offer several advantages over bacterial probiotics, including resilience to gastrointestinal conditions, tolerance to antibiotics, and the ability to exert immunomodulatory, anti-inflammatory, and pathogen-inhibitory effects. This chapter explores the future perspectives of yeast-based probiotics, emphasizing advancements in strain selection, genetic engineering, and delivery technologies that enhance their efficacy and stability. It also considers their emerging role in addressing global health challenges such as antibiotic resistance, gastrointestinal disorders, and metabolic syndromes. The integration of omics technologies and systems biology is paving the way for a more precise understanding of yeast-host interactions, facilitating the development of next-generation probiotic formulations. Furthermore, the chapter discusses regulatory considerations, commercial viability, and the potential of yeast probiotics in personalized nutrition and functional food development. As the field progresses, yeast probiotics are poised to complement or even surpass bacterial probiotics in diverse therapeutic and nutritional applications.

Keywords: probiotics, yeast, *Saccharomyces cerevisiae*, *Saccharomyces boulardii*

9.1 Introduction

Probiotics are characterized as "live microorganisms that provide a health benefit to the host when given in sufficient quantities" [9], and they are mainly linked to bacte-

Corresponding author: R. Nithyatharani, Department of Microbiology, Shrimati Indira Gandhi College, Tiruchirappalli, Tamil Nadu, India, e-mail:nithyatharanir@gmail.com
R. Manikandan, Department of Biochemistry, Shrimati Indira Gandhi College, Tiruchirappalli, Tamil Nadu, India
M. Vinoth, Department of Microbiology, Government Arts and Science College, Perambalur, Tamil Nadu, India
R. Malathi, Department of Biochemistry, Enathi Rajappa Arts and Science College, Pattukkottai, Tamil Nadu, India

https://doi.org/10.1515/9783112205150-009

rial species, especially those from the genera *Lactobacillus* and *Bifidobacterium*. However, in recent decades, yeasts have emerged as promising candidates for probiotic applications due to their unique physiological traits, ability to survive harsh gastrointestinal (GI) conditions, and their health-promoting properties. The most extensively studied probiotic yeast is *Saccharomyces cerevisiae* var. *boulardii*, which has shown efficacy in managing GI disorders and enhancing immune functions [4]. As scientific understanding deepens, the potential of non-*Saccharomyces* yeasts and genetically engineered strains is being explored, broadening the scope of yeast-based probiotics. This chapter explores the current status of yeast as probiotics and their future potential in health, nutrition, and functional food sectors, while addressing associated challenges and innovations.

The probiotic properties of yeasts are primarily attributed to their resilience in the GI environment. Unlike many bacterial probiotics that may be susceptible to stomach acidity and bile salts, yeast cells possess robust cell walls composed of β-glucans, mannoproteins, and chitin, which confer resistance to GI stressors [18]. These structural components not only aid survival but also interact beneficially with the host's immune system. For instance, β-glucans have been shown to modulate immune responses by activating macrophages and dendritic cells [33]. Moreover, yeasts do not transfer antibiotic resistance genes, a growing concern with bacterial probiotics, which further enhances their safety profile for therapeutic use.

The most extensively studied probiotic yeast, *S. cerevisiae* var. *boulardii*, has been successfully used in the treatment of various GI disorders, including antibiotic-associated diarrhea, *Clostridium difficile* infections, and inflammatory bowel diseases [19]. It exerts its effects through multiple mechanisms: enhancing intestinal barrier function, modulating immune responses, inhibiting pathogen adherence, and secreting proteases that neutralize bacterial toxins [14]. Beyond its therapeutic roles, *S. boulardii* also contributes to gut microbiota balance, thereby promoting long-term GI health. Its safety record, substantiated by clinical trials and over-the-counter use, provides a foundation for the development of new yeast-based probiotics. Table 9.1 shows the comparison of characteristics of yeast and bacterial probiotics.

The exploration of nonconventional yeasts as potential probiotics is an emerging area of interest. Species such as *Kluyveromyces marxianus* (formerly known as *Candida kefyr*), *Pichia kudriavzevii*, and *Yarrowia lipolytica* have shown promising probiotic traits, including acid and bile tolerance, adhesion to intestinal epithelial cells, antioxidant activity, and antimicrobial properties [12]. These yeasts also offer advantages in terms of metabolic versatility and fermentative capabilities, which can be harnessed for synbiotic applications and bioactive compound production. For example, *K. marxianus* has been identified as a potential probiotic due to its thermotolerance and ability to metabolize a wide range of carbohydrates, making it suitable for applications in tropical and resource-limited settings [10]. This chapter reviews the future perspective of probiotic yeast.

Table 9.1: Comparative characteristics of yeast versus bacterial probiotics.

Characteristics	Yeast probiotics (e.g., S. boulardii)	Bacterial probiotics (e.g., Lactobacillus spp.)
Antibiotic resistance	Naturally resistant	Susceptible
Survival in gastrointestinal tract	High	Moderate to high
Tolerance to bile and acid	Strong	Variable
Immunomodulatory potential	Strong (β-glucans in cell wall)	Moderate
Genetic stability	High	Moderate to low
Shelf life	Long	Shorter
Common applications	Diarrhea, inflammatory bowel disease, and inflammation	Gut health, lactose digestion, and immunity

9.2 Future Perspective: Yeast as Probiotics

9.2.1 Exploration of Novel Yeast Strains

The future of probiotic research is poised to expand beyond the traditionally used *S. cerevisiae* var. *boulardii*, with increasing attention directed toward the exploration of nonconventional yeast strains. These lesser-known yeasts, including species such as *K. marxianus* (formerly known as *C. kefyr*), *P. kudriavzevii*, *Y. lipolytica*, and *Torulaspora delbrueckii*, exhibit unique physiological traits that may render them suitable for specialized probiotic applications. Unlike bacterial probiotics, these yeasts often possess innate resistance to antibiotics and environmental stressors, making them robust candidates for future functional food and pharmaceutical products [12, 25].

Advances in high-throughput screening, genomics, and metabolomics are expected to accelerate the identification and characterization of such yeasts, enabling the discovery of novel health-promoting mechanisms, such as specific immunomodulatory or anti-inflammatory actions [24]. Moreover, the ability of some novel yeast strains to ferment diverse substrates, synthesize bioactive metabolites, and survive under harsh GI conditions further supports their future use in personalized probiotic therapies [10]. As consumer demand grows for targeted, sustainable, and nonbacterial alternatives, the inclusion of diverse yeast strains in probiotic formulations could lead to more inclusive and effective microbiome management strategies. Nonetheless, future research must also focus on ensuring safety, regulatory compliance, and long-term stability of these novel strains, especially those derived from nontraditional environments such as fermented foods or extreme habitats [27]. In conclusion, the exploration of novel probiotic yeast strains represents a promising frontier in microbial biotechnology, with the potential to diversify and enhance the functional capacities of next-generation probiotics.

9.2.2 Genetic Engineering and Synthetic Biology

The application of genetic engineering and synthetic biology to probiotic yeasts represents a transformative frontier in microbial biotechnology, offering unprecedented opportunities to enhance and customize their probiotic functionalities. Yeasts, particularly *S. cerevisiae*, are well-characterized eukaryotic systems that are amenable to precise genetic modifications, making them ideal chassis for synthetic biology applications [7]. Future advances are expected to leverage CRISPR-Cas9 and other genome-editing tools to engineer probiotic yeasts capable of performing targeted therapeutic functions, such as delivering immunomodulatory molecules, degrading gut toxins, or synthesizing essential micronutrients in situ [36].

Engineered yeasts could also be designed to sense environmental cues in the GI tract and respond dynamically by producing bioactive compounds, positioning them as smart, living therapeutics for diseases such as inflammatory bowel disease, metabolic syndrome, and even cancer [21]. Moreover, synthetic biology enables the introduction of metabolic pathways that are not native to yeast, allowing the production of novel compounds such as antimicrobial peptides, short-chain fatty acids, or anti-inflammatory agents. In the context of probiotic development, this approach could also enhance traits such as acid and bile tolerance, adhesion to intestinal mucosa, or competitive exclusion of pathogens [35]. However, regulatory, ethical, and biosafety considerations surrounding the use of genetically modified organisms in food and medicine remain critical and must be addressed through transparent risk assessments and stakeholder engagement. As synthetic biology continues to evolve, engineered yeast probiotics are poised to play a central role in precision medicine, personalized nutrition, and sustainable healthcare solutions.

9.2.3 Development of Yeast-Based Biotherapeutics

The development of yeast-based biotherapeutics marks a promising and innovative frontier in probiotic and biomedical research, leveraging the intrinsic safety, robustness, and metabolic versatility of yeast cells to deliver therapeutic benefits beyond conventional probiotic effects. Yeasts such as *S. cerevisiae* and *S. boulardii* have already demonstrated utility in modulating gut microbiota and host immunity; however, future advancements in molecular engineering and systems biology are expected to elevate their role as platforms for delivering biotherapeutic agents directly within the host [35]. Engineered yeast strains could be tailored to secrete enzymes that degrade GI toxins, produce anti-inflammatory cytokines, or release therapeutic peptides and metabolites in response to specific physiological cues [36]. This precision targeting allows for site-specific therapeutic action, minimizing systemic side effects and maximizing efficacy.

Additionally, yeast-based systems offer unique advantages over bacterial systems, such as a lower likelihood of horizontal gene transfer and resistance gene dissemination, which are critical safety considerations in clinical applications [1]. Their ability to survive through the GI tract, adhere to epithelial cells, and interact with immune components further enhances their therapeutic potential. Notably, *S. boulardii* has been explored for its role in treating inflammatory bowel disease, *Clostridium difficile* infection, and even in modulating the gut-brain axis [14]. The emerging field of yeast biotherapeutics is also expanding into oncology and metabolic disease, where yeasts are being designed to deliver checkpoint inhibitors, insulin analogs, or tumor-suppressing compounds. While regulatory and scalability challenges remain, the convergence of synthetic biology, personalized medicine, and probiotic technology is expected to solidify yeast-based biotherapeutics as integral tools in future healthcare paradigms.

9.2.4 Integration into Functional Foods

The incorporation of probiotic yeasts into functional foods represents a compelling future direction in both nutritional science and food biotechnology, aligning with the growing consumer demand for health-promoting, sustainable, and naturally fermented products. Unlike bacterial probiotics, yeasts offer several technological advantages, including enhanced tolerance to acidic and high-temperature environments, natural resistance to antibiotics, and the ability to ferment complex carbohydrates and produce flavor-enhancing metabolites [12, 18]. As food systems evolve, probiotic yeasts such as *S. boulardii*, *K. marxianus*, and *P. kudriavzevii* are being explored for incorporation into a wide range of functional products, including dairy alternatives, bakery goods, cereal bars, kombucha, and plant-based beverages [15].

Future innovations will likely see the development of tailored food matrices designed to optimize the viability and efficacy of yeast strains, utilizing microencapsulation, prebiotic fortification, and controlled fermentation technologies. These formulations can enhance shelf stability and targeted delivery in the GI tract, thereby improving probiotic functionality and consumer appeal [28]. Additionally, the integration of probiotic yeasts in traditionally fermented foods – such as kefir, sourdough, and fermented vegetables – provides an opportunity to merge cultural culinary practices with modern health science, thereby appealing to both traditional and health-conscious markets. Moreover, the application of omics technologies and sensory science will facilitate the design of functional foods that are not only health-enhancing but also organoleptically pleasing. As regulatory frameworks evolve to accommodate novel probiotic carriers, yeast-based functional foods are poised to become a cornerstone of personalized nutrition and preventive health in the coming decades.

9.2.5 Personalized Nutrition and Microbiome Modulation

The integration of probiotic yeasts into the emerging field of personalized nutrition and microbiome modulation holds immense promise for the development of targeted, individualized health interventions. With increasing recognition that the human gut microbiota exhibits considerable interindividual variability, there is growing interest in tailoring probiotic supplementation based on a person's unique microbial composition, dietary habits, and genetic predispositions [39]. Probiotic yeasts such as *S. boulardii* and *K. marxianus* offer unique benefits in this context due to their ability to survive GI transit, interact dynamically with the host immune system, and resist antibiotic disruption – traits that make them strong candidates for integration into precision nutrition frameworks [10, 14].

Future strategies are expected to incorporate high-throughput microbiome profiling, metabolomics, and machine learning to predict individual responses to specific yeast strains and formulate bespoke dietary interventions. Additionally, engineered yeast strains may be programmed to perform person-specific therapeutic actions, such as metabolizing dietary components, restoring dysbiotic microbial communities, or modulating host gene expression [36]. This personalized approach not only enhances therapeutic efficacy but also minimizes adverse reactions and improves long-term adherence to dietary regimes. Furthermore, probiotic yeasts could play a crucial role in managing noncommunicable diseases, such as obesity, diabetes, and inflammatory disorders, which are closely linked to gut microbiota imbalances. As scientific understanding of host-microbiome interactions deepens, yeast-based probiotics are likely to become integral tools in next-generation personalized healthcare and nutrition systems.

9.2.6 Synbiotic Product Innovations

The development of synbiotic products – formulations that combine probiotics and prebiotics – has gained considerable momentum in recent years, and future innovations are increasingly expected to incorporate probiotic yeasts as vital components of these advanced therapeutic systems. Unlike traditional bacterial strains, yeast probiotics such as *S. boulardii*, *P. kudriavzevii*, and *K. marxianus* exhibit enhanced resilience under GI and industrial processing conditions, making them ideal candidates for next-generation synbiotic formulations [12, 15].

In the synbiotic framework, carefully selected prebiotics – nondigestible food ingredients that promote the growth and activity of beneficial microbes – can be tailored to specifically support the metabolic activity and colonization efficiency of yeast strains in the host gut. For instance, oligosaccharides such as inulin, fructooligosaccharides, and beta-glucans may serve as targeted substrates for yeast proliferation, enhancing their probiotic efficacy and functional integration within the gut microbiome [32]. Future product innovations are likely to employ precision-based design, incorporating multi-omics and bioinformatics tools to match specific yeast strains with optimal prebiotic partners, leading to customized synbiotic solutions for individual health profiles.

Moreover, synbiotic foods that combine yeast probiotics with plant-based ingredients rich in natural prebiotics such as chicory root, oats, or bananas can serve as functional foods for managing gut health, metabolic syndrome, and immune modulation. Advanced delivery systems like encapsulation and nanoformulation are also expected to improve the bioavailability and stability of synbiotic yeast products [29]. As regulatory frameworks evolve and consumer awareness of gut health increases, yeast-based synbiotics are poised to become a significant segment of the nutraceutical and functional food industries, offering robust, sustainable, and scientifically backed health benefits.

9.2.7 Expansion in Animal Health and Sustainable Agriculture

The future of yeast-based probiotics is increasingly promising within the domains of animal health and sustainable agriculture, offering innovative solutions to improve livestock productivity, animal welfare, and ecological sustainability. Yeast strains such as *S. cerevisiae* and *S. boulardii* have long been used in animal feed to enhance gut health, improve digestion, and bolster immune responses, and future research will likely expand their application through more targeted and sustainable interventions [6, 40]. Probiotic yeasts can contribute to the modulation of the gut microbiota in animals, promoting a balanced microbial community that reduces the incidence of GI diseases, such as colibacillosis and enteritis, and enhances feed efficiency [16]. Table 9.2 shows the applications of yeast as a probiotic in human and animal health.

Moreover, these yeasts can produce bioactive compounds like enzymes, vitamins, and antimicrobial peptides, which may further enhance livestock health while reducing the need for antibiotics and chemical additives, in line with global efforts to combat antimicrobial resistance [6]. In the context of sustainable agriculture, yeast-based probiotics offer potential benefits beyond animal health, contributing to soil microbiome management and plant growth promotion. Yeasts have been shown to exert beneficial effects on plant root systems, enhancing nutrient uptake and resistance to pathogens, which could aid in reducing the reliance on synthetic fertilizers and pesticides in crop production [34].

Table 9.2: Emerging applications of yeast probiotics in human and animal health.

Application area	Specific potential uses	Advantages of yeast probiotics
Human gastrointestinal health	Prevention of antibiotic-associated diarrhea, treatment of irritable bowel syndrome and inflammatory bowel disease	Nonpathogenic and promotes gut microbiota balance
Immunotherapy	Modulation of innate and adaptive immune responses	Contains β-glucans with immunostimulatory properties
Metabolic health	Management of obesity, type 2 diabetes, and hyperlipidemia	Anti-inflammatory and antioxidant effects
Veterinary medicine	Growth promotion, immune modulation, and disease resistance in livestock	Safe alternative to antibiotics in animal feed
Functional foods	Incorporation into dairy, beverages, and baked products	High survivability in food processing

Furthermore, as part of integrated pest management strategies, yeast-based probiotics may help in suppressing harmful microorganisms and promoting soil health, leading to more resilient farming systems [17]. Moving forward, advancements in genomics and biotechnology will likely enable the development of custom-designed yeast strains tailored for specific animal species, crop types, or environmental conditions, further driving the expansion of yeast-based probiotics in both animal and plant health sectors. This growing trend aligns with the broader goals of achieving food security, reducing environmental impact, and fostering sustainable agricultural practices.

9.2.8 Environmental and Industrial Biotechnology Applications

Yeast-based probiotics are poised to play a significant role in the fields of environmental and industrial biotechnology, offering innovative solutions for sustainability and efficiency in various industrial processes. Yeasts such as *S. cerevisiae* and *Pichia pastoris* have been widely utilized in industrial fermentation for the production of biofuels, biochemicals, and enzymes [41]. In the future, probiotic yeasts are likely to be engineered to improve the productivity of these processes by enhancing tolerance to environmental stressors, such as high temperatures, extreme pH, and the presence of toxic compounds [8]. For instance, yeast strains with enhanced stress resistance can optimize the fermentation of lignocellulosic biomass into biofuels, contributing to more sustainable and cost-effective renewable energy production [38]. Additionally, the bioremediation potential of probiotic yeasts offers another exciting frontier for environmental applications.

Yeasts are capable of absorbing and metabolizing pollutants such as heavy metals, pesticides, and petroleum hydrocarbons from contaminated environments [37]. Future research will likely focus on engineering yeast strains to enhance their pollutant-degrading capabilities and their ability to function in challenging environmental conditions, thus contributing to the cleanup of contaminated soil and water in a sustainable manner [23]. Furthermore, in the realm of waste management, yeast-based probiotics could be employed in the conversion of organic waste into valuable products such as biodegradable plastics, organic acids, and other bioproducts, promoting a circular economy [3]. By harnessing the natural metabolic diversity of yeasts and integrating them with industrial processes, yeast-based probiotics are expected to be at the forefront of developing environmentally friendly and economically viable technologies across various sectors, from biofuel production to waste treatment and environmental remediation.

9.2.9 Advanced Delivery Systems

The effectiveness of probiotic yeasts in promoting health is highly dependent on their ability to survive the harsh conditions of the GI tract, including acidic pH, digestive enzymes, and bile salts. As such, one of the most promising areas for future development in yeast-based probiotics is the design of advanced delivery systems that enhance the stability, viability, and targeted release of these microorganisms. Innovations in delivery technologies such as microencapsulation, nanoencapsulation, and controlled-release systems are expected to significantly improve the bioavailability of probiotic yeasts and optimize their health benefits [26, 30]. Microencapsulation techniques, for instance, involve enclosing probiotic yeast cells in biopolymer matrices, which protect the cells from environmental stressors and facilitate their gradual release in the gut [2]. These systems can be further enhanced by selecting specific materials, such as alginate, chitosan, or pectin, which are resistant to gastric acid but dissolve under the more neutral pH of the small intestine [13]. In addition to protecting the probiotics, advanced delivery systems can also be tailored to release the yeast strains at specific locations within the GI tract, such as the colon, where they can exert their therapeutic effects more effectively [11].

Nanoencapsulation, which involves the use of nanomaterials to encapsulate yeast cells, offers additional benefits by improving the dispersion, stability, and targeted release of probiotics, as well as reducing the size of the encapsulated particles to enhance cellular uptake [13]. Furthermore, smart delivery systems that respond to changes in pH, temperature, or enzymatic activity could provide a more dynamic and controlled release of yeast-based probiotics, thus increasing their therapeutic potential. Such systems may also enable synergistic interactions with prebiotics or other bioactive compounds, creating functional combinations that support gut health, immune modulation, and metabolic function. As the demand for personalized nutrition and microbiome-based therapeutics grows, the development of advanced delivery technologies will be key in ensuring that yeast probiotics can reach their intended sites of action, maximizing their clinical and functional benefits [11].

9.2.10 Clinical Research and Evidence-Based Applications

The future of yeast-based probiotics in clinical practice hinges on robust clinical research and the accumulation of evidence to substantiate their therapeutic efficacy and safety across various health conditions. While yeast strains such as *S. boulardii* have shown promise in the management of GI disorders, including diarrhea, irritable bowel syndrome, and inflammatory bowel diseases, future clinical trials are expected to provide deeper insights into their long-term effectiveness, dosage regimens, and mechanisms of action [5, 19, 20]. As the field progresses, clinical research will likely expand beyond gut health to explore the broader therapeutic potential of yeast-based probiotics in conditions such as metabolic syndrome, obesity, diabetes, and even mental health disorders, particularly in relation to the gut-brain axis [22]. For instance, recent studies have suggested that *S. boulardii* may have a positive impact on glucose metabolism and insulin sensitivity, making it a potential candidate for inclusion in the management of type 2 diabetes [31]. Table 9.3 shows the future research direction and technological advancements in yeast as a probiotics.

Table 9.3: Future research directions and technological advancements in yeast probiotics.

Research area	Description	Potential impact
Genetic engineering	Use of CRISPR/Cas and other tools to enhance probiotic traits	Development of tailored probiotic strains
Omics-based profiling	Application of genomics, proteomics, and metabolomics to study yeast functions	Precision understanding of host-microbe interactions
Encapsulation and delivery systems	Microencapsulation, freeze-drying, and nanocarriers	Improved viability and targeted delivery in the gut
Strain-specific efficacy studies	Clinical trials assessing strain-specific outcomes	Regulatory approval and consumer confidence
Synbiotic formulations	Combination with prebiotics to enhance probiotic performance	Enhanced colonization and efficacy

The development of evidence-based clinical guidelines for the use of yeast-based probiotics will require large-scale, multicenter randomized controlled trials that rigorously assess their safety, efficacy, and mechanisms in diverse patient populations. Additionally, the integration of microbiome analyses into clinical studies will allow for a more precise understanding of how probiotic yeasts interact with the host microbiota to exert their health benefits. This personalized approach to clinical research will help optimize the selection of yeast strains and their corresponding therapeutic applications based on individual microbiome profiles [39]. Moreover, as yeast-based probiotics are integrated into mainstream healthcare practices, further studies are needed to explore their role in preventing and managing chronic diseases, reducing antibiotic use, and enhancing overall patient outcomes. Ultimately, clinical evidence will form the foundation for regulatory approvals and the establishment of standardized treatment protocols, enabling healthcare providers to confidently prescribe yeast-based probiotics as part of evidence-based clinical care.

9.3 Conclusion

Yeast-based probiotics, particularly strains such as *S. cerevisiae* and *S. boulardii,* have emerged as promising alternatives or complements to conventional bacterial probiotics. Their unique physiological attributes such as acid and bile tolerance, robust cell wall structure, and natural resistance to antibiotics make them highly suitable for survival and functionality in the GI tract. Beyond their established roles in managing GI disorders, yeast probiotics are increasingly recognized for their broader immunological, metabolic, and antimicrobial properties. Advances in molecular biology, functional genomics, and fermentation technology are paving the way for the development of tailor-made probiotic yeast strains with enhanced health benefits and targeted therapeutic potential. Looking forward, the integration of yeast probiotics into personalized medicine, precision nutrition, and functional food innovation holds immense promise. Cutting-edge technologies such as CRISPR-based gene editing and multi-omics approaches will enable deeper insights into yeast-host interactions and optimize strain performance for specific health outcomes. Moreover, their application in sustainable agriculture, veterinary medicine, and biotherapeutics further expands their potential beyond human health. However, to fully harness the benefits of yeast probiotics, challenges related to regulatory approval, large-scale production, strain standardization, and long-term safety must be addressed through rigorous scientific validation and interdisciplinary collaboration. In summary, the future of yeast probiotics is not only promising but also essential in the context of growing global health concerns, such as antibiotic resistance, chronic inflammatory diseases, and the need for more resilient and diverse probiotic options. With continued research and innovation, yeasts are poised to play a transformative role in the next generation of probiotic science.

Conflicts of Interests: The authors do not declare any conflict of interest.

References

[1] Aggarwal J, Adholeya A, Prakash A. Yeasts as novel probiotics: Current status and future perspectives. World Journal of Microbiology and Biotechnology. 2022;38(2):30. https://doi.org/10.1007/s11274-022-03162-0.

[2] Almeida GR, Costa FC, Almeida RF. Advanced encapsulation strategies for probiotic yeasts: Challenges and opportunities. International Journal of Food Science & Technology. 2021;56(9):4449–4462. https://doi.org/10.1111/ijfs.14898.

[3] Bommarius AS, Sievers C. Yeast in industrial biotechnology: Applications in biofuels and biochemicals. Industrial Biotechnology. 2021;17(2):84–97. https://doi.org/10.1089/ind.2021.0018.

[4] Czerucka D, Piche T, Rampal P. Review article: Yeast as probiotics – *Saccharomyces boulardii*. Alimentary Pharmacology & Therapeutics. 2007;26(6):767–778.

[5] Czerucka D, Piche T, Rampal P. *Saccharomyces boulardii* in gastrointestinal diseases: A review. American Journal of Gastroenterology. 2007;102(11):2411–2427. https://doi.org/10.1111/j.1572-0241.2007.01409.x.

[6] Deng Z, Zhao Y, Zhang Y. Application of probiotics in animal husbandry: A new opportunity for sustainable livestock production. Frontiers in Veterinary Science. 2019;6:1–14. https://doi.org/10.3389/fvets.2019.00001.

[7] Duina AA, Miller ME, Keeney JB. Budding yeast for budding geneticists: A primer on the *Saccharomyces cerevisiae* model system. Genetics. 2014;197(1):33–48. https://doi.org/10.1534/genetics.114.163188.

[8] Dulermo R, Gabarra JR, Molina J. Engineering stress-tolerant yeast strains for industrial fermentation processes. Applied Microbiology and Biotechnology. 2020;104(8):3413–3424. https://doi.org/10.1007/s00253-020-10649-4.

[9] FAO/WHO. Guidelines for the Evaluation of Probiotics in Food. 2002, London, Ontario.

[10] Fonseca GG, Heinzle E, Wittmann C, Gombert AK. The yeast *Kluyveromyces marxianus* and its biotechnological potential. Applied Microbiology and Biotechnology. 2008;79(3):339–354. https://doi.org/10.1007/s00253-008-1458-6.

[11] Gänzle MG, Eim V, Eijck L. Advanced delivery systems for probiotics: Methods, applications, and challenges. Food Research International. 2019;123:612–626. https://doi.org/10.1016/j.foodres.2019.04.041.

[12] Hatoum R, Labrie S, Fliss I. Antimicrobial and probiotic properties of yeasts: From fundamental to novel applications. Frontiers in Microbiology. 2012;3:421. https://doi.org/10.3389/fmicb.2012.00421.

[13] Jafari SM, Sheikh Z, Zeynizadeh B. Encapsulation technologies for probiotics: A review. Critical Reviews in Food Science and Nutrition. 2017;57(12):2352–2366. https://doi.org/10.1080/10408398.2016.1163680.

[14] Kelesidis T, Pothoulakis C. Efficacy and safety of the probiotic *Saccharomyces boulardii* for the prevention and therapy of gastrointestinal disorders. Therapeutic Advances in Gastroenterology. 2012;5(2):111–125. https://doi.org/10.1177/1756283X11428502.

[15] Khalil R, Darwish H, Hassan S. Probiotic yeasts and their potential in dairy and non-dairy functional food development: Current status and future outlook. Probiotics and Antimicrobial Proteins. 2022;14(1):1–15. https://doi.org/10.1007/s12602-021-09821-z.

[16] Kim M, Lee K, Ryu S. Effect of *Saccharomyces cerevisiae* and *Saccharomyces boulardii* on gastrointestinal health and immune function in livestock. Animal Feed Science and Technology. 2021;275:114786. https://doi.org/10.1016/j.anifeedsci.2021.114786.

[17] Köhl J, Latz M, Amossé C. Yeasts as biocontrol agents in integrated pest management: Opportunities and challenges. Pest Management Science. 2019;75(5):1336–1348. https://doi.org/10.1002/ps.5455.

[18] Kumura H, Tanoue Y, Tsukahara M, Tanaka T, Shimazaki K. Screening of dairy yeast strains for probiotic applications. Journal of Dairy Science. 2004;87(12):4050–4056. https://doi.org/10.3168/jds. S0022-0302(04)73547-9.

[19] McFarland LV. *Saccharomyces boulardii* in the prevention and treatment of gastrointestinal disorders. Clinical Infectious Diseases. 2010;51(Suppl 1):S12–S15. https://doi.org/10.1086/653645.

[20] McFarland LV. Systematic review and meta-analysis of *Saccharomyces boulardii* in adult patients. World Journal of Gastroenterology. 2010;16(18):2202.

[21] Mimee M, Tucker AC, Voigt CA, Lu TK. Programming a human commensal bacterium, *Bacteroides thetaiotaomicron*, to sense and respond to stimuli in the murine gut microbiota. Cell Systems. 2015;1 (1):62–71. https://doi.org/10.1016/j.cels.2015.06.001.

[22] Möller AH, West KL, Yang RY. Exploring the gut-brain axis: The role of probiotics in mental health and neuroinflammation. Neurotherapeutics. 2021;18(4):2102–2113. https://doi.org/10.1007/s13311-021-01032-2.

[23] Mroczek A, Wegrzyn G, Ziolkowski W. Bioremediation potential of yeast-based probiotics in the removal of environmental pollutants. Microorganisms. 2021;9(1):99. https://doi.org/10.3390/microor ganisms9010099.

[24] Oliveira RPDS, Perego P, Converti A, de Oliveira MN. Novel probiotic yeasts with potential therapeutic applications. Current Opinion in Food Science. 2021;38:1–7. https://doi.org/10.1016/j. cofs.2020.10.002.

[25] Parapouli M, Vasileiadis A, Afendra AS, Hatziloukas E. *Saccharomyces cerevisiae* and its industrial applications. AIMS Microbiology. 2020;6(1):1–31. https://doi.org/10.3934/microbiol.2020001.

[26] Patel S, Kaur G, Chawla P. Nanoencapsulation of probiotics: A future perspective for gut health. Food Research International. 2017;100:107–119. https://doi.org/10.1016/j.foodres.2017.07.014.

[27] Pinto C, Pinho D, Egas C, Gomes AC. Unravelling the diversity of grapevine microbiome. Microbial Ecology. 2015;70(2):425–436. https://doi.org/10.1007/s00248-015-0591-9.

[28] Ranadheera CS, Evans CA, Adams MC, Baines SK. Probiotic viability and physicochemical and sensory properties of plain and stirred fruit yogurts made from goat's milk. Food Science & Nutrition. 2017;5(4):865–872. https://doi.org/10.1002/fsn3.489.

[29] Ranadheera CS, Evans CA, Adams MC, Baines SK. In vitro analysis of gastrointestinal tolerance and colonic adhesion of probiotics in goat's milk yogurt. Food Science & Nutrition. 2017;5(3):626–635. https://doi.org/10.1002/fsn3.447.

[30] Sadeghi P, Salehi M, Najafi M. Advanced probiotic delivery systems: Recent innovations and future perspectives. Current Opinion in Food Science. 2020;32:66–75. https://doi.org/10.1016/j.cofs.2019. 12.003.

[31] Salem MA, Amrouche T, Louet A. Probiotic effects of *Saccharomyces boulardii* on glucose metabolism and insulin sensitivity in type 2 diabetes. Journal of Clinical Endocrinology & Metabolism. 2021;106 (5):e1887–e1895. https://doi.org/10.1210/clinem/dgab091.

[32] Saulnier DM, Kolida S, Gibson GR. Microbiology of the human intestinal tract and approaches for its dietary modulation. Current Pharmaceutical Design. 2008;14(13):1344–1355. https://doi.org/10.2174/ 138161208784139756.

[33] Stier H, Ebbeskotte V, Gruenwald J. Immune-modulatory effects of dietary yeast β-glucans. Nutrition Journal. 2014;13:38.

[34] Turan M, Yildirim E, Erdal S. Beneficial effects of probiotic yeasts on plant growth and productivity. Biocontrol Science. 2021;26(2):157–166. https://doi.org/10.1016/j.biocontrol.2020.12.003.

[35] van der Hoek SA, Darbani B, Zugaj KE, Prabhala BK, Biron MB, Borodina I. Engineering probiotic yeast *Saccharomyces boulardii* for the delivery of therapeutic proteins in the gut. ACS Synthetic Biology. 2021;10(1):69–81. https://doi.org/10.1021/acssynbio.0c00438.

[36] Wijsman M, van der Meer JR, Sasso S. Synthetic biology approaches for the development of next-generation probiotics. Current Opinion in Biotechnology. 2022;73:219–226. https://doi.org/10.1016/j.copbio.2021.09.003.

[37] Wong CM, Lee Y, Lee SH. Biodegradation of toxic compounds by yeasts in environmental remediation. Fungal Biology Reviews. 2018;32(1):12–20. https://doi.org/10.1016/j.fbr.2017.12.001.

[38] Yang Y, Liu Y, Gao W. Enhancing the biofuel production process through engineered yeast strains for lignocellulosic biomass conversion. Bioresource Technology. 2022;346:126462. https://doi.org/10.1016/j.biortech.2021.126462.

[39] Zeevi D, Korem T, Zmora N, Israeli D, Rothschild D, Weinberger A, . . . Segal E. Personalized nutrition by prediction of glycemic responses. Cell. 2015;163(5):1079–1094. https://doi.org/10.1016/j.cell.2015.11.001.

[40] Zhang H, Zong X, Zhang L. Potential of *Saccharomyces boulardii* in animal health and its impact on gut microbiota. Veterinary Microbiology. 2022;267:109338. https://doi.org/10.1016/j.vetmic.2021.109338.

[41] Zhang Y, Liu S, Li S. Industrial applications of yeast in biofuel production. Frontiers in Bioengineering and Biotechnology. 2021;9:630663. https://doi.org/10.3389/fbioe.2021.630663.

Sachin Kumar*, Ashwani Kumar, Ashish Kumar Singh,
Nirmala Sehrawat, Mukesh Yadav

Chapter 10
Diet Patterns and Lifestyle Changes for Improved Gut Health

Abstract: Gut health encompasses the optimal functioning of the gastrointestinal system and the balanced relationship between the host and its microbiota. This chapter explores how diet and lifestyle collectively influence gut microbial composition, immune function, metabolic health, and disease risk. Beneficial dietary patterns – such as fiber-rich diets, the Mediterranean diet, and consumption of fermented foods – enhance microbial diversity, support the production of short-chain fatty acids (SCFAs), and promote intestinal and systemic health. In contrast, Western diets, artificial sweeteners, and low-fiber regimens disrupt microbial balance, reduce SCFA levels, and increase inflammation. Lifestyle factors such as stress, sleep, physical activity, and antibiotic use also play crucial roles in modulating gut microbiota. This comprehensive analysis highlights that dietary quality, coupled with supportive lifestyle practices, is central to fostering a healthy and resilient gut ecosystem.

Keywords: gut health, host-probiotic interaction, dietary pattern, lifestyle

10.1 Introduction

Gut health refers to the optimal functioning and balance of the gastrointestinal tract, particularly the relationship between the host and its gut microbiota.

Gut health is not only about proper digestion but also about the symbiotic relationship between humans and their gut flora. A healthy gut supports:

–   Nutrient absorption and metabolism [1, 2]
–   Immune system development [3–6]
–   Protection against pathogens [7, 8]
–   Mood regulation and brain function via the gut-brain axis [9–11]

*Corresponding author: Sachin Kumar, Department of Bioinformatics, Janta Vedic College, Baraut, Baghpat, Uttar Pradesh, India, e-mail: sachinsuryan@gmail.com
Ashwani Kumar, Department of Biotechnology, KVSCOS, Swami Vivekanand Subharti University, Meerut, Uttar Pradesh, India
Ashish Kumar Singh, Biomolecular Toxicology Lab, CSIR-Indian Institute of Toxicology Research, Lucknow 206001, Uttar Pradesh, India
Nirmala Sehrawat, Mukesh Yadav, Department of Bio-Sciences and Technology, MMEC, Maharishi Markandeshwar (Deemed to be University), Mullana-Ambala 133207, Haryana, India

https://doi.org/10.1515/9783112205150-010

Dysbiosis (imbalance in gut bacteria) has been linked to diseases such as obesity, type 2 diabetes, inflammatory bowel disease (IBD), cardiovascular disease and mental health disorders like depression and anxiety [12].

There are various factors related to diet and lifestyle that significantly affect the gut health. A diverse, fiber-rich, and fermented plant-based diet plays a key role in supporting beneficial gut bacteria and overall microbial diversity [13, 14]. On the contrary, the overuse of antibiotics and certain medications can disrupt the gut microbiome, reducing the population of helpful microbes and potentially leading to long-term health issues [15]. Chronic psychological stress has also been found to alter the gut microbiota composition and impair gut-brain communication, contributing to various gastrointestinal and mood disorders [16]. Additionally, poor sleep quality and a sedentary lifestyle have been associated with negative changes in gut microbial balance [17, 18]. Lastly, frequent infections or overly sterile environments, especially during early childhood, may interfere with the development of a healthy and diverse gut microbiome [19].

This chapter thoroughly examines how dietary choices and lifestyle interventions can influence gut microbiota composition, and ultimately, improve health outcomes.

10.2 Role of Diet in Shaping the Gut Microbiome

Diet is the most influential factor affecting gut microbial diversity and composition. There are various diets that benefit the gut health while various diets harm as well. The chapter highlights both beneficial and harmful dietary patterns.

10.2.1 Beneficial Diets

A gut-friendly diet typically includes fiber-rich foods, such as fruits, vegetables, whole grains, and legumes, which support healthy digestion and promote the growth of beneficial gut bacteria [20]. Fermented foods such as yogurt, kefir, sauerkraut, and kimchi introduce probiotics, which can enhance the microbial balance [12]. Limiting processed foods, added sugars, and excessive fats also helps maintain gut integrity and reduce inflammation.

10.2.1.1 High-Fiber Diet: Fiber and Prebiotics

Among various dietary components, dietary fiber plays a crucial role in shaping the gut environment and promoting microbial diversity. Dietary fiber refers to the indigestible portion of plant foods that resists digestion in the upper gastrointestinal tract but is fermented by the microbiota in the colon [21]. A high-fiber diet includes two main types of fiber: soluble fiber and insoluble fiber. Soluble fiber dissolves in water to form a gel-like material. Insoluble fiber does not dissolve in water:

- **Insoluble Fiber:** Adds bulk to stool and supports regular bowel movement. Examples: whole grains, wheat bran, and vegetables.
- **Soluble Fiber:** Forms gel-like substance, lowers cholesterol, and helps digestion. Examples: oats, fruits, and legumes.

Prebiotics (Table 10.1) are a specific type of dietary fiber that nourishes the good bacteria in your gut. Unlike regular fibers, prebiotics are not digested by human enzymes but are instead fermented by gut bacteria, promoting the growth of beneficial microorganisms such as bifidobacteria and *Lactobacillus* [22]. This process helps to improve gut health, digestive function, and immune system response. Including prebiotics in your diet not only aids in digestion but also supports the overall well-being by maintaining a healthy balance of gut microbiota [23].

10.2.1.1.1 Role of Fiber in Gut Microbiota Modulation

The gut microbiota relies heavily on dietary fiber as a source of fermentable substrate. When fiber reaches the colon, it undergoes fermentation by resident bacteria, producing short-chain fatty acids (SCFAs) such as acetate, propionate, and butyrate [24, 25]. These SCFAs have multiple health-promoting functions:

- **Butyrate** serves as the primary energy source for colonocytes and helps maintain the intestinal barrier integrity [26].
- **Propionate** influences gluconeogenesis in the liver and has anti-inflammatory properties.
- **Acetate** plays a role in lipid metabolism and appetite regulation.

These metabolites not only enhance gut health but also exert systemic benefits on metabolism and immune function.

10.2.1.1.2 Fiber and Microbial Diversity

A high-fiber diet is associated with increased microbial diversity, which is considered a hallmark of a healthy gut ecosystem [27]. Diverse microbial populations are more resilient to pathogenic invasions and environmental stressors. Low fiber intake, common in Western diets, has been linked to dysbiosis – an imbalance in microbial populations – contributing to conditions such as irritable bowel syndrome (IBS), IBD, and even obesity [28, 29].

This way, you are showing that prebiotics are an important part of a high-fiber diet, while also highlighting their unique benefit: feeding good bacteria in the gut. The daily recommended fiber intake is approximately 25–30 g for adults, but most populations consume significantly less than this [30, 31].

10.2.1.1.3 Clinical Evidence and Implications

Several clinical trials have demonstrated the benefits of high-fiber diets in managing gastrointestinal disorders. For instance, a randomized controlled trial showed that supplementation with soluble fiber (psyllium) improved symptoms in patients with IBS [32]. Additionally, dietary interventions that increase fiber intake have been shown to shift microbial composition toward beneficial taxa such as bifidobacteria and *Lactobacillus* species [33].

Table 10.1: Different types of prebiotics, their sources, and functions.

Prebiotic fibers	Food sources	Functions	References
Inulin	Chicory root, garlic, onion, and banana	Feeds bifidobacteria	[34]
FOS (fructo-oligosaccharides)	Asparagus, leeks, and chicory	Boosts good gut flora	[34, 35]
GOS (galacto-oligosaccharides)	Legumes and lentils	Stimulates the growth of *Lactobacillus* and bifidobacteria	[36]
Resistant starch	Cooked/cooled potatoes and green banana	Improves gut and insulin sensitivity	[37]
Beta-glucan	Oats and barley	Lowers cholesterol and supports microbiota	[38]
Pectin	Apples and citrus fruits	Improves digestion and gels in intestine	[39, 40]
Mucilage	Flaxseeds and okra	Soothes digestion and forms viscous gel	[41]
Arabinoxylans	Whole grains (wheat and rye)	Modulates microbiota and gut health	[42]
Lignans	Flaxseeds and sesame seeds	Acts as antioxidant and mild prebiotic	[43]

10.2.1.2 Mediterranean Diet

The Mediterranean diet (MD) is a dietary pattern traditionally followed in countries bordering the Mediterranean Sea, such as Greece, Italy, and Spain.

It is known for its benefits to heart health, longevity, and chronic disease prevention [44].

10.2.1.2.1 Key Components of the Mediterranean Diet

The MD is characterized by:

- **High intake** of fruits, vegetables, legumes, nuts, seeds, whole grains, and olive oil
- **Moderate intake** of fish, poultry, eggs, and dairy products
- **Low consumption** of red meat, processed foods, and added sugars
- **Regular use** of herbs and spices instead of salt
- **Moderate consumption** of red wine with meals (optional)

Beyond cardiovascular and metabolic health, as the components of MD are rich in dietary fiber, polyphenols, omega-3 fatty acids, and antioxidants, this diet also supports gut health by positively influencing the gut microbiota [45, 46].

10.2.1.2.2 Impact on Gut Microbiota

The MD fosters a gut environment conducive to beneficial bacteria. Studies have shown increased levels of bifidobacteria, lactobacilli, and SCFA-producing bacteria such as *Faecalibacterium prausnitzii* in individuals adhering to this diet [45, 47].

SCFAs, especially butyrate, are key metabolites that improve gut barrier function, reduce inflammation, and enhance immunity [48]. The MD's high fiber and polyphenol content promotes the growth of these SCFA-producing microbes.

10.2.1.2.3 Anti-inflammatory Effects

Chronic low-grade inflammation is a common feature of gut-related disorders such as IBD and IBS. The MD has been shown to reduce systemic inflammation markers like C-reactive protein and interleukin-6, likely due to its high antioxidant and anti-inflammatory nutrient profile [49].

10.2.1.2.4 Disease Prevention and Management

Adherence to MD is associated with a lower risk of gastrointestinal diseases and metabolic conditions that have gut microbiota involvement. For example:

- **Colorectal cancer:** The MD reduces the risk by limiting red/processed meat and enhancing microbial diversity [50, 51].
- **Obesity and diabetes:** Improvements in insulin sensitivity and reduced inflammation via microbiome-mediated pathways have been observed [52].
- **Neurodegenerative diseases:** Gut-brain axis modulation via the MD may influence the progression of Alzheimer's and Parkinson's diseases [53].

10.2.1.2.5 Comparative Studies

A study by De Filippis et al. [45] found that individuals following the MD had a significantly higher abundance of fiber-fermenting bacteria and lower levels of potentially harmful bacteria compared to those following a Western diet. Another intervention study reported increased microbial richness and improved metabolic profiles after only 8 weeks of adopting the MD [54].

Hence, MD supports gut health through its rich content of fiber, polyphenols, and healthy fats. By encouraging microbial diversity and SCFA production, it not only maintains gastrointestinal integrity but also contributes to the prevention of a wide range of inflammatory and metabolic diseases. The MD stands as a model dietary pattern for both gut and systemic health.

10.2.1.3 Fermented Foods and Probiotics

Fermented foods have been consumed for centuries across various cultures for their flavor, preservation benefits, and health-promoting properties. In recent years, there has been growing scientific interest in their role in modulating the gut microbiota and promoting gastrointestinal and systemic health. Fermentation not only enhances the nutrient bioavailability but also introduces live microorganisms – commonly known as probiotics – that may benefit gut health [55, 56].

10.2.1.3.1 What Are Fermented Foods?

Fermented foods are those that have been transformed by the action of microorganisms such as bacteria, yeasts, or fungi. Common examples include:
- Dairy products: Yogurt, kefir, and buttermilk
- Vegetables: Sauerkraut, kimchi, and pickles (naturally fermented)
- Legume-based products: Miso, tempeh, and natto
- Beverages: Kombucha and kvass
- Others: Fermented soy sauces and sourdough bread

These foods are rich in beneficial microbes, enzymes, and bioactive compounds that can have direct or indirect effects on human health [57]. They may also contain or be consumed alongside prebiotic fibers – such as inulin, resistant starch, and oligosaccharides – which serve as food for gut microbes and enhance their survival and function.

10.2.1.3.2 Microbial Content and Probiotic Potential

The main microbial strains found in fermented foods include *Lactobacillus, Bifidobacterium, Leuconostoc, Saccharomyces*, and *Pediococcus* species. When consumed, these microorganisms may survive gastrointestinal transit and temporarily colonize the gut, where they can compete with pathogens, support gut barrier function, and modulate the immune response [58].

However, it is important to distinguish between fermented foods and probiotic supplements. Not all fermented foods contain live microbes at the point of consumption, especially those that have been pasteurized. Only those with viable microorganisms in sufficient quantities can be considered sources of probiotics [22, 59].

10.2.1.3.3 Mechanisms Supporting Gut Health

Fermented foods support gut health through multiple mechanisms:

- **Restoration of Microbial Balance**: Regular consumption may help correct dysbiosis – a microbial imbalance linked with diseases such as IBD and obesity [60, 61].
- **Production of Beneficial Metabolites**: Fermentation enhances the production of SCFAs and bioactive peptides that contribute to colon health, reduce inflammation, and regulate immune responses [62].
- **Improvement of Gut Barrier Integrity**: Fermented foods have been shown to reduce intestinal permeability ("leaky gut") and strengthen the mucosal barrier [63].
- **Inhibition of Pathogenic Microbes**: Probiotic strains from fermented foods produce antimicrobial substances (e.g., lactic acid and bacteriocins) that inhibit harmful bacteria in the gut [64].
- **Synergistic Effects with Prebiotics**: When consumed alongside prebiotic-rich foods, such as garlic, onions, bananas, and whole grains, fermented foods may exert synergistic effects – known as synbiotics – further enhancing the gut microbial diversity and health [65].

10.2.1.3.4 Clinical and Population-Based Evidence

Several clinical studies have demonstrated the benefits of fermented foods in gut-related conditions. Yogurt and kefir have shown to alleviate symptoms in lactose-intolerant individuals and reduce inflammation in patients with IBD [66, 67]. Kimchi and sauerkraut have been associated with increased microbial diversity and reduced markers of inflammation [68, 69].

In a landmark study, fermented food consumption was linked with increased gut microbiome diversity and decreased inflammatory markers in healthy adults [70].

10.2.1.3.5 Safety and Considerations

While fermented foods are generally safe, some caution is needed for individuals with histamine intolerance or compromised immune systems. Additionally, not all fermented products offer equal benefits – some commercially produced foods are pasteurized, destroying beneficial microbes. Therefore, choosing raw, unpasteurized, or homemade versions may be more effective for gut health.

10.2.2 Harmful Diets and Their Impact on Gut Health

Diets high in ultra-processed foods, artificial ingredients, and low in fiber can lead to dysbiosis, increased intestinal permeability, and systemic inflammation – factors implicated in metabolic, gastrointestinal, and neurological disorders. This chapter examines three diet patterns shown to adversely affect gut health, incorporating recent evidence.

10.2.2.1 Western Diet

The Western diet is characterized by high consumption of processed foods, red meats, added sugars, saturated fats, and a lack of fiber-rich plant foods. This pattern promotes a pro-inflammatory gut microbiota and decreases the abundance of beneficial bacteria such as *Faecalibacterium prausnitzii* and *Akkermansia muciniphila* [12, 71].

Increased intake of saturated fats and refined sugars is associated with greater intestinal permeability ("leaky gut") and systemic low-grade inflammation, which underpins many chronic diseases, including obesity, diabetes, and cardiovascular disorders [12, 72]. The review by Valdes and colleagues concluded that habitual Western dietary patterns consistently impair gut microbiota composition and function, accelerating the risk of inflammatory diseases [12].

10.2.2.1.1 Artificial Sweeteners
Common nonnutritive sweeteners such as aspartame, saccharin, and sucralose are marketed as sugar alternatives but have been shown to disrupt gut microbial balance. Early studies, including Suez et al. [73], demonstrated that saccharin altered microbial composition and induced glucose intolerance in both mice and humans.

Recent work by Suez et al. [74] extended these findings by showing that different artificial sweeteners have distinct and individualized effects on the human microbiome and glucose metabolism. In this randomized controlled trial, saccharin and sucralose altered both microbial composition and gene expression, suggesting microbiome-mediated metabolic consequences even with modest intake.

10.2.2.1.2 High-Protein, Low-Carbohydrate Diets (e.g., Ketogenic Diet)
Diets such as the ketogenic diet, which emphasize high fat and protein intake with minimal carbohydrates, may provide short-term weight loss and glycemic control benefits. However, their long-term impact on gut health is less favorable. Due to the extremely low fiber content, ketogenic diets limit substrates for fermentation, reducing beneficial SCFA production – especially butyrate, a critical anti-inflammatory metabolite [48].

Recent microbiome research [75] found that sustained low-carbohydrate diets led to a decline in microbial diversity, particularly in SCFA-producing bacteria like *Roseburia* and *Faecalibacterium*. Furthermore, an increase in bile-tolerant species such as *Bilophila wadsworthia* was observed, which is associated with gut inflammation and metabolic endotoxemia.

Hence, the modern dietary patterns, particularly the Western diet, high-protein low-carb regimens, and the use of artificial sweeteners, have demonstrable negative effects on gut microbial ecology. These diets often reduce microbial diversity, decrease SCFA production, and compromise gut barrier function, leading to inflammation and increased disease risk. Recognizing and mitigating these risks through informed dietary choices are key to preserving the gut and overall health.

10.3 Lifestyle Changes for Gut Health

While diet plays a foundational role in shaping the gut microbiome, it is far from the only influence. Several lifestyle factors also significantly impact microbial diversity, function, and gut health. By optimizing these aspects of daily life, we can support the balance of our gut microbiota and improve overall well-being.

The composition and function of the gut microbiome are heavily influenced by lifestyle factors. Positive changes in habits such as physical activity, stress management, sleep, and responsible antibiotic use can promote a diverse and resilient gut microbiota, which in turn supports better digestion, immune function, and even neurological health. The following sections explore how these lifestyle components contribute to gut health, as supported by recent scientific research.

10.3.1 Physical Activity

Regular exercise has been shown to beneficially alter the gut microbiota by increasing microbial diversity and the abundance of beneficial bacteria. These changes support metabolic health, reduce inflammation, and strengthen the gut barrier. Monda et al. [76] demonstrated that both aerobic and resistance training can enhance gut microbial composition, which may help in preventing chronic diseases. Additionally, synchronized circadian rhythms influenced by consistent activity patterns can further enhance the gut microbiome's role in regulating metabolism [77].

10.3.2 Stress Management

Chronic psychological stress can disrupt the gut-brain axis, leading to gut dysbiosis and increased intestinal permeability. This imbalance may contribute to mood disorders and neurodegenerative conditions. Loh et al. [16] emphasized that stress can modulate gut microbial populations through neural, hormonal, and immune pathways. Managing stress through mindfulness, relaxation techniques, and physical activity can help maintain a healthier gut microbiome, protecting both gastrointestinal and neurological health.

10.3.3 Sleep Quality

Sleep plays a critical role in maintaining a balanced microbiome. Disrupted or insufficient sleep can lead to gut microbiota imbalances, inflammation, and impaired immunity. According to Sun et al. [78], sleep deprivation alters the diversity and stability of the gut microbiota, potentially increasing the risk of metabolic and cognitive disorders. Lin et al. [79] further noted that the gut and sleep systems are interconnected in a feedback loop, with microbial metabolites influencing sleep patterns and quality. Prioritizing consistent, restful sleep is therefore essential for microbiome health.

10.3.4 Avoiding Unnecessary Antibiotics

Antibiotics, while lifesaving in many situations, can significantly disrupt the gut microbiome by reducing microbial diversity and allowing opportunistic pathogens to flourish. Studies by Dethlefsen and Relman [80], Jernberg et al. [81], and Patangia et al. [82] show that the recovery of the gut microbiome after antibiotic use is often incomplete and highly individualized. Overuse or misuse of antibiotics may lead to long-term health consequences such as increased susceptibility to infections, allergies, and metabolic disorders. Lathakumari et al. [83] emphasize the importance of using antibiotics judiciously, and only when prescribed by a healthcare professional, to minimize unnecessary harm to the microbiome.

10.4 Emerging Trends and Future Perspectives

As the importance of the gut microbiome in human health becomes more evident, the field of gut health research and interventions is quickly advancing, with a growing focus on personalized nutrition, probiotic and prebiotic supplements, and fecal microbiota transplantation.

10.4.1 Personalized Nutrition

Individual responses to foods vary significantly due to differences in gut microbiota composition, genetics, and lifestyle factors. The future of gut health is moving toward precision nutrition, where dietary recommendations are tailored using microbiome testing to optimize health outcomes [84]. Personalized nutrition, tailored to an individual's gut microbiota, has shown promising potential in promoting gut health and metabolic balance [85]. Recent advancements using artificial intelligence-driven dietary recommendations have further demonstrated measurable improvements in microbiome diversity [86].

10.4.2 Probiotic and Prebiotic Supplements

While a diverse, fiber-rich diet remains the most effective way to support gut microbiota, probiotic and prebiotic supplements are being increasingly explored to fill nutritional gaps. However, variability in product quality and strain-specific effects highlights the need for more standardized clinical research to determine appropriate dosages and efficacy [87, 88].

10.4.3 Fecal Microbiota Transplantation

Fecal microbiota transplantation involves the transfer of fecal material from a healthy donor to a patient with established success in treating recurrent *Clostridioides difficile* infections. Ongoing research is exploring its potential for metabolic disorders, IBDs, and even neurological conditions, though long-term safety and regulatory standards remain areas of active investigation [89–91].

10.5 Conclusion

This chapter highlights the strong connection between gut health and dietary, lifestyle choices. A balanced, fiber-rich diet with fermented and plant-based foods, along with regular physical activity, stress management, and adequate sleep, can significantly enhance the gut microbiome and overall health. The chapter emphasizes the importance of gut health for long-term well-being and underscores that simple, sustainable lifestyle changes, such as avoiding processed foods, unnecessary antibiotics, and artificial sweeteners, are key strategies. As research on gut microbiota advances, the integration of personalized nutrition and microbiome-targeted therapies shows promise for preventing and managing chronic diseases.

Author contributions: All the authors have accepted responsibility for the entire content of this submitted manuscript and approved submission.

Conflicts of interest: The authors do not declare any conflicts of interest.

References

[1]   Wu GD, Chen J, Hoffmann C, Bittinger K, Chen YY, Keilbaugh SA, et al. Linking long-term dietary patterns with gut microbial enterotypes. Science. 2011;334(6052):105–108. doi: 10.1126/science.1208344.

[2]   Fujisaka S, Watanabe Y, Tobe K. The gut microbiome: A core regulator of metabolism. Journal of Endocrinology. 2023;256(3):e220111. doi: 10.1530/JOE-22-0111.

[3]   Belkaid Y, Hand TW. Role of the microbiota in immunity and inflammation. Cell. 2014;157(1):121–141. doi: 10.1016/j.cell.2014.03.011.

[4]   Yang Q, Cai Y, Guo S, Wang Z, Wang Y, Yu X, Sun W, Qiu S, Zhang A. Decoding immune interactions of gut microbiota for understanding the mechanisms of diseases and treatment. Frontiers in Microbiology. 2023;14:1238822. doi: 10.3389/fmicb.2023.1238822.

[5]   Ardeshir A, Gensollen T, Zeng M, Al Nabhani Z, Blumberg R. Editorial: The early life window of opportunity: Role of the microbiome on immune system imprinting. Frontiers in Immunology. 2024;15:1417060. doi: 10.3389/fimmu.2024.1417060.

[6]   Zhu J, He M, Li S, Lei Y, Xiang X, Guo Z, Wang Q. Shaping oral and intestinal microbiota and the immune system during the first 1,000 days of life. Frontiers in Pediatrics. 2025;13:1471743. doi: 10.3389/fped.2025.1471743.

[7]   Kamada N, Chen GY, Inohara N, Núñez G. Control of pathogens and pathobionts by the gut microbiota. Nature Immunology. 2013;14(7):685–690. doi: 10.1038/ni.2608.

[8]   Spragge F, Bakkeren E, Jahn MT, Araujo BN, Pearson CF, Wang X, et al. Microbiome diversity protects against pathogens by nutrient blocking. Science. 2023;382(6676):eadj3502. doi: 10.1126/science.adj3502.

[9]   Cryan JF, Dinan TG. Mind-altering microorganisms: The impact of the gut microbiota on brain and behaviour. Nature Reviews Neuroscience. 2012;13(10):701–712. doi: 10.1038/nrn3346.

[10] Tan HE. The microbiota-gut-brain axis in stress and depression. Frontiers in Neuroscience. 2023;17:1151478. doi: 10.3389/fnins.2023.1151478.

[11] Verma A, Inslicht SS, Bhargava A. Gut-Brain Axis: Role of Microbiome, Metabolomics, Hormones, and Stress in Mental Health Disorders. Cells. 2024;13(17):1436. doi: 10.3390/cells13171436.

[12] Valdes AM, Walter J, Segal E, Spector TD. Role of the gut microbiota in nutrition and health. BMJ. 2018;361:2179. doi: 10.1136/bmj.k2179

[13] Soldán M, Argalášová Ľ, Hadvinová L, Galileo B, Babjaková J. The effect of dietary types on gut microbiota composition and development of non-communicable diseases: A narrative review. Nutrients. 2024;16(18):3134. doi: 10.3390/nu16183134.

[14] Rinninella E, Tohumcu E, Raoul P, Fiorani M, Cintoni M, Mele MC, et al. The role of diet in shaping human gut microbiota. Best Practice & Research Clinical Gastroenterology. 2023;62–63:101828. https://doi.org/10.1016/j.bpg.2023.101828.

[15] Garg K, Mohajeri MH. Potential effects of the most prescribed drugs on the microbiota-gut-brain axis: A review. Brain Research Bulletin. 2024;207:110883. doi: 10.1016/j.brainresbull.2024.110883.

[16] Loh JS, Mak WQ, Tan LKS, Lim WL, Lee SC, Lim SY. Microbiota–gut–brain axis and its therapeutic applications in neurodegenerative diseases. Signal Transduction and Targeted Therapy. 2024;9:37. doi: 10.1038/s41392-024-01743-1.

[17]   Wu J, Zhang B, Zhou S, Huang Z, Xu Y, Lu X, Zheng X, Ouyang D. Associations between gut microbiota and sleep: A two-sample, bidirectional Mendelian randomization study. Frontiers in Microbiology. 2023;14:1236847. doi: 10.3389/fmicb.2023.1236847.

[18]   Farré N, Gozal D. Sleep and the microbiome: A two-way relationship. Archivos de Bronconeumología. 2019;55(1):7–8. doi: 10.1016/j.arbres.2018.04.007.

[19]   Roslund M, Puhakka R, Grönroos M, Nurminen N, Oikarinen S, Haahtela T, et al. Biodiversity intervention enhances immune regulation and health-associated commensal microbiota among daycare children. Science Advances. 2020;6(42):eaba2578. https://doi.org/10.1126/sciadv.aba2578.

[20]   Sonnenburg J, Sonnenburg E. The Good Gut: Taking Control of Your Weight, Your Mood, and Your Long-term Health. 2015, Penguin Publishing Group.

[21]   Stephen AM, Champ MM, Cloran SJ, et al. Dietary fibre in Europe: Current state of knowledge on definitions, sources, recommendations, intakes and relationships to health. Nutrition Research Reviews. 2017;30(2):149–190. doi: 10.1017/S095442241700004X.

[22]   Hill C, Guarner F, Reid G, Gibson GR, Merenstein DJ, Pot B, et al. Expert consensus document: The International Scientific Association for Probiotics and Prebiotics consensus statement on the scope and appropriate use of the term probiotic. Nature Reviews Gastroenterology & Hepatology. 2014;11 (8):506–514. doi: 10.1038/nrgastro.2014.66.

[23]   Sanders ME. Probiotics: Definition, sources, selection, and uses. Clinical Infectious Diseases. 2008;46 (Supplement_2):S58–S61. https://doi.org/10.1086/523341.

[24]   Flint HJ, Duncan SH, Scott KP, Louis P. Interactions and competition within the microbial community of the human colon: Links between diet and health. Environmental Microbiology. 2007;9(5):1101– 1111. doi: 10.1111/j.1462-2920.2007.01281.x.

[25]   Vinelli V, Biscotti P, Martini D, Del Bo' C, Marino M, Meroño T, Nikoloudaki O, Calabrese FM, Turroni S, Taverniti V, et al. Effects of dietary fibers on short-chain fatty acids and gut microbiota composition in healthy adults: a systematic review. Nutrients. 2022;14(13):2559. doi: 10.3390/ nu14132559.

[26]   Canani RB, Costanzo MD, Leone L, Pedata M, Meli R, Calignano A. Potential beneficial effects of butyrate in intestinal and extraintestinal diseases. World Journal of Gastroenterology. 2011;17 (12):1519. doi: 10.3748/wjg.v17.i12.1519.

[27]   Mosca A, Leclerc M, Hugot JP. Gut microbiota diversity and human diseases: Should we reintroduce key predators in our ecosystem?. Frontiers in Microbiology. 2016;7:455. https://doi.org/10.3389/ fmicb.2016.00455.

[28]   De Filippo C, Cavalieri D, Di Paola M, et al. Impact of diet in shaping gut microbiota revealed by a comparative study in children from Europe and rural Africa. PNAS. 2010;107(33):14691–14696. doi: 10.1073/pnas.1005963107.

[29]   Nguyen TL, Zhao J. Editorial: Role of prebiotics in managing obesity and metabolic disorders. Frontiers in Endocrinology (Lausanne). 2024;15:1432530. https://doi.org/10.3389/fendo.2024. 1432530.

[30]   Slavin J. Fiber and prebiotics: Mechanisms and health benefits. Nutrients. 2013;5(4):1417–1435. doi: 10.3390/nu5041417.

[31]   Correa AC, Lopes MS, Perna RF, Silva EK. Fructan-type prebiotic dietary fibers: Clinical studies reporting health impacts and recent advances in their technological application in bakery, dairy, meat products and beverages. Carbohydrate Polymers. 2024;323:121396. doi: 10.1016/j. carbpol.2023.121396.

[32]   Bijkerk CJ, de Wit NJ, Muris JW, Whorwell PJ, Knottnerus JA, Hoes AW. Soluble or insoluble fibre in irritable bowel syndrome in primary care? Randomised placebo controlled trial. BMJ. 2009;339: b3154. doi: 10.1136/bmj.b3154.

[33]   Conlon MA, Bird AR. The impact of diet and lifestyle on gut microbiota and human health. Nutrients. 2015;7(1):17–44. doi: 10.3390/nu7010017.

[34] Cui L, Zhang H. Prebiotics: Recent advances in understanding their health effects. Foods. 2024;13 (3):446. doi: 10.3390/foods13030446.

[35] Roberfroid MB, Gibson GR, Hoyles L, McCartney AL, Rastall R, Rowland I, et al. Prebiotic effects: Metabolic and health benefits. British Journal of Nutrition. 2010;104(S2):S1–S63. https://doi.org/10.1017/S0007114510003363.

[36] Wang S, Xiao Y, Tian F, Zhao J, Zhang H, Zhai Q, Chen W. Rational use of prebiotics for gut microbiota alterations: Specific bacterial phylotypes and related mechanisms. Journal of Functional Foods. 2020;66:103838. doi: 10.1016/j.jff.2020.103838.

[37] Bindels LB, Segura Munoz RR, Gomes-Neto JC, Mutemberezi V, Martínez I, Salazar N, et al. Resistant starch can improve insulin sensitivity independently of the gut microbiota. Microbiome. 2017;5(1):12. doi: 10.1186/s40168-017-0230-5.

[38] Jaeger JW, Brandt A, Gui W, Yergaliyev T, Hernández-Arriaga A, Muthu MM, et al. Microbiota modulation by dietary oat beta-glucan prevents steatotic liver disease progression. JHEP Reports. 2024;6(3):100987. doi: 10.1016/j.jhepr.2023.100987.

[39] Chung WSF, Meijerink M, Zeuner B, Holck J, Louis P, Meyer AS, et al. Prebiotic potential of pectin and pectic oligosaccharides to promote anti-inflammatory commensal bacteria in the human colon. FEMS Microbiology Ecology. 2017;93(11):fix127. doi: 10.1093/femsec/fix127.

[40] Zhang W, Luo H, Keung W, Chan Y, Chan K, Xiao X, Li F, Lyu A, Dong C, Xu J. Impact of pectin structural diversity on gut microbiota: A mechanistic exploration through in vitro fermentation. Carbohydrate Polymers. 2025;355:123367. doi: 10.1016/j.carbpol.2025.123367.

[41] Sungatullina A, Petrova T, Kharina M, Mikshina P, Nikitina E. Effect of flaxseed mucilage on the probiotic, antioxidant, and structural-mechanical properties of the different Lactobacillus cells. Fermentation. 2023;9(5):486. doi: 10.3390/fermentation9050486.

[42] Njoku EN, Mottawea W, Hassan H, et al. Bioengineered wheat arabinoxylan – Fostering next-generation prebiotics targeting health-related gut microbes. Plant Foods for Human Nutrition. 2023;78(5):698–703. https://doi.org/10.1007/s11130-023-01120-3.

[43] Gao Z, Cao Q, Deng Z. Unveiling the power of flax lignans: From plant biosynthesis to human health benefits. Nutrients. 2024;16(20):3520. doi: 10.3390/nu16203520.

[44] Willett WC, Sacks F, Trichopoulou A, Drescher G, Ferro-Luzzi A, Helsing E, Trichopoulos D. Mediterranean diet pyramid: A cultural model for healthy eating. The American Journal of Clinical Nutrition. 1995;61(6 Suppl):1402S–1406S. doi: 10.1093/ajcn/61.6.1402S.

[45] De Filippis F, Pellegrini N, Vannini L, et al. High-level adherence to a Mediterranean diet beneficially impacts the gut microbiota and associated metabolome. Gut. 2016;65(11):1812–1821. doi: 10.1136/gutjnl-2015-309957.

[46] Tosti V, Bertozzi B, Fontana L. Health benefits of the Mediterranean diet: Metabolic and molecular mechanisms. The Journals of Gerontology. Series A, Biological Sciences and Medical Sciences. 2018;73(3):318–326. doi: 10.1093/gerona/glx227.

[47] Singh RK, Chang HW, Yan D, et al. Influence of diet on the gut microbiome and implications for human health. Journal of Translational Medicine. 2017;15(1):73. doi: 10.1186/s12967-017-1175-y.

[48] Koh A, De Vadder F, Kovatcheva-Datchary P, Bäckhed F. From dietary fiber to host physiology: Short-chain fatty acids as key bacterial metabolites. Cell. 2016;165(6):1332–1345. doi: 10.1016/j.cell.2016.05.041.

[49] Estruch R, Ros E, Salas-Salvadó J, et al. Primary prevention of cardiovascular disease with a Mediterranean diet. The New England Journal of Medicine. 2013;368(14):1279–1290. doi: 10.1056/NEJMoa1200303.

[50] Shivappa N, Godos J, Hébert JR, Wirth MD, Piuri G, Speciani AF, et al. Dietary Inflammatory Index and colorectal cancer risk – A meta-analysis. Nutrients. 2017;9(9):1043. https://doi.org/10.3390/nu9091043.

[51] Tabung FK, Steck SE, Ma Y, Liese AD, Zhang J, Caan B, et al. The association between dietary inflammatory index and risk of colorectal cancer among postmenopausal women: Results from the Women's Health Initiative. Cancer Causes & Control. 2015;26(3):399–408. doi: 10.1007/s10552-014-0515-y.

[52] Telle-Hansen VH, Holven KB, Ulven SM. Impact of a healthy dietary pattern on gut microbiota and systemic inflammation in humans. Nutrients. 2018;10(11):1783. doi: 10.3390/nu10111783.

[53] Lourida I, Soni M, Thompson-Coon J, Purandare N, Lang IA, Ukoumunne OC, et al. Mediterranean diet, cognitive function, and dementia: A systematic review. Epidemiology. 2013;24(4):479–489. doi: 10.1097/EDE.0b013e3182944410.

[54] Meslier V, Laiola M, Roager HM, De Filippis F, Roume H, Quinquis B, et al. Mediterranean diet intervention in overweight and obese subjects lowers plasma cholesterol and causes changes in the gut microbiome and metabolome independently of energy intake. Gut. 2020;69(7):1258–1268. https://doi.org/10.1136/gutjnl-2019-320438.

[55] Marco ML, Heeney D, Binda S, Cifelli CJ, Cotter PD, Foligné B, et al. Health benefits of fermented foods: Microbiota and beyond. Current Opinion in Biotechnology. 2017;44:94–102. https://doi.org/10.1016/j.copbio.2016.11.010.

[56] Abbaspour N. Fermentation's pivotal role in shaping the future of plant-based foods: An integrative review of fermentation processes and their impact on sensory and health benefits. Applied Food Research. 2024;4(2):100468. doi: 10.1016/j.afres.2024.100468.

[57] Tamang JP, Watanabe K, Holzapfel WH. Review: Diversity of microorganisms in global fermented foods and beverages. Frontiers in Microbiology. 2016;7:377. doi: 10.3389/fmicb.2016.00377.

[58] Rezac S, Kok CR, Heermann M, Hutkins R. Fermented foods as a dietary source of live organisms. Frontiers in Microbiology. 2018;9:1785. https://doi.org/10.3389/fmicb.2018.01785.

[59] World Gastroenterology Organisation. Probiotics and prebiotics. 2023. Available from: https://www.worldgastroenterology.org/UserFiles/file/guidelines/probiotics-and-prebiotics-english-2023.pdf.

[60] Kahleova H, Fleeman R, Hlozkova A, Holubkov R, Barnard ND. A plant-based diet in overweight individuals in a 16-week randomized clinical trial: Metabolic benefits of plant protein. Nutrition & Diabetes. 2018 Nov 2;8(1):58. doi: 10.1038/s41387-018-0067-4.

[61] Landry A, Masini F, Sanmarchi M, Fiore A, Coa AA, Castagna G, et al. Cardiovascular health and cancer risk associated with plant-based diets: An umbrella review. PLoS One. 2024;19(5):e0300711. doi: 10.1371/journal.pone.0300711.

[62] Pavlidou E, Fasoulas A, Mantzorou M, Giaginis C. Clinical evidence on the potential beneficial effects of probiotics and prebiotics in cardiovascular disease. International Journal of Molecular Sciences. 2022;23(24):15898. https://doi.org/10.3390/ijms232415898.

[63] Suez J, Zmora N, Segal E, Elinav E. The pros, cons, and many unknowns of probiotics. Nature Medicine. 2019;25(5):716–729. doi: 10.1038/s41591-019-0439-x.

[64] Arqués JL, Rodríguez E, Langa S, Landete JM, Medina M. Antimicrobial activity of lactic acid bacteria in dairy products and gut: Effect on pathogens. BioMed Research International. 2015;2015:584183. doi: 10.1155/2015/584183.

[65] Fuloria S, Mehta J, Talukdar MP, Sekar M, Gan SH, Subramaniyan V, Rani NNIM, Begum MY, Chidambaram K, Nordin R, Maziz MNH, Sathasivam KV, Lum PT, Fuloria NK. Synbiotic effects of fermented rice on human health and wellness: A natural beverage that boosts immunity. Frontiers in Microbiology. 2022;13:950913. doi: 10.3389/fmicb.2022.950913.

[66] Santos A, San Mauro M, Sanchez A, Torres JM, Marquina D. The antimicrobial properties of different strains of Lactobacillus spp. isolated from kefir. Systematic and Applied Microbiology. 2003;26 (3):434–437. https://doi.org/10.1078/072320203322497464.

[67] Tingirikari JMR, Sharma A, Lee HJ. Kefir: A fermented plethora of symbiotic microbiome and health. Journal of Ethnic Foods. 2024;11:35. doi: 10.1186/s42779-024-00252-4.

[68] Park KY, Jeong JK, Lee YE, Daily JW III. Health benefits of kimchi (Korean fermented vegetables) as a probiotic food. Journal of Medicinal Food. 2014;17(1):6–20. https://doi.org/10.1089/jmf.2013.3083.

[69] Nugroho D, Thinthasit A, Surya E, Hartati JS, Jang JG, Benchawattananon R, et al. Immunoenhancing and antioxidant potentials of kimchi, an ethnic food from Korea, as a probiotic and postbiotic food. Journal of Ethnic Foods. 2024;11:12. https://doi.org/10.1186/s42779-024-00232-8.

[70] Wastyk HC, Fragiadakis GK, Perelman D, Dahan D, Merrill BD, Yu FB, et al. Gut-microbiota-targeted diets modulate human immune status. Cell. 2021;184(16):4137–4153.e14. doi: 10.1016/j.cell.2021.06.019.

[71] Zinöcker MK, Lindseth IA. The Western diet–microbiome-host interaction and its role in metabolic disease. Nutrients. 2018;10(3):365. doi: 10.3390/nu10030365.

[72] Cani PD, et al. Metabolic endotoxemia initiates obesity and insulin resistance. Diabetes. 2008;57 (7):1470–1481. doi: 10.2337/db07-1403.

[73] Suez J, et al. Artificial sweeteners induce glucose intolerance by altering the gut microbiota. Nature. 2014;514(7521):181–186. doi: 10.1038/nature13793.

[74] Suez J, et al. Personalized microbiome-driven effects of non-nutritive sweeteners on human glucose tolerance. Cell. 2022;185(18):3307–3328.e19. doi: 10.1016/j.cell.2022.07.016.

[75] Li L, Zhao X, Abdugheni R, Yu F, Zhao Y, Ma BF, et al. Gut microbiota changes associated with low-carbohydrate diet intervention for obesity. Open Life Sciences. 2024;19(1):20220803. doi: 10.1515/biol-2022-0803.

[76] Monda V, Villano I, Messina A, Valenzano A, Esposito T, Moscatelli F, et al. Exercise modifies the gut microbiota with positive health effects. Oxidative Medicine and Cellular Longevity. 2017;2017:3831972. https://doi.org/10.1155/2017/3831972.

[77] Gutierrez Lopez DE, Lashinger LM, Weinstock GM, Bray MS. Circadian rhythms and the gut microbiome synchronize the host's metabolic response to diet. Cell Metabolism. 2021;33(5):873–887. doi: 10.1016/j.cmet.2021.03.015.

[78] Sun J, Fang D, Wang Z, Liu Y. Sleep Deprivation and Gut Microbiota Dysbiosis: Current Understandings and Implications. International Journal of Molecular Sciences. 2023;24(11):9603. doi: 10.3390/ijms24119603.

[79] Lin Z, Jiang T, Chen M, Ji X, Wang Y. Gut microbiota and sleep: Interaction mechanisms and therapeutic prospects. Open Life Sciences. 2024 Jul 18;19(1):20220910. doi: 10.1515/biol-2022-0910.

[80] Dethlefsen L, Relman DA. Incomplete recovery and individualized responses of the human distal gut microbiota to repeated antibiotic perturbation. Proceedings of the National Academy of Sciences of the United States of America. 2010;108(Suppl 1):4554–4561. doi: 10.1073/pnas.1000087107.

[81] Jernberg C, Lofmark S, Edlund C, Jansson JK. Long-term impacts of antibiotic exposure on the human intestinal microbiota. Microbiology. 2010;156(11):3216–3223. doi: 10.1099/mic.0.040618-0.

[82] Patangia DV, Ryan CA, Dempsey E, Ross RP, Stanton C. Impact of antibiotics on the human microbiome and consequences for host health. MicrobiologyOpen. 2022;11(1):e1260. https://doi.org/10.1002/mbo3.1260.

[83] Lathakumari RH, Vajravelu LK, Satheesan A, Ravi S, Thulukanam J. Antibiotics and the gut microbiome: Understanding the impact on human health. Medical Microecology. 2024;20:100106. doi: 10.1016/j.medmic.2024.100106.

[84] Zeevi D, Korem T, Zmora N, Israeli D, Rothschild D, Weinberger A, et al. Personalized nutrition by prediction of glycemic responses. Cell. 2015;163(5):1079–1094. doi: 10.1016/j.cell.2015.11.001.

[85] Iacomino G, Rufián Henares JÁ, Lauria F. Editorial: Personalized nutrition and gut microbiota: Current and future directions. Frontiers in Nutrition. 2024;11:1375157. doi: 10.3389/fnut.2024.1375157.

[86] Rouskas K, Guela M, Pantoura M, et al. The influence of an AI-driven personalized nutrition program on the human gut microbiome and its health implications. Nutrients. 2025;17(7):1260. https://doi.org/10.3390/nu17071260.

[87] Sanders ME, Lenoir-Wijnkoop I, Salminen S, Merenstein DJ, Gibson GR, Petschow BW, et al.
      Probiotics and prebiotics: Prospects for public health and nutritional recommendations. Annals of
      the New York Academy of Sciences. 2014;1309(1):19–29. https://doi.org/10.1111/nyas.12377.

[88] Sarita B, Samadhan D, Hassan MZ, Kovaleva EG. A comprehensive review of probiotics and human
      health – Current prospective and applications. Frontiers in Microbiology. 2025;15:1487641.
      https://doi.org/10.3389/fmicb.2024.1487641.

[89] Cammarota G, Ianiro G, Tilg H, Rajilić-Stojanović M, Kump P, Satokari R, et al. European consensus
      conference on faecal microbiota transplantation in clinical practice. Gut. 2017;66(4):569–580. doi:
      10.1136/gutjnl-2016-313017.

[90] Zhang Q, Bi Y, Zhang B, Jiang Q, Mou CK, Lei L, Deng Y, Li Y, Yu J, Liu W, Zhao J. Current landscape of
      fecal microbiota transplantation in treating depression. Frontiers in Immunology. 2024;15:1416961.
      doi: 10.3389/fimmu.2024.1416961.

[91] Cao Z, Gao T, Bajinka O, Zhang Y, Yuan X. Fecal microbiota transplantation – Current perspective on
      human health. Frontiers in Medicine. 2025;12:1523870. doi: 10.3389/fmed.2025.1523870.

P. Saranraj*, K. Gayathri, B. Lokeshwari, L. Charlie Jelura,
K. Kesavardhini, U. Subhalakshmi

Chapter 11
Gut Microbiome in Connection with Various Organ Axes: Role and Regulation

Abstract: Emerging research underscores the gut microbiome's integral function as a mediator of human health, orchestrating complex interactions with distant organ systems via specialized communication pathways. These include the gut-brain, gut-heart, gut-lung, gut-liver, gut-skin, and gut-kidney axes. Communication along these axes is facilitated by microbial metabolites, immune regulation, and neuroendocrine signaling, which collectively influence host physiology in states of both health and pathology. Metabolites such as short-chain fatty acids, produced through bacterial fermentation, exhibit potent anti-inflammatory properties and are crucial for immune homeostasis. Conversely, a state of dysbiosis, or microbial imbalance, is increasingly implicated in the pathogenesis of a spectrum of conditions, including cardiovascular, neurological, and metabolic disorders. Cutting-edge meta-omics technologies are rapidly expanding our comprehension of these host-microbe dynamics. Furthermore, modifiable factors such as diet, probiotic intake, and lifestyle choices are recognized as powerful determinants of microbial composition, presenting promising avenues for therapeutic intervention in diseases like inflammatory bowel disease, asthma, and certain mental health conditions. The microbiome's capacity to produce neurotransmitters and hormones further solidifies its status as a vital "virtual endocrine organ." This chapter consolidates evidence advocating for the maintenance of microbial equilibrium through increased dietary fiber consumption, incorporation of fermented foods, and judicious antibiotic use to support systemic health. Future investigative efforts should prioritize the development of personalized, microbiome-targeted strategies to ameliorate specific diseases and enhance the symbiotic relationship between host and microbiota.

Keywords: dysbiosis, short-chain fatty acids, gut-organ axes, gut microbiome and microbial metabolites

*Corresponding author: **P. Saranraj**, PG and Research Department of Microbiology, Sacred Heart College (Autonomous), Tirupattur, Tamil Nadu, India, e-mail: microsaranraj@gmail.com
K. Gayathri, B. Lokeshwari, L. Charlie Jelura, K. Kesavardhini, U. Subhalakshmi, PG and Research Department of Microbiology, Sacred Heart College (Autonomous), Tirupattur, Tamil Nadu, India

https://doi.org/10.1515/9783112205150-011

11.1 Introduction

For billions of years, microorganisms have been integral to life on the Earth, inhabiting virtually every environment, including the human body. Within their host, these microbes form complex communities that engage in a dynamic exchange of genetic and metabolic products with human cells. Through these sophisticated interactions, microorganisms participate in essential signaling and metabolic activities that have played a foundational role in human evolution [1]. The human gastrointestinal (GI) tract contains the body's most concentrated and diverse microbial population, known as the gut microbiota, consisting of bacteria, viruses, fungi, and archaea [2]. Research over the past two decades has dramatically advanced, uncovering the profound influence of this symbiotic ecosystem on daily physiological functions. Emerging evidence now positions the gut microbiota as a master regulator of host physiology, significantly impacting both health and disease trajectories [3].

Modern high-throughput DNA sequencing and advanced bioinformatics have transformed our ability to characterize the microbiota's structure and function. These studies reveal a remarkable fact: microbial genes represent approximately 99% of the unique genetic material found in the human body, totaling over 100 million genes [4]. This vast genetic repository highlights a deep evolutionary partnership, indicating that the microbiome is instrumental in developing and refining all human physiological systems. Many modern chronic diseases appear linked to contemporary lifestyles; the widespread adoption of highly processed diets and increased use of pharmaceuticals may have gradually reduced our innate resilience to illness [5]. The distribution of gut microbes is not uniform but varies significantly along the digestive tract. The small intestine shows higher concentrations of Bacilli (a class within Firmicutes) and Actinobacteria, while the colon is dominated by Bacteroidetes and Lachnospiraceae (a family of Firmicutes) [6]. This geographical variation extends to specific niches within the gut itself, with distinct communities occupying the intestinal lumen, mucosal layer, and epithelial surface, often existing in relative isolation [7].

Although less numerous than bacteria, fungi contribute significantly to the gut's ecological diversity. Recent investigations have identified hundreds of fungal species residing in the gut; some show high variability between individuals while others appear more consistent [8]. The foundation of this microbial ecosystem is established at birth. An infant's gut microbiota closely mirrors the maternal vaginal microbiota, emphasizing the critical importance of the birthing process [9]. Consequently, infants delivered by cesarean section develop a substantially different initial microbial community compared to those born vaginally. From this earliest stage, these microorganisms hold considerable potential to influence human physiology in both beneficial and detrimental ways [10].

The collective genetic potential housed within our gut microbiome far surpasses that of the human genome. While the human genome contains approximately 20,000–25,000 genes, the microbial community contributes millions of unique genes. This dynamic genetic reservoir is constantly changing, shaped by the daily influx of new microorganisms and the continuous coevolution of resident microbes with their host. A crucial mechanism enabling this adaptation is horizontal gene transfer, the ability of microbes to exchange genetic material with each other and their environment. This profound interdependence has given rise to the "hologenome" concept, which proposes that an individual's genetic identity should not be considered in isolation. Rather, humans function as superorganisms, comprising both the host genome and the collective genomes of all associated microorganisms, all subject to evolutionary pressures. This symbiotic relationship is now recognized as a fundamental driver of core biological processes. This perspective transforms our understanding of the gut microbiome from passive inhabitants to an essential, integrated organ system. Its intricate connection to host genetics serves as a major determinant of health and disease. Understanding how these microbial communities have coevolved with humans to establish mutually beneficial relationships is crucial to appreciating their vital role in maintaining overall well-being [5].

11.2 Gut Microbiome

The community of microorganisms residing in our digestive system, known as the gut microbiota, represents a complex and dynamic ecosystem that plays a decisive role in our overall health. While including archaea, viruses, and fungi, this community is predominantly bacterial, with the phyla Firmicutes and Bacteroidetes comprising the majority in healthy individuals. A primary function of these microbial populations is the breakdown of dietary fibers that human enzymes cannot digest. Through this process, they generate beneficial short-chain fatty acids (SCFAs), notably butyrate, propionate, and acetate, which exert powerful anti-inflammatory effects and help regulate immune responses [11].

The diversity within this gut ecosystem is increasingly recognized as a key indicator of health and is associated with longevity. Nutritional approaches, including prebiotics and probiotics, can support a robust microbiota and potentially extend the health span. Conversely, microbial imbalance, known as dysbiosis, has been connected to numerous disease states. Our ability to study this hidden organ has been revolutionized by meta-omics technologies. By integrating insights from metagenomics (revealing genetic potential), metatranscriptomics (showing active genes), and metaproteomics (identifying functional proteins), researchers can now build comprehensive models of how these microbes interact with each other and their human host, clarifying their precise roles in health and disease [12].

Scientists employ two primary methodological approaches to study microbial communities: shotgun metagenomics and 16S rRNA sequencing. While shotgun metagenomics captures a broader range of organismal diversity by sequencing all genetic material in a sample, it typically offers lower taxonomic resolution at genus and family levels compared to 16S rRNA sequencing. Conversely, 16S sequencing targets a specific conserved genomic region, providing higher resolution for assessing bacterial diversity within individual samples. The choice between these techniques depends primarily on research questions and available resources [13]. Growing evidence indicates that the gut microbiome significantly influences the brain function and may contribute to various neurological disorders. One proposed mechanism involves microbially produced SCFAs, which may affect cognitive processes by regulating inflammatory pathways in the brain [14].

Early gut microbiome research highlighted the influence of host genetics, showing how variations in human genes shape microbial community function. These studies identified specific genetic variations, particularly single-nucleotide polymorphisms, linked to various microbiome characteristics, health outcomes, and diseases, demonstrating the substantial impact of human genetics on gut microbiome dynamics [15]. Analyzing these extensive datasets now requires advanced computational tools, including artificial intelligence and bioinformatics, to identify patterns and correlations between microbial composition and diverse health conditions.

The microbiota is essential for regulating immune activity and maintaining host health. Dysbiosis, or microbial imbalance, correlates with various pathologies, including cardiovascular disease, cancers, and respiratory illnesses. Probiotic supplementation has shown potential as an adjunct therapy for managing colorectal tumors, intestinal inflammatory diseases, and certain cardiovascular conditions. However, their efficacy in treating other disorders such as inflammatory bowel diseases, rotavirus diarrhea, NSAID-induced enteropathy, and irritable bowel syndrome remains debated. Modulating the gut microbiota with probiotics also offers novel therapeutic approaches for lung diseases and may help reduce hyperinflammation in COVID-19 [16]. Despite this potential, the safety of microbiota-targeted therapies requires careful evaluation, especially since modifiable lifestyle factors such as diet powerfully influence both gut microbiome composition and gut-brain communication. The Mediterranean diet has demonstrated benefits for neurodegenerative disorders, psychological diseases, cancers, and atherosclerosis. The ketogenic diet alters microbiota composition and provides protective antiepileptic effects against acute seizures. Additionally, various pharmaceuticals, including both antibiotic and nonantibiotic drugs, significantly impact the gut microbial community [17].

11.3 Gut-Organ Axis: Elucidating Interorgan Connections

Humans and their gut microorganisms have coevolved over millennia, resulting in a sophisticated, bidirectional relationship. The gut ecosystem represents a dynamic system formed through collaborative interaction between host and resident microbes. As noted, the human gut contains billions of microorganisms that metabolize dietary nutrients to produce various bioactive compounds. These microbial metabolites can interact with distant organs, influencing hormonal signaling, immune responses, host metabolism, and other physiological functions, thereby enabling the gut microbiota to modulate metabolic homeostasis and organ function [18]. The complex relationships between gut microbiota and the host immune system affect physiological processes that create functional axes connecting the gut to other organs. This cross-talk occurs through multiple signaling channels and direct chemical relationships between microorganisms and host cells. The host-microbe metabolic axis represents a network of bidirectional communication between various microbial species and host cellular pathways [19]. This axis involves numerous bacterial species that successively influence metabolic processes through the synthesis of bile acids, choline, and SCFAs, essential for host health [19]. The production of these metabolites significantly impacts host metabolism and susceptibility to disease development. The dynamic elements constituting the gut microbiome, when shaped by dietary changes or environmental pressures, may alter disease or health risks by modifying species diversity or composition. Thus, bacterial components and signaling molecules, including endotoxin or DNA, generated and delivered by intestinal epithelial cells or dendritic cells, profoundly influence the host's physiological and pathological processes [20].

11.4 Gut-Brain Axis

The human body's first encounter with microbes occurs during birth, serving as the initial inoculation for the infant's developing microbiome. The delivery method – vaginal or cesarean – plays a fundamental role in determining which bacteria first colonize the infant. This early microbial community is then shaped by numerous factors, including maternal health and diet (including breast feeding), host genetics, environmental exposures, and maternal conditions such as obesity or stress. Early interventions, particularly antibiotic use, can profoundly alter this delicate developmental phase [21]. Although initial composition varies, the gut microbiota stabilizes in adulthood into a unique, fingerprint-like profile. This vast microbial community is a central component of the gut-brain axis (GBA), a complex bidirectional communication system connecting the gut and brain. This dialogue involves neural pathways, hormones, and immune signals. Accumulating research confirms that the gut microbiota actively contributes to this process, influencing brain function, behavior, and cognitive processes [22]. The varied ecology of a healthy adult gut microbiota primarily consists of bacteria from the Firmicutes and Bacteroidetes families. Though less abundant, members of the Actinobacteria, Proteobacteria, Fusobacteria, and Verrucomicrobia phyla remain vital components. However, this complex community is not static; its delicate balance can be disrupted by factors such as aging, dietary changes, infections, illness, and medications with significant consequences for overall health.

11.4.1 Neuronal Pathways for Gut-Brain Axis Interactions

Two neuroanatomical pathways connect the gut and brain [2]: the gut's enteric nervous system (ENS) and the spinal cord's vagus nerve (VN) and autonomic nervous system (ANS). By stimulating ENS afferent neurons and the VN, bacteria establish a direct neurological link between the brain and GI microbiota [23]. The gut microbiota and probiotics benefit from vagal stimulation, which also exhibits anti-inflammatory properties. Preclinical research indicates that gut microbial imbalances, or "dysbiosis," have been linked to neurological illnesses, intestinal disorders, and mental ailments, including anxiety, depression, autism spectrum disorder (ASD), Alzheimer's disease (AD), multiple sclerosis (MS), and Parkinson's disease (PD) [24].

11.4.2 Gut Microbes and Communication with the Nervous System

Host-dependent factors that are difficult to modify, such as the immune system, age, sex, and genetic background, influence gut microbiota composition and activity [10]. Gut microbiota can be directly impacted by medical practices such as gastric bypass surgery and the use of antidiabetic, antacid, and antibiotic medications [25]. Nutrition and dietary patterns have significantly influenced the organization of the gut microbiota over the past two decades.

11.4.3 Gut Microbiome and Neurodevelopment

The gut microbiome, comprising over 100 trillion microorganisms, plays a vital role in brain development. The first 3 years of life are crucial for brain synaptogenesis and GI microbiota development [26]. The gut microbiota contributes to establishing brain neural networks, as shown in germ-free animal models. It can directly or indirectly activate the GI tract's ENS and send information to the brain via the VN. Although probiotic administration affects behavioral features through the VN, operational changes in neuronal circuits require further investigation. In BALB/c mouse brains, *Lactobacillus rhamnosus* (JB-1) significantly reduced anxiety, corticosterone levels, and γ-aminobutyric acid (GABA) receptor upregulation, suggesting potential applications for psychobiotics in personalized anxiety disorder treatment [27]. Mouse gut flora influences the behavior without requiring ANS and GI neurotransmitters and increases cerebral neurotrophic factors. Additionally, by stimulating immune cells and influencing brain function, gut symbionts aid in establishing the host immune system. From infancy to adulthood, GI microbiota and their metabolites modify brain microglia, the primary macrophages, and can influence inflammatory responses in the central nervous system (CNS) [28]. Recent clinical data show that brain disorders correlate with GI problems, including dysbiosis in autistic individuals who experience persistent constipation, increased intestinal permeability, abdominal discomfort, and disrupted intestinal flora [29].

11.4.4 Gut Microbiota-Brain Signaling Through the Immune System

The CNS and immune system are complex systems regulating numerous physiological processes. The brain produces innate immunity molecules that control the brain development, including Toll-like receptors (TLRs), cytokines, the complement family, antibody receptors, and the major histocompatibility complex [30]. Microglia and lymphocytes influence cognition and connect neuronal networks. The tissue-specific, selective blood-brain barrier (BBB) found in most brain vasculature permits molecule passage while blocking dangerous chemicals and cells. Microglia constitute approximately 10% of brain cells and are essential for active immune responses. Disease and CNS function can be altered by immune cell or microglia infiltration [31].

Brain microglia serve physiological and immunological roles and can proliferate from embryonic progenitor cells [32]. Immune cells such as neutrophils, macrophages, T cells, and natural killer cells can infiltrate the brain from the bloodstream. Once in the CNS, these cells can modulate behavioral functions and contribute to neurodegenerative disorder pathogenesis. The gut microbiome is critical to brain immunity because it influences early microglial cell development and adult microglia require gut bacteria. Autoimmunity in the CNS requires brain-infiltrating immune cells; gut microorganisms are also involved in autoimmune disorders [33]. Gut microbiota plays a significant role in both MS and EAE, and transplanting gut bacteria increases the experimental autoimmune encephalomyelitis (EAE) risk compared to healthy donors. By influencing T cells, SCFAs from the gut microbiome reduce axon damage associated with EAE [34]. The immune system and microbiota often collaborate to elicit sufficient immunological responses. AD acceleration involves "leaky gut syndrome" and GI bacterial dysbiosis, which push neurotoxic compounds from the microbiome across the BBB. Gut microbial populations enlist various immune cells to regulate gut immune responses in different immunological applications [35].

11.4.5 Gut-Brain Axis (GBA): Gut as "Second Brain" Influencing Mood and Behavior

The GBA represents a bidirectional communication system between the brain and GI system [11] that can mutually influence the function through humoral, immunological, endocrine, or neurological connections [36]. GBA associations have been demonstrated in studies using germ-free model organisms, probiotics, antimicrobial agents, dysbiosis linked to CNS diseases, and functional GI disorders. Genetically modified (GM) organisms can impact the brain functions, including behavior, hunger control, intestinal gluconeogenesis, and serotonin metabolism [37]. Neurological conditions such as anxiety, MS, ASD, and PD have been connected to GM alterations. AD patients show less prevalent Firmicutes and *Bifidobacterium* with more common Bacteroidetes. PD patients exhibit altered gut flora structures with increased Enterobacteriaceae and Ruminococcaceae and reduced Prevotellaceae levels [38].

11.4.6 Microbiota and Mental Health: Exploring the Gut-Brain Axis

The GBA represents a fascinating bidirectional communication system between digestive tract microbes and the brain [39]. Increasing research connects it to conditions such as anxiety, depression, and ASDs, showing that this continuous dialogue significantly influences brain chemistry including emotional states. This influence mechanism involves gut bacteria producing neuroactive substances. Certain species generate essential neurotransmitters, including serotonin and GABA, fundamental to mood regulation and anxiety reduction [40]. Furthermore, bacterial by-products, such as SCFAs can travel through the blood stream, helping to maintain the BBB and modulate neuroinflammation. Brain imaging studies are beginning to link individual gut microbiome composition with activity in emotion-processing brain regions like the amygdala and insula. This emerging science is advancing new mental health treatment approaches. Therapeutic strategies aimed at nurturing beneficial gut ecosystems through tailored diets, prebiotics, or specific probiotic regimens are being explored as innovative complementary approaches for managing and potentially alleviating psychiatric symptoms.

11.4.7 Role of Gut-Microbiome in Brain Physiology

Both internal and external variables impact human CNS development. In germ-free and antibiotic-treated animals, neurochemistry and microbiota can affect the CNS function [41]. Neuropsychiatric and gastrointestinal disorders such as anxiety, depression, and autism demonstrate interaction. The CNS state is impacted by GBA connecting gut microbiota to the brain. Gut microbial dysbiosis can trigger atypical immunological signaling, host homeostasis imbalance, and CNS illness progression. Environmental variables can also influence CNS neurogenesis. For example, maternal-fetal interface permeability allows gut microbiota regulatory elements to trigger TLR 2, promoting fetal neural development and influencing adult cognitive performance [42]. Polyunsaturated fatty acids from gut microbiota are necessary for developing structural elements and vasculature in the highly selective, semipermeable BBB.

11.5 Gut-Heart Axis

Bacterial metabolites supporting intestinal health and other physiological systems, including the circulatory system, remain healthy when gut flora is balanced. SCFAs represent critical gut flora by-products with numerous cardioprotective benefits. Beneficial substances such as butyrate are among many compounds that increase or decrease when equilibrium is disrupted by substances such as trimethylamine-N-oxide (TMAO), lipopolysaccharide (LPS), phenylacetylglutamine (PAGln), indoxyl sulfate, and p-cresyl sulfate. This damages the intestinal barrier and allows substantial toxic substances to enter the blood, a known risk factor for atherosclerosis and other cardiac issues [43]. Atherosclerosis, among the most prevalent causes of cardiovascular disease, contributes to heart failure, stroke, myocardial infarction, and claudication [44].

11.5.1 Gut Microbiome and Cardiovascular Complications

Cardiovascular and metabolic diseases have emerged as the primary cause of global mortality in recent decades, with projections indicating forthcoming increased prevalence. This illness group shows wide symptom variability, making them quite diverse. Over recent decades, experimental research has examined GI microbial composition's direct and indirect effects on several cardiovascular and cardiometabolic disorders' development and progression. Atherosclerosis is a chronic inflammatory disease damaging the inner lining of artery, leading to lipid accumulation and plaque formation [45]. Several risk factors cause it, including obesity, diabetes, high blood pressure, high cholesterol, and physical inactivity. Microbial DNA discovery in atherosclerotic deposits raised possibilities that gut microbiota might influence plaque development [46]. Earlier studies showed that individuals with symptomatic atherosclerosis exhibited higher *Collinsella* sp. frequency than healthy individuals. Enterobacteriaceae, including *Klebsiella* spp., *Enterobacter aerogenes*, *Escherichia coli*, and certain *Streptococcus* spp., were more prevalent in the gut microbiome of 218 atherosclerosis patients compared to 187 healthy controls in a recent metagenomic study [47].

11.5.2 Microbial Metabolites in Cardiovascular Pathogenesis

Among the various compounds produced by gut bacteria that influence systemic physiological processes, the methylamines TMAO and its precursor trimethylamine (TMA) have emerged as particularly significant. Multiple bacterial species generate TMA from dietary sources rich in carnitine and choline. Following its production in the gut, TMA crosses the intestinal epithelium and undergoes hepatic oxidation to form TMAO [48]. Epidemiological studies have consistently demonstrated a positive association between circulating TMAO levels and cardiovascular disease risk, suggesting its potential utility as a biomarker for risk stratification in both general populations and patients with established cardiovascular conditions, including heart failure, coronary artery disease, and peripheral artery disease [49]. Despite numerous investigations attempting to establish causality between TMAO and cardiovascular pathogenesis [50], the evidence remains inconclusive regarding whether it functions as a direct toxin, represents an adaptive response marker, or simply serves as a confounding variable. Notably, research has disproportionately focused on TMAO while neglecting its precursor TMA, despite historical documentation of its toxic properties, including encephalopathy, behavioral disturbances, developmental impairments, and mucosal irritation, leading to its classification as a uremic toxin [51]. Recent investigations have revealed that TMA, rather than TMAO, demonstrates direct cardiovascular toxicity through mechanisms, including increased mortality rates, blood pressure elevation, and cytotoxic effects on cardiac muscle cells [52].

11.5.3 Bidirectional Signaling in the Gut-Heart Network

Accumulating evidence supports the existence of a sophisticated bidirectional communication system connecting cardiac and intestinal functions. In heart failure patients, characteristic pathological changes including peripheral vasoconstriction, diminished cardiac output, and systemic congestion collectively compromise GI integrity. These hemodynamic alterations reduce intestinal perfusion, promote wall thickening throughout the colonic and ileal regions, and stimulate collagen deposition in small intestinal tissues [53]. The resulting microcirculatory impairment causes hypoxia-induced damage to intestinal epithelial cells, disrupting their normal function. Impaired nutrient absorption secondary to intestinal ischemia can precipitate both malnutrition and cachexia. Furthermore, compromised barrier integrity facilitates translocation of microorganisms and their metabolic products into systemic circulation. This microbial translocation initiates a systemic inflammatory cascade characterized by elevated circulating cytokines, including tumor necrosis factor (TNF)-α, contributing to structural tissue damage, reduced cardiac contractility, and increased mortality. LPS additionally stimulates catecholamine release from granulocytes and phagocytes, creating complex effects on intestinal perfusion. The gut microbiota itself undergoes significant compositional changes in heart failure, with patients over 60 showing reduced abundance of Bacteroidetes and *Faecalibacterium* species. Patients with chronic heart failure exhibiting intestinal wall thickening demonstrate increased proportions of *Lactobacillus* and *Proteobacteria*, accompanied by enhanced intestinal permeability (evidenced by lactulose/mannitol testing) and impaired absorptive function (demonstrated by D-xylose testing). Bacterial infiltration has been observed within sigmoid colon mucosal biofilms in these patients. Collectively, these changes establish a vicious cycle of intestinal ischemia, nutritional deficiency, and chronic inflammation. Investigators have documented significantly increased colonization by enteropathogens such as *Salmonella* sp., *Candida* sp., *Shigella* sp., *Campylobacter* sp., and *Yersinia enterocolitica* in congestive heart failure patients compared to healthy controls [53].

Gut microbiota-derived metabolites contribute significantly to disease progression. Uremic toxins such as TMAO, *p*-cresyl sulfate, and indoxyl sulfate originate from microbial fermentation of dietary proteins. Indoxyl sulfate demonstrates pro-hypertrophic properties, while TMAO shows promise as a predictive biomarker for cardiovascular risk. A substantial cohort study revealed that elevated plasma TMAO levels correlated with increased risk of mortality, myocardial infarction, and stroke in patients undergoing elective coronary angiography [54]. Heart failure patients additionally exhibit markedly elevated TMAO concentrations compared to healthy individuals. The inflammatory and metabolic processes fundamental to cardiovascular disease pathogenesis, particularly atherosclerosis, appear substantially influenced by gut microbial activity. Emerging research suggests an alternative pathway connecting microbial metabolism, atherogenesis, and dietary lipid intake. The gut microbiome processes dietary phosphatidylcholine (lecithin) to generate TMAO, choline, and betaine, with TMAO demonstrating particularly strong positive correlation with cardiovascular risk.

Experimental evidence indicates that TMAO suppresses bile acid synthesis and inhibits reverse cholesterol transport processes mechanistically linked to atherosclerosis development, though precise mechanisms require further elucidation [55]. Additionally, persistent infections with pathogens, such as *Chlamydia pneumoniae* and *Helicobacter pylori*, and their consequent immune activation may contribute to atherosclerosis formation in certain models. Multiple studies support associations between elevated circulating endotoxin, LPS levels, and atherosclerotic disease. Individuals with compromised intestinal barrier function, including those with hepatic cirrhosis or inflammatory bowel disease, exhibit increased LPS concentrations and higher atherosclerosis prevalence. LPS interacts with lipoprotein metabolism, affecting low-density lipoprotein (LDL) particles and promoting endothelial damage through superoxide anion release during LDL oxidation [54]. Oxidized LDL stimulates cytokine production (including TNF-α and interleukin (IL)-1) from macrophages, driving their transformation into foam cells. These mechanisms collectively contribute to atherosclerosis initiation and progression [55].

11.6 Pulmonary Microbiome and Systemic Connections

11.6.1 Gut-Lung Axis

Traditional medical teaching described the pulmonary system as a sterile environment, based primarily on negative bacterial culture results from healthy lung samples. This perception has been radically revised with contemporary understanding that the lungs maintain a low biomass but functionally important microbial community, primarily inoculated through microaspiration of oropharyngeal secretions into the lower respiratory tract [56]. Consequently, the lung microbiome demonstrates taxonomic similarities with oral microbial communities, featuring predominant genera, including *Streptococcus*, *Prevotella*, and *Veillonella*. At the phylum level, healthy lungs predominantly contain Bacteroidetes and Firmicutes, reflecting oropharyngeal composition. However, the pulmonary environment constitutes a distinct ecological niche from the oral cavity, shaped by unique selective pressures including oxygen tension, redox potential, and continuous immune monitoring. These conditions create notable compositional differences, with the lung environment often supporting greater relative abundance of certain Proteobacteria, including Enterobacteriaceae, *Ralstonia* species, and *Haemophilus* species, while *Prevotella*-related taxa appear less prevalent compared to oral communities.

11.6.1.1 Gut Microbiota and Respiratory Immunity

Disruption of gut-lung communication has been associated with increased susceptibility to infectious and inflammatory airway diseases, including allergic conditions [57]. Murine models have demonstrated that gut microbiota regulates both innate and adaptive immune mechanisms, providing protection against bacterial and viral pulmonary infections alongside allergic airway diseases [58]. Maintenance of immune homeostasis and proper immune system development depends on this gut-lung interconnection. Intestinal and respiratory mucosal surfaces provide physical barriers against pathogen invasion, while commensal colonization confers resistance against pathogens. Typical gut commensals, including segmented filamentous bacteria, *Bifidobacterium* species, and colonic *Bacteroides* species, stimulate the production of immunomodulatory compounds such as lactic acid [59]. Given the intestinal tract's role as the body's largest immune organ, gut microbiome composition significantly influences immune responses both locally and in distant sites, including the lungs [60]. Multiple mechanisms facilitate gut-lung communication, including systemic dissemination of microbial metabolites, particularly SCFAs and other immunomodulatory molecules derived from bacterial metabolic activity [61].

11.6.1.2 Immunological Mechanisms in Asthma and Allergy

Asthmatic and allergic conditions frequently feature dysregulated T-helper type 2 (Th2) cell responses [61]. Th2 cells are characterized by their capacity to produce inflammatory cytokines, including IL-4, IL-5, IL-9, and IL-13. Substantial evidence indicates that modifications to gut immune responses directly affect the development of allergic conditions in the respiratory system [62]. For example, antibiotic-treated mice receiving oral *Candida albicans* developed gut dysbiosis and subsequently exhibited enhanced CD4$^+$ T-cell-mediated pulmonary inflammation following aerosolized allergen challenge compared to mice with intact gut microbiota. This suggests that altered gut microbiome composition may increase susceptibility to respiratory allergies. Research attention is increasingly focusing on the contributions of additional T-cell subsets, including Th9 and Th17 cells, to allergic and asthmatic disease pathogenesis [63].

11.6.1.3 Microbiota and Respiratory Infection Response

The gut microbiota significantly influences host immune responses to respiratory viral infections such as influenza [64]. In infected murine models, intestinal microbiota directly shapes CD4$^+$ and CD8$^+$ T-cell population dynamics. Furthermore, intact gut flora appears necessary for optimal production of pro-inflammatory cytokines, including pro-IL-1β and pro-IL-18, which are essential for effective viral clearance. This suggests that gut microbiota provide essential microbial signals required for proper immune system education and preparation against viral pneumonia pathogens [65, 66]. Similar findings have emerged regarding bacterial respiratory infections using germ-free animal models. These animals exhibit elevated IL-10 levels and reduced neutrophil recruitment, increasing susceptibility to *Klebsiella pneumoniae* infections due to impaired pathogen containment [67, 68].

11.7 Hepatic-Microbial Interactions

11.7.1 Gut-Liver Communication Mechanisms

11.7.1.1 Intestinal Barrier Function

The GI barrier consists of tightly opposed epithelial cells that prevent pathogenic microorganisms from entering host systems from the gut lumen while permitting selective nutrient absorption. Gut microbiota contributes to barrier integrity through metabolite production (including SCFAs) and maintenance of immunological signaling networks [69, 70]. Consequently, disruption of these processes can increase intestinal permeability. For example, inflammatory conditions or consumption of high-fat diets, alcohol, or antibiotics compromise gut barrier function, leading to microbial loss. Barrier impairment enables translocation of microorganisms and their toxic products into the portal system, potentially reaching the liver and distant organs to trigger inflammatory damage [71].

11.7.1.2 Portal Circulation Transfer

Intestinal dysbiosis features compromised the barrier function and the presence of pathogen-associated molecular patterns in portal circulation, which activate pattern recognition receptors on hepatic cells to initiate pro-inflammatory signaling cascades [72]. Pathogens or their products reaching the liver trigger TLR signaling, inducing the expression of cytokines including IL-1β and TNF-α that target infectious agents [73]. Persistent activation of these pathways through enhanced TLR signaling and sustained cytokine production can promote hepatic injury in various liver diseases.

Elevated systemic LPS levels correlate with liver conditions such as non-alcoholic fatty liver disease (NAFLD) and non-alcoholic steatohepatitis (NASH). Murine models of nonalcoholic fatty liver disease show increased TLR4 expression and proinflammatory cytokine production in steatohepatitis [74]. Bacterial DNA, found in elevated concentrations in NASH patients, drives TLR9 overexpression in experimental models. TLR2 binding to gram-positive bacterial cell wall components (peptidoglycan and lipoteichoic acids) also contributes to inflammation. TLR2-deficient animals show reduced cytokine production and resistance to diet-induced steatohepatitis [75]. TLR5, which recognizes bacterial flagellin, helps maintain the gut ecological balance, and its disruption contributes to obesity and hepatic steatosis through microbiota alterations.

11.8 Dermal-Microbial Interactions

11.8.1 Gut-Skin Axis Fundamentals

The GI system and skin jointly contribute to physiological equilibrium, with skin homeostasis and adaptive responses intimately connected to gut microbiota status [76]. Skin regeneration processes are essential for thermoregulation, protection, and hydration maintenance. Diet and GI health shape the gut microbiome, dominated by four primary bacterial phyla: Bacteroidetes, Firmicutes, Proteobacteria, and Actinobacteria. Gut microbiota influence the skin microbiology by modulating SCFA production, which regulates cutaneous immune responses. Skin regulatory cells affect physiological processes, including hair follicle development and wound healing. Thus, gut microbiota plays a crucial role in skin health regulation through immune modulation and homeostasis promotion [77]. For example, gut microbiome disruption participates in atopic dermatitis pathogenesis characterized by immune dysfunction and epidermal barrier defects and can trigger psoriasis and other systemic inflammatory conditions in 7–11% of inflammatory bowel disease patients.

11.8.2 Gut Microbiota in Skin Homeostasis Regulation

Gut microbiota significantly influences skin homeostasis (Table 11.1), affecting physiology and pathology of cutaneous commensals through microbial translocation, metabolic activity, and immune modulation. Compounds such as polysaccharide A, commensal *Clostridium* clusters IV and XI bacteria, and retinoic acid promote regulatory T-cell differentiation, tempering excessive inflammatory responses [78]. Homeostasis disruption occurs when intestinal barrier function is compromised, allowing gut bacteria and their metabolites to enter circulation and accumulate in skin tissues. Psoriasis patients demonstrate plasma enriched with microbial DNA from intestinal bacteria, suggesting direct gut-skin communication. SCFAs produced from dietary fiber fermentation in the gut are critical for maintaining healthy skin microbiota and modulating cutaneous immunity. Propionic acid exhibits antibiotic activity against methicillin-resistant *Staphylococcus aureus*. Notably, commensal skin bacteria such as *Staphylococcus epidermidis* and *Cutibacterium acnes* show greater propionic acid resistance, suggesting their metabolic activities help maintain skin microbial balance.

11.8.3 Gut Flora in Dermatological Conditions

The gut-skin connection extends to numerous dermatological conditions, including atopic dermatitis, psoriasis, acne vulgaris, rosacea, hidradenitis suppurativa, and chronic spontaneous urticaria, all associated with specific gut microbiota alterations [79].

11.8.3.1 Acne Pathogenesis and Intestinal Health

Acne vulgaris development involves complex interactions between gut microbiota and the mTOR signaling pathway, which regulates lipid metabolism and cellular proliferation. Compromised intestinal barrier function and microbial imbalance can initiate metabolic inflammation that contributes significantly to acne pathophysiology. Psychological stress, GI disorders, and conditions like depression and anxiety appear to exacerbate both systemic and intestinal inflammatory processes. The neuropeptide substance P, which participates in both gut dysbiosis and acne formation, may intensify inflammatory signaling cascades [80]. The relationship between acne and GI dis-

Table 11.1: Association of the gut microbiome with various skin disorders.

S. no.	Skin condition	Gut microbiome findings
1	General impact of Western diet	Western dietary patterns disrupt gut microbiota equilibrium, leading to a decrease in the production of short-chain fatty acids.
2	General microbiota disruption	Beneficial microbes such as *Faecalibacterium prausnitzii* and *Bifidobacterium* are reduced.
3	Atopic dermatitis	The presence of *Escherichia coli* correlates with a higher risk of eczema.
4	Atopic dermatitis	*Clostridioides difficile* is linked with increased susceptibility to atopic diseases.
5	General microbiota disruption	Diminished populations of *Akkermansia muciniphila*, *Bacteroides*, Proteobacteria, and *Faecalibacterium prausnitzii*.
6	General microbiota disruption	Elevated levels of Actinobacteria and Firmicutes were observed.
7	Psoriasis	Microbial DNA originating from the gut has been detected in the bloodstream of individuals experiencing active psoriasis.
8	UV damage and skin aging	Orally consumed *Lactobacillus* strains exhibit anti-inflammatory properties and help mitigate ultraviolet-induced skin aging.
9	Hypersensitivity reactions	*Lactobacillus johnsonii* supplementation has been shown to reduce hypersensitive skin responses.
10	UV damage and skin aging	*Lactobacillus johnsonii* La1 helps maintain a healthy skin response under UV exposure.
11	UV damage and skin aging	Oral intake of *Lactobacillus plantarum* boosts skin moisture and elasticity and reduces the expression of matrix metalloproteinase-1, a marker of skin aging.
12	Hidradenitis suppurativa	Increased presence of *Ruminococcus gnavus* and *Clostridium ramosum* has been noted.
13	Chronic spontaneous urticaria	Greater abundance of *Lactobacillus*, *Turicibacter*, and *Lachnobacterium*, and reduced levels of *Phascolarctobacterium*.

turbances may involve small intestinal bacterial overgrowth (SIBO) mediated by toxic metabolite production, enhanced intestinal permeability, and systemic inflammatory responses. The Table 11.1 summarizes the specific alterations in gut microbiome composition linked to various skin disorders, demonstrating that conditions like atopic dermatitis, psoriasis, and acne are associated with a reduction in beneficial bacteria and an increase in potentially harmful microbes.

11.8.3.2 Atopic Dermatitis and Microbial Interactions

Current understanding suggests that dietary patterns influence allergic predisposition through modifications of gut bacterial communities. Research on diet-microbiome relationships demonstrates that Western dietary patterns characterized by high fat and low fiber consumption, promote dysbiosis and disrupt microbial equilibrium. This imbalance reduces the production of crucial microbial metabolites, particularly SCFAs, which are essential for proper immune system development and function. Diminished SCFA levels impair regulatory T-cell activity, weakening the body's capacity to maintain immunological tolerance. Investigations show that atopic dermatitis patients typically harbor reduced populations of beneficial SCFA-producing bacteria while displaying increased colonization by potentially pathogenic microorganisms, including *E. coli* and *Clostridium difficile*. The confluence of impaired intestinal barrier function and deficient SCFA protection establishes a self-perpetuating inflammatory cycle. Barrier compromise facilitates allergen penetration, initiating immune activation that eventually manifests as cutaneous damage and characteristic atopic dermatitis symptoms [81].

11.8.3.3 Psoriasis and Systemic Microbial Influence

Intestinal microbiota significantly impacts T-cell differentiation and immune tolerance mechanisms. Psoriasis patients demonstrate reduced abundance of protective species, such as *Akkermansia muciniphila, Bacteroides*, Proteobacteria, and *Faecalibacterium prausnitzii*, while showing increased representation of Firmicutes and Actinobacteria [82]. Gut dysbiosis can produce systemic effects since microbial DNA and metabolic products can influence distant organs, including the skin and joints. Circulating bacterial DNA of gastric origin has been detected in blood samples from patients with active psoriasis.

11.8.3.4 Photodamage and Microbial Protection

Oral administration of specific *Lactobacillus* strains demonstrates protective effects against ultraviolet-induced skin damage and exhibits anti-inflammatory properties. *Lactobacillus johnsonii* supplementation reduces UV-induced hypersensitivity in murine models by increasing IL-10 production and decreasing Langerhans cell populations [83]. This strain also promotes appropriate cutaneous immune responses following UV exposure. Oral supplementation with *Lactobacillus plantarum* suppresses matrix metalloproteinase-1 expression in dermal fibroblasts, a primary factor in skin aging while improving skin hydration and elasticity.

11.8.3.5 Additional Dermatological Conditions

Therapeutic interventions targeting SIBO with rifaximin correlate with improved cutaneous symptoms in Korean women with rosacea [84]. Microbial analysis reveals increased abundance of *Ruminococcus gnavus* and *Clostridium ramosum* in hidradenitis suppurativa patients, while chronic spontaneous urticaria patients show higher prevalence of *Lactobacillus*, *Turicibacter*, and *Lachnobacterium* species.

11.9 Microbial Endocrinology

The field of microbial endocrinology examines how gut bacteria produce hormone-like compounds and regulate hormonal secretion (Table 11.2). This discipline investigates how microbial community structure and distribution, along with stress-mediated effects on the microbiome, influence endocrine function [85]. Elevated norepinephrine levels stimulate the proliferation of gram-negative bacteria, including *E. coli*, potentially altering the microbiome's hormonal signaling and contributing to behavioral modifications and disease states in the host. The complexity of host-microbiota interactions necessitates careful consideration of each partner's individual contributions.

Intestinal microorganisms constitute a vital component of host health, performing essential functions including nutrient processing, polysaccharide digestion, pathogen inhibition, and vitamin biosynthesis. This collection of microorganisms functions as a "virtual endocrine organ" capable of modulating other bodily systems and responding to secretions from host organs. The biochemical versatility of gut microbiota enables the synthesis of numerous neuroactive compounds, including monoamine neurotransmitters such as serotonin, dopamine, and norepinephrine, which act within the CNS. The remarkable metabolic flexibility arises from the tremendous diversity of microbial cells inhabiting the human organism [86].

The endocrine capabilities of gut bacteria substantially influence both metabolic processes and immune function. Microbial fermentation of dietary fibers generates SCFAs that play critical roles in regulating hormonal secretion and immune responses. These fatty acids activate G-protein-coupled receptors on enteroendocrine cells, modulating appetite regulation and glucose homeostasis. Future research should focus on elucidating the sophisticated interactions between microbiota and host endocrine regulation, which will be essential for understanding chronic conditions, including obesity, diabetes, and neurodegenerative disorders. The Table 11.2 catalogs the diverse hormone-like metabolites produced by the gut microbiota, including short-chain fatty acids, neurotransmitters, and bile acids, which play critical roles in regulating energy, metabolism, immunity, and brain function, solidifying the microbiome's role as a "virtual endocrine organ."

Table 11.2: Hormones of the gut microbiota.

S. no.	Class	Metabolites	Physiological functions	Properties	References
1	Short-chain fatty acids (SCFAs)	Acetate, butyrate, and propionate	Energy, metabolic regulation, cell signaling, and anti-inflammatory activity	Bacterial fermentation products, histone deacetylation, influence neural function and behavior	[87]
2	Neurotransmitters	Serotonin, dopamine, norepinephrine, and GABA	Emotional regulation. cognitive function, reward processing (CNS), and digestive function (ENS)	Microbial synthesis capability and host-microbe co-regulation pathways	[88]
3	Neurotransmitter precursors	Tryptophan, kynurenine, and L-DOPA	Serotonin synthesis, neuroactive metabolite production, and dopaminergic pathway support	Gut microbiota regulates tryptophan metabolism and modulates the kynurenine pathway	[89]
4	Bile acid metabolites	Secondary bile acids (e.g., deoxycholic acid)	Microbial growth inhibition and metabolic regulation	Nuclear receptor activation (FXR and TGR5)	[90]
5	Choline derivatives	Trimethylamine (TMA)	Lipid metabolism regulation	Hepatic conversion to TMAO and cardiovascular risk associations	[91]
6	Stress response mediators	Cortisol	Stress adaptation and metabolic control	Gut-brain axis modulation and hypothalamic-pituitary-adrenal axis influence	[92]
7	Gastrointestinal peptides	Ghrelin, leptin, GLP-1, and PYY	Nutrient absorption and intestinal function	Microbial metabolite-mediated regulation and enteroendocrine cell activation	[93]

11.10 Gut-Kidney Axis: Coordinating Waste Management

The kidneys and GI tract maintain a reciprocal, synergistic relationship crucial for systemic homeostasis. One pathway involves microbial metabolism producing uremic toxins, including TMAO, p-cresyl sulfate, and indoxyl sulfate. Alternatively, the uremic state itself can disrupt gut microbiota composition and metabolic activity. These co-metabolites demonstrate both vascular and renal toxicity, representing products of host-microbiota collaboration. Consequently, disruption of this bidirectional communication can precipitate severe outcomes, including acute kidney injury (AKI), end-stage renal disease (ESRD), and chronic kidney disease (CKD) [94].

Colonic microbiome composition is substantially determined by dietary intake and small intestinal absorption efficiency. Dietary fibers that escape digestion in the upper GI tract serve as the primary carbohydrate source for colonic flora, undergoing microbial fermentation to produce SCFAs. Similarly, dietary proteins that resist upper gut digestion provide nitrogen sources for colonic microbiota [95]. Subsequent processing of protein fermentation products generates uremic toxins such as p-cresyl sulfate and indoxyl sulfate. These compounds exhibit strong albumin binding affinity through noncovalent interactions, facilitating their circulation in bloodstream until renal tubular secretion eliminates them. Accumulation of these uremic retention solutes accelerates renal disease progression and glomerular sclerosis, making their plasma concentrations useful indicators of renal functional capacity [96].

Dietary lipid components including lecithin, carnitine, and choline require specialized enzymatic processing. Mammals lack enzymes to cleave the cyanide bonds in these compounds, but colonic microbiota produce TMA lyases that perform this function [97]. Collaborative action between microbial TMA lyases and hepatic enzymes (particularly flavin-containing monooxygenases) converts carnitine and choline to TMAO, which enters systemic circulation and undergoes renal elimination like other uremic toxins. Elevated TMAO levels directly correlate with CKD progression [98]. ESRD patients demonstrate TMAO concentrations up to 20-fold higher than healthy individuals. TMAO's adverse effects include enhanced platelet activity, increased thrombosis risk, renal tubulointerstitial fibrosis, and atherosclerosis development [99].

Colonic microbiota composition influences uremic toxin elimination kinetics; proteolytic bacteria proliferate with reduced intestinal transit time. Furthermore, carbohydrate deficiency in the colon promotes excessive bacterial protein fermentation [100, 101]. Patients with septic AKI exhibit altered microbial ecology potentially resulting from compromised epithelial integrity or inflammatory mediators. Gut microbiota and their metabolic products can modify the expression of bacterial receptors in GI tissues, leading to microbial community relocation [102, 103]. In septic patients, the gut microbiome experiences disruption through multiple physiological alterations arising from both intrinsic factors (increased intestinal permeability and systemic inflammation) and extrinsic factors (medications and parenteral nutrition). Administration of microbially produced SCFAs to septic AKI patients has demonstrated improved renal function [104, 105].

11.11 Gut-Bone Axis: Microbial Influence on Skeletal Health

The human skeleton provides structural support, protects vital organs, maintains calcium homeostasis, supports hematopoiesis and fat storage in bone marrow, and stores cytokines and growth factors. Interestingly, gut microbiota and their derivatives significantly influence bone health. Recent investigations of the "gut-bone axis" have revealed complex relationships between GI health and skeletal integrity [106]. Intrauterine and early postnatal environmental exposures regulate growth patterns, bone mineralization, and body composition alongside gut microbiota development. Multiple preclinical studies have demonstrated gut microbiota impacts on bone strength and mineral density, with *Lactobacillus* species identified as particularly influential [107, 108].

The relationship between gut microbiota and bone health is further supported by observations of bone loss accompanying GI bacterial overgrowth. Gut microbiota regulates the bone formation through several potential mechanisms: immune system maturation, nutrient absorption efficiency, production of various metabolites, alterations in gut permeability, gut-derived serotonin signaling, and LPS-mediated systemic inflammation. Microbial communities significantly affect nutrient absorption; for instance, elevated *Lactobacillus reuteri* and *Bifidobacterium longum* concentrations in the GI tract correlate with improved bone mineral density through enhanced calcium, phosphate, and magnesium absorption. Gut microbiota also contributes substantially to the production of vitamins B and K, which are essential nutrients for bone health regulation and bile acid metabolism [109, 110].

11.12 Strategies for Optimizing Gut Microbial Health

Plant-based diets abundant in dietary fiber, vitamins, minerals, and phytochemicals incorporating diverse fruits, vegetables, whole grains, legumes, and nuts enhance the gut microbial richness [111, 112]. Dietary diversity is recommended to support a healthier gut microbiome. Fermented foods containing probiotics (kefir, yogurt, sauerkraut, kimchi, and miso) help maintain balanced gut environments and increase microbial diversity [113]. Reducing consumption of highly processed foods is crucial, as these typically lack the fiber and nutrients necessary for healthy gut microbiota while

promoting inflammation and microbial alterations that may transmit across genera-
tions through epigenetic mechanisms [114]. Consumption of prebiotic fiber-rich foods
(whole grains, onions, garlic, bananas, and asparagus) fosters balanced gut flora [115].
Adequate sleep duration benefits gut microbiome health, as sleep deprivation correlates with microbial alterations. Finally, antibiotic administration disrupts gut microbial equilibrium and promotes resistant bacterial strain proliferation [116].

11.13 Conclusion

The gut microbiome plays a pivotal role in maintaining human health by influencing
various organ systems through bidirectional communication axes, such as the gut-
brain, gut-heart, gut-lung, gut-liver, gut-skin, and gut-kidney axes. These interactions
are mediated by microbial metabolites, immune modulation, and neuroendocrine signaling, impacting both health and disease states. Dysbiosis in the gut microbiota is
linked to numerous conditions, including cardiovascular diseases, neurodegenerative
disorders, and metabolic syndromes. Advances in metagenomics and metatranscriptomics have deepened our understanding of these complex interactions, highlighting
the potential for personalized microbiome-based interventions. Dietary choices, probiotics, and lifestyle modifications are critical in shaping gut microbiota composition,
offering therapeutic avenues for conditions like inflammatory bowel disease, asthma,
and mental health disorders. Maintaining gut microbial balance through fiber-rich
diets, fermented foods, and judicious antibiotic use is essential for promoting overall
health and preventing disease. Future research should focus on optimizing host-
microbiome symbiosis to target specific diseases and enhance well-being.

Conflicts of interests: The authors do not declare any conflicts of interest.

References

[1]   Nicholson JK, Holmes E, Kinross J, Burcelin R, Gibson G, Jia W, Pettersson S. Host-gut microbiota
      metabolic interactions. Science. 2012;336:1262–1267.
[2]   Qin Y, Wade PA. Crosstalk between the microbiome and epigenome: Messages from bugs. Journal
      of Biochemistry. 2018;163:105–112.
[3]   Shapira M. Gut microbiotas and host evolution: Scaling up symbiosis. Trends in Ecology and
      Evolution. 2016;31:539–549.
[4]   Falony G, Vandeputte D, Caenepeel C, Vieira-Silva S, Daryoush T, Vermeire S, Raes J. The human
      microbiome in health and disease: Hype or hope. Acta Clinica Belgica. 2019;74:53–64.
[5]   Ding RX, Goh WR, Wu RN, Yue XQ, Luo X, Khine WWT, Wu JR, Lee YK. Revisit gut microbiota and its
      impact on human health and disease. Journal of Food and Drug Analysis. 2019;27:623–631.

[6]   Tierney BT, Yang Z, Luber JM, Beaudin M, Wibowo MC, Baek C, Mehlenbacher E, Patel CJ, Kostic AD. The landscape of genetic content in the gut and oral human microbiome. Cell Host & Microbe. 2019;26:283–295.

[7]   Park W. Gut Microbiomes and Their Metabolites Shape Human and Animal Health. 2018, New York, NY: Springer.

[8]   Thursby E, Juge N. Introduction to the human gut microbiota. Biochemical Journal. 2017;474:1823–1836.

[9]   Sekirov I, Russell SL, Caetano M, Antunes L, Finlay BB. Gut microbiota in health and disease. Physiological Reviews. 2010;90:859–904.

[10]   Hawrelak JA, Myers SP. The causes of intestinal dysbiosis: A review. Alternative Medicine Review. 2004;9:180–197.

[11]   Jandhyala SM, Talukdar R, Subramanyam C, Vuyyuru H, Sasikala M, Nageshwar Reddy D. Role of the normal gut microbiota. World Journal of Gastroenterology. 2015;21(29):8787–8803.

[12]   Kho ZY, Lal SK. The human gut microbiome – A potential controller of wellness and disease. Frontiers in Microbiology. 2018;9:1835.

[13]   Wensel CR, Pluznick JL, Salzberg SL, Sears CL. Next-generation sequencing: Insights to advance clinical investigations of the microbiome. Journal of Clinical Investigation. 2022;132(7):e154944.

[14]   Zhang X, Li L, Butcher J, Stintzi A, Figeys D. Advancing functional and translational microbiome research using meta-omics approaches. Microbiome. 2019;7(1):154–166.

[15]   Bergamaschi M, Tiezzi F, Howard J, Huang YJ, Gray KA, Schillebeeckx C, McNulty NP, Maltecca C. Gut microbiome composition differences among breeds impact feed efficiency in swine. Microbiome. 2020;8(1):110–119.

[16]   De Oliveira GLV, Oliveira CNS, Pinzan CF, De Salis LVV, Cardoso CRB. Microbiota modulation of the gut-lung axis in COVID-19. Frontiers in Immunology. 2021;12:635471.

[17]   Fan Y, Wang H, Liu X, Zhang J, Liu G. Crosstalk between the ketogenic diet and epilepsy: From the perspective of gut microbiota. Mediators of Inflammation. 2019;8373060. https://onlinelibrary.wiley.com/doi/epdf/10.1155/2019/8373060

[18]   Forkosh E, Ilan Y. The heart-gut axis: New target for atherosclerosis and congestive heart failure therapy. Open Heart. 2019;6(1):e000993.

[19]   Nicholson JK, Holmes E, Kinross J, Burcelin R, Gibson G, Jia W, Pettersson S. Host-gut microbiota metabolic interactions. Science. 2012;336(6086):1262–1267.

[20]   Rogler G, Rosano G. The heart and the gut. European Heart Journal. 2014;35(7):426–430.

[21]   Athanasopoulou K, Adamopoulos PG, Scorilas A. Unveiling the human gastrointestinal tract microbiome: The past, present, and future of metagenomics. Biomedicines. 2023;11(3):827–835.

[22]   Campbell C, Kandalgaonkar M, Golonka RM, Yeoh BS, Vijay-Kumar M, Saha P. Crosstalk between gut microbiota and host immunity: Impact on inflammation and immunotherapy. Biomedicines. 2023;11 (2):294–310.

[23]   Chelakkot C, Choi Y, Kim D, Park HT, Ghim J, Kwon Y, Jeon J, Kim MS, Jee Y, Gho YS, Park H, Kim Y, Ryu SH. *Akkermansia muciniphila*-derived extracellular vesicles influence gut permeability by regulating tight junctions. Experimental & Molecular Medicine. 2018;50(2):e450.

[24]   Emencheta SC, Olovo CV, Eze OC, Kalu CF, Berebon DP, Onuigbo E, Vila MMDC, Balcão VM, Attama AA. The role of bacteriophages in the gut microbiota: Implications for human health. Pharmaceutics. 2023;15(10):2416–2430.

[25]   Gong Y, Chen A, Zhang G, Shen Q, Zou L, Li J, Miao Y, Liu W. Cracking brain diseases from gut microbes-mediated metabolites for precise treatment. International Journal of Biological Sciences. 2023;19(10):2974–2990.

[26]   Kasarełło K, Cudnoch-Jędrzejewska A, Czarzasta K. Communication of gut microbiota and brain *via* immune and neuroendocrine signaling. Frontiers in Microbiology. 2023;14:1118529.

[27] Khan I, Bai Y, Zha L, Ullah N, Ullah H, Shah SRH, Sun H, Zhang C. Mechanism of the gut microbiota colonization resistance and enteric pathogen infection. Frontiers in Cellular and Infection Microbiology. 2021;11:716299.

[28] Kumpunya S, Kawang K, Pollapong K, Nilaratanakul V. The effects of repeated fecal transplantation and activated charcoal treatment on gut dysbiosis induced by concurrent ceftriaxone administration in mice. Research Square. 2025;Preprint.

[29] Li H, Jiaojiao H, Wang J. The influence of gut microbiota on drug metabolism and toxicity. Expert Opinion on Drug Metabolism & Toxicology. 2015;12(1):31–40.

[30] Liu Y, Lau HCH, Yu J. Microbial metabolites in colorectal tumorigenesis and cancer therapy. Gut Microb. 2023;15(1):2203968.

[31] Lv J, Guo L, Liu J, Zhao H, Zhang J, Wang JH. Alteration of the esophageal microbiota in Barrett's esophagus and esophageal adenocarcinoma. World Journal of Gastroenterology. 2019;25 (18):2149–2164.

[32] Miri S, Yeo J, Abubaker S, Hammami R. Neuromicrobiology, an emerging neurometabolic facet of the gut microbiome?. Frontiers in Microbiology. 2023;14:1098412.

[33] O'Riordan KJ, Moloney GM, Keane L, Clarke G, Cryan JF. The gut microbiota-immune-brain axis: Therapeutic implications. Cell Reports Medicine. 2025;6(3):101982.

[34] Paul JK, Azmal M, Haque ASNB, Meem M, Talukder OF, Ghosh A. Unlocking the secrets of the human gut microbiota: Comprehensive review on its role in different diseases. World Journal of Gastroenterology. 2024;31(5):99913.

[35] Politi C, Mobrici M, Parlongo RM, Spoto B, Tripepi G, Pizzini P, Cutrupi S, Franco D, Tino R, Farruggio G, Failla C, Marino F, Pioggia G, Testa A. Role of gut microbiota in overweight susceptibility in an adult population in Italy. Nutrients. 2023;15(13):2834–2845.

[36] Portincasa P, Bonfrate L, Vacca M, De Angelis M, Farella I, Lanza E, Khalil M, Wang DQH, Sperandio M, Di Ciaula A. Gut microbiota and short chain fatty acids: Implications in glucose homeostasis. International Journal of Molecular Sciences. 2022;23(3):1105–1115.

[37] Rusch J, Layden BT, Dugas LR. Signalling cognition: The gut microbiota and hypothalamic-pituitary-adrenal axis. Frontiers in Endocrinology. 2023;14:1130689.

[38] Tabrizi E, Tabrizi FPF, Khaled GM, Sestito M, Jamie S, Boone BA. Unraveling the gut microbiome's contribution to pancreatic ductal adenocarcinoma: Mechanistic insights and therapeutic perspectives. Frontiers in Immunology. 2024;15:1434771.

[39] Di Vincenzo F, Di Gaudio A, Petito V, Lopetuso LR, Scaldaferri F. Gut microbiota, intestinal permeability, and systemic inflammation: A narrative review. Internal and Emergency Medicine. 2023;19(2):275–286.

[40] Yassin LK, Nakhal MM, Alderei A, Almehairbi A, Mydeen AB, Akour A, Hamad MIK. Exploring the microbiota-gut-brain axis: Impact on brain structure and function. Frontiers in Neuroanatomy. 2025;19:1504065.

[41] Sharon G, Sampson TR, Geschwind DH, Mazmanian SK. The central nervous system and the gut microbiome. Cell. 2016;167(4):915–932.

[42] Mohajeri MH, La Fata G, Steinert RE, Weber P. Relationship between the gut microbiome and brain function. Nutrition Reviews. 2018;10(7):1025–1036.

[43] Li YH, Chen JW, Lin TH, Wang YC, Wu CC, Yeh HI. A performance guide for major risk factors control in patients with atherosclerotic cardiovascular disease in Taiwan. Journal of the Formosan Medical Association. 2020;119:674–684.

[44] Nemet I, Saha PP, Gupta N, Zhu W, Romano KA, Skye SM. A cardiovascular disease-linked gut microbial metabolite acts *via* adrenergic receptors. Cell. 2020;180:862–877.

[45] Gurung M, Li Z, You H, Rodrigues R, Jump DB, Morgun A. Role of gut microbiota in type 2 diabetes pathophysiology. EBioMedicine. 2020;51:102590.

[46] Sanna S, Van Zuydam NR, Mahajan A, Kurilshikov A, Vich Vila A, Vosa U. Causal relationships among the gut microbiome, short-chain fatty acids and metabolic diseases. Nature Genetics. 2019;51:600–605.

[47] Li Q, Chang Y, Zhang K, Chen H, Tao S, Zhang Z. Implication of the gut microbiome composition of type 2 diabetic patients from northern China. Scientific Reports. 2020;10:5450.

[48] Yamamoto H, Yamanashi Y, Takada T, Mu S, Tanaka Y, Komine T. Hepatic expression of Niemann-Pick C1-like 1, a cholesterol reabsorber from bile, exacerbates Western diet–induced atherosclerosis in LDL receptor mutant mice. Molecular Pharmacology. 2019;96:47–55.

[49] Ferrell JM, Boehme S, Li F, Chiang JYL. Cholesterol 7α-hydroxylase-deficient mice are protected from high-fat/high-cholesterol diet-induced metabolic disorders. Journal of Lipid Research. 2016;57:1144–1154.

[50] Jaworska K, Bielinska K, Gawrys-Kopczynska M, Ufnal M. TMA (trimethylamine), but not its oxide TMAO (trimethylamine-N-oxide), exerts haemodynamic effects: Implications for interpretation of cardiovascular actions of gut microbiome. Cardiovascular Research. 2019;115(14):1948–1949.

[51] Jaworska K, Hering D, Mosieniak G, Bielak-Zmijewska A, Pilz M, Konwerski M. TMA, a forgotten uremic toxin, but not TMAO, is involved in cardiovascular pathology. Toxins (Basel). 2019;11(9):490–498.

[52] Jaworska K, Konop M, Hutsch T, Perlejewski K, Radkowski M, Grochowska M. Trimethylamine but not trimethylamine N-oxide increases with age in rat plasma and affects smooth muscle cells viability. Journal of Gerontology: Series A – Biological Sciences and Medical Sciences. 2020;75(7):1276–1283.

[53] Battson ML, Lee DM, Weir TL, Gentile CL. The gut microbiota as a novel regulator of cardiovascular function and disease. Journal of Nutritional Biochemistry. 2018;56:1–15.

[54] Jie Z, Xia H, Zhong SL, Feng Q, Li S, Liang S, Zhong H, Liu Z. The gut microbiome in atherosclerotic cardiovascular disease. Nature Communications. 2017;8:845–856.

[55] Jin M, Qian Z, Yin J, Xu W, Zhou X. The role of intestinal microbiota in cardiovascular disease. Journal of Cellular and Molecular Medicine. 2019;23:2343–2350.

[56] Bassis CM, Dickson RP, Freeman CM, Schmidt TM, Young VB. Analysis of the upper respiratory tract microbiotas as the source of the lung and gastric microbiotas in healthy individuals. Molecular Biology. 2015;6(2):7–15.

[57] Hasegawa K, Linnemann RW, Mansbach JM, Ajami NJ, Espinola JA, Petrosino JF. The fecal microbiota profile and bronchiolitis in infants. Pediatrics. 2016;138(1):e20160218.

[58] Russell SL, Gold MJ, Hartmann M, Willing BP, Thorson L, Wlodarska M. Early life antibiotic-driven changes in microbiota enhance susceptibility to allergic asthma. EMBO Reports. 2012;13:440–447.

[59] Sigurs N, Aljassim F, Kjellman B, Robinson PD, Sigurbergsson F, Bjarnason R. Asthma and allergy patterns over 18 years after severe RSV bronchiolitis in the first year of life. Thorax. 2010;65:1045–1052.

[60] Bacharier LB, Cohen R, Schweiger T, Yin-Declue H, Christie C, Zheng J. Determinants of asthma after severe respiratory syncytial virus bronchiolitis. Journal of Allergy and Clinical Immunology. 2012;130:91–100.

[61] Mathieu E, Escribano-Vazquez U, Descamps D, Cherbuy C, Langella P, Riffault S. Paradigms of lung microbiota functions in health and disease, particularly in asthma. Frontiers in Physiology. 2018;9:1168–1176.

[62] Jung JW, Choi JC, Shin JW, Kim JY, Park IW, Choi BW. Lung microbiome analysis in steroid-naïve asthma patients by using whole sputum. Tuberculosis and Respiratory Diseases. 2016;79:165–178.

[63] Bruzzese E, Callegari ML, Raia V, Viscovo S, Scotto R, Ferrari S. Disrupted intestinal microbiota and intestinal inflammation in children with cystic fibrosis and its restoration with *Lactobacillus GG*: A randomised clinical trial. PLOS One. 2014;9(2):e87796.

[64] McLoughlin RM, Mills KH. Influence of gastrointestinal commensal bacteria on the immune responses that mediate allergy and asthma. Journal of Allergy and Clinical Immunology. 2011;127 (5):1097–1107.

[65] Noverr MC, Huffnagle GB. Role of antibiotics and fungal microbiota in driving pulmonary allergic responses. Infection and Immunity. 2004;72(9):4996–5003.

[66] Fagundes CT, Amaral FA, Souza AL, Vieira AT, Xu D, Liew FY. Transient TLR activation restores inflammatory response and ability to control pulmonary bacterial infection in germfree mice. Journal of Immunology. 2012;188(3):1411–1420.

[67] Molyneaux PL, Cox MJ, Wells AU, Kim HC, Ji W, Cookson WOC. Changes in the respiratory microbiome during acute exacerbations of idiopathic pulmonary fibrosis. Respiratory Research. 2017;18:10–22.

[68] Ramírez-Labrada AG, Isla D, Artal A, Arias M, Rezusta A, Pardo. The influence of lung microbiota on lung carcinogenesis, immunity, and immunotherapy. Trends in Cancer. 2020;6:86–97.

[69] Loverdos K, Bellos G, Kokolatou L, Vasileiadis I, Giamarellos E, Pecchiari M. Lung microbiome in asthma: Current perspectives. Journal of Clinical Medicine. 2019;8(11):1967.

[70] Rohr MW, Narasimhulu CA, Rudeski-Rohr TA, Parthasarathy S. Negative effects of a high-fat diet on intestinal permeability: A review. Advances in Nutrition. 2020;11:77–91.

[71] Nakanishi K. Exogenous administration of low-dose lipopolysaccharide potentiates liver fibrosis in a choline-deficient l-amino-acid-defined diet-induced murine steatohepatitis model. International Journal of Molecular Sciences. 2019;20:2724–2732.

[72] Himes RW, Smith CW. Tlr2 is critical for diet-induced metabolic syndrome in a murine model. FASEB Journal. 2010;24:731–739.

[73] Belkaid Y, Hand TW. Role of the microbiota in immunity and inflammation. Cell. 2014;157(1):121–141.

[74] O'Neill CA, Monteleone G, McLaughlin JT, Paus R. The gut-skin axis in health and disease: A paradigm with therapeutic implications. BioEssays. 2016;38(11):1167–1176.

[75] Zambanini T, Kleineberg W, Sarikaya E, Buescher JM, Meurer G, Wierckx N. Enhanced malic acid production from glycerol with high-cell density *Ustilago trichophora* TZ1 cultivations. Biotechnology for Biofuels. 2016;9:135–144.

[76] Salem I, Ramser A, Isham N, Ghannoum MA. The gut microbiome as a major regulator of the gut-skin axis. Frontiers in Microbiology. 2018;9:1459–1466.

[77] Berni Canani R, Sangwan N, Stefka AT, Nocerino R, Paparo L, Aitoro R, Segata N. *Lactobacillus rhamnosus* GG-supplemented formula expands butyrate-producing bacterial strains in food allergic infants. ISME Journal. 2016;10(3):742–750.

[78] Ring HC, Theut Riis P, Zacho M, Miller IM, Prens E, Saunte DM, Jemec GB. The microbiome of hidradenitis suppurativa: A systematic review. British Journal of Dermatology. 2017;177(5):1117–1127.

[79] Tremaroli V, Bäckhed F. Functional interactions between the gut microbiota and host metabolism. Nature. 2012;489(7415):242–249.

[80] Yano JM, Yu K, Donaldson GP, Shastri GG, Ann P, Ma L. Indigenous bacteria from the gut microbiota regulate host serotonin biosynthesis. Cell. 2015;161(2):264–276.

[81] Cryan JF, Dinan TG. Mind-altering microorganisms: The impact of the gut microbiota on brain and behaviour. Nature Reviews. Neuroscience. 2012;13(10):701–712.

[82] Mayer EA. Gut feelings: The emerging biology of gut–brain communication. Nature Reviews. Neuroscience. 2011;12(8):453–466.

[83] Cani PD, Delzenne NM. The gut microbiome as therapeutic target. Pharmacology & Therapeutics. 2011;130(2):202–212.

[84] Cryan JF, O'Riordan KJ, Sandhu K, Peterson V, Dinan TG. The gut microbiome in neurological disorders. The Lancet Neurology. 2020;19(2):179–194.

[85] Wang Z, Klipfell E, Bennett BJ, Koeth R, Levison BS, DuGar B, Hazen SL. Gut flora metabolism of phosphatidylcholine promotes cardiovascular disease. Nature. 2011;472(7341):57–63.

[86] Frost G, Sleeth ML, Sahuri-Arisoylu M, Lizarbe B, Cerdan S, Brody L, Bell JD. The short-chain fatty acid acetate reduces appetite via a central homeostatic mechanism. Nature Communications. 2014;5:3611–3621.

[87] Dalile B, Van Oudenhove L, Vervliet B, Verbeke K. The role of short-chain fatty acids in microbiota–gut–brain communication. Nature Reviews Gastroenterology & Hepatology. 2019;16 (8):461–478.

[88] Silva YP, Bernardi A, Frozza RL. The role of short-chain fatty acids from gut microbiota in gut-brain communication. Frontiers in Endocrinology. 2020;11:25–36.

[89] Ridlon JM, Harris SC, Bhowmik S, Kang DJ, Hylemon PB. Consequences of bile salt biotransformations by intestinal bacteria. Gut Microbes. 2016;7(1):22–39.

[90] Wang Z, Roberts AB, Buffa JA, Levison BS, Zhu W, Org E, Hazen SL. Non-lethal inhibition of gut microbial trimethylamine production for the treatment of atherosclerosis. Cell. 2015;163 (7):1585–1595.

[91] Everard A, Belzer C, Geurts L, Ouwerkerk JP, Druart C, Bindels LB, Cani PD. Cross-talk between *Akkermansia muciniphila* and intestinal epithelium controls diet-induced obesity. Proceedings of the National Academy of Sciences of the United States of America. 2013;110(22):9066–9071.

[92] Johnson T, Gomez B, McIntyre M, Dubick M, Christy R, Nicholson S, Burmeister D. The cutaneous microbiome and wounds: New molecular targets to promote wound healing. International Journal of Molecular Sciences. 2018;19:2699–2706.

[93] Karlsson FH, Fak F, Nookaew I, Tremaroli V, Fagerberg B, Petranovic D, Bäckhed F, Nielsen J. Symptomatic atherosclerosis is associated with an altered gut metagenome. Nature Communications. 2012;3:1245–1256.

[94] Zhang J, Ankawi G, Sun J, Digvijay K, Yin Y, Rosner MH, Ronco C. Gut–kidney crosstalk in septic acute kidney injury. Critical Care. 2018;22(1):117–125.

[95] Evenepoel P, Poesen R, Meijers B. The gut kidney axis. Pediatric Nephrology. 2017;32(11):2005–2014.

[96] Leboucher A, Rath M, Kleinridders A. Increased uremic toxins in cerebrospinal fluid of obese mice cause insulin resistance. Diabetologie Und Stoffwechsel. 2018;13(S 01):59–66.

[97] Tang WH, Wang Z, Kennedy DJ, Wu Y, Buffa JA, Agatisa-Boyle B, Li XS, Levison BS. Gut microbiota-dependent trimethylamine N-oxide (TMAO) pathway contributes to both development of renal insufficiency and mortality risk in chronic kidney disease. Circulation Research. 2015;116(3):448–455.

[98] Jazani NH, Savoj J, Lustgarten M, Lau WL, Vaziri ND. Impact of gut dysbiosis on neurohormonal pathways in chronic kidney disease. Diseases. 2019;7(1):21–30.

[99] Hiippala K, Jouhten H, Ronkainen A, Hartikainen A, Kainulainen V, Jalanka J, Satokari R. The potential of gut commensals in reinforcing intestinal barrier function and alleviating inflammation. Nutrients. 2018;10(8):988–1002.

[100] Barrios C, Beaumont M, Pallister T, Villar J, Goodrich JK, Clark A, Pascual J, Ley RE. Gut microbiota–metabolite axis in early renal function decline. PLOS One. 2015;10(8):e0134311.

[101] Schefold JC, Filippatos G, Hasenfuss G, Anker SD, von Haehling S. Heart failure and kidney dysfunction: Epidemiology, mechanisms and management. Nature Reviews Nephrology. 2016;12 (10):610–623.

[102] Zampini A, Nguyen AH, Rose E, Monga M, Miller AW. Defining dysbiosis in patients with urolithiasis. Scientific Reports. 2019;9(1):5425–5433.

[103] Medina-Gomez C. Bone and the gut microbiome: A new dimension. Journal of Laboratory and Precision Medicine. 2018;3:96–106.

[104] Chen YC, Greenbaum J, Shen H, Deng HW. Association between gut microbiota and bone health: Potential mechanisms and prospective. Journal of Clinical Endocrinology & Metabolism. 2017;102 (10):3635–3646.

[105] Conlon MA, Bird AR. The impact of diet and lifestyle on gut microbiota and human health. Nutrients. 2015;7(1):17–44.

[106]   McGinty T, Mallon PW. Fractures and the gut microbiome. Current Opinion in HIV and AIDS. 2018;13 (1):28–37.

[107]   Frost M, Andersen T, Gossiel F, Hansen S, Bollerslev J, Van Hul W, Eastell R, Kassem M. Levels of serotonin, sclerostin, bone turnover markers as well as bone density and microarchitecture in patients with high bone-mass phenotype due to a mutation in Lrp5. Journal of Bone and Mineral Research. 2011;26(7):1721–1728.

[108]   Pietschmann P, Mechtcheriakova D, Meshcheryakova A, Föger-Samwald U, Ellinger I. Immunology of osteoporosis: A mini-review. Gerontology. 2016;62(2):128–137.

[109]   Lucas S, Omata Y, Hofmann J, Böttcher M, Iljazovic A, Sarter K, Albrecht O, Schulz O. Short-chain fatty acids regulate systemic bone mass and protect from pathological bone loss. Nature Communications. 2018;9(1):55–69.

[110]   Wu X, Liu J, Xiao L, Lu A, Zhang G. Alterations of gut microbiome in rheumatoid arthritis. Osteoarthritis and Cartilage. 2017;25:S287–S288.

[111]   Costello ME, Ciccia F, Willner D, Warrington N, Robinson PC, Gardiner B, Marshall M, Kenna TJ. Intestinal dysbiosis in ankylosing spondylitis. Arthritis & Rheumatology. 2015;67(3):686–691.

[112]   Leeuwendaal NK, Stanton C, O'Toole PW. Fermented foods, health and the gut microbiome. Nutrients. 2022;14(8):1527–1540.

[113]   Shi Z. Gut microbiota: An important link between western diet and chronic diseases. Nutrients. 2019;11(9):2287–2299.

[114]   Fu J, Zheng Y, Gao Y. Dietary fiber intake and gut microbiota in human health. Microorganisms. 2022;10(12):2507–2516.

[115]   Karl JP, Whitney CC, Wilson MA, Fagnant HS, Radcliffe PN, Chakraborty N, Campbell R, Hoke A, Gautam A, Hammamieh R, Smith TJ. Severe, short-term sleep restriction reduces gut microbiota community richness but does not alter intestinal permeability in healthy young men. Scientific Reports. 2023;13:213–225.

[116]   Patangia DV, Anthony Ryan C, Dempsey E, Ross RP, Stanton C. Impact of antibiotics on the human microbiome and consequences for host health. Microbiology and Biotechnology Open. 2022;11: e1260.

Index

https://doi.org/10.1515/9783112205150-012